Orchids as Aphrodisiac, Medicine or Food

Dendrobium nobile flowering in April on a tree in Sikkim, India. Pseudobulbs of this beautiful, popular orchid are employed medicinally as *shihu* in China. Although the species is widely distributed, its existence in nature is now under threat because of over-collection from the forests of southern China and the adjacent countries. (©Teoh Eng Soon 2019. All Rights Reserved.)

Eng Soon Teoh

Orchids as Aphrodisiac, Medicine or Food

Eng Soon Teoh
Singapore, Singapore

ISBN 978-3-030-18254-0   ISBN 978-3-030-18255-7   (eBook)
https://doi.org/10.1007/978-3-030-18255-7

This Springer imprint is published by the registered company Springer Nature Switzerland AG.
The registered company address is: Gewerbestrasse 11, 6330 Cham, Switzerland

for Phaik Khuan, John, Kristine, Chrissie and Ning

Preface

Orchids are more than pretty exotic flowers. For thousands of years, some orchid species have played an important role in traditional herbal medicine in China, India and Europe. Even today several hundred orchid species are employed medicinally to treat injuries and disease or as food and delicacies all over the world. Vanilla, a favourite flavour with the Aztecs and now ubiquitous in Western confectionary, is derived from an orchid fruit.

Recently, the modernization of China has wrought a paradigm shift in the development and practice of Traditional Chinese Medicine. Not only are new modalities being employed for diagnosis and treatment, scientists are scrutinizing ancient remedies at the molecular level to determine whether they actually contain useful compounds and, if they do, their modes of action. In the process, new potential uses are being discovered. Numerous compounds present in orchids act against viruses, bacteria, fungi, protozoa and worms. Other compounds are toxic to cancer cells, causing programmed cell death (apoptosis); depriving malignant cells of their blood supply; or preventing their spread. Some orchid phytochemicals protect against liver damage, brain damage and ultraviolet damage to skin; lower blood sugar; promote fetal lung maturation; and prevent osteoporosis. Every week, new information is appearing in scientific journals. However, most of the excitement is confined within the laboratory. Clinical trials are few and far between. They need to be properly designed and performed.

Globalization, rapid communication and a fast pace of life are rapidly eliminating distinctive cultures and creating homogenous stereotypic communities. Whether in cities or in tribal settlements, people no longer receive detailed knowledge of their past and sometimes they do not make an effort to look after and preserve nature's bounty. Knowledge of ethnomedicine is fast disappearing. Thus, there is a need to discover and record all that was known in the past about medicinal orchid usage before orchids become mere toys for gardening hobbyists or flowers for decoration. I do not deny the importance of the latter, being a fancier of orchids myself, but we should always look beyond horticulture and attempt to realize every potential in the orchids. This transitional period when traditional practices meet science offers many opportunities to the scientist and to traditional healers.

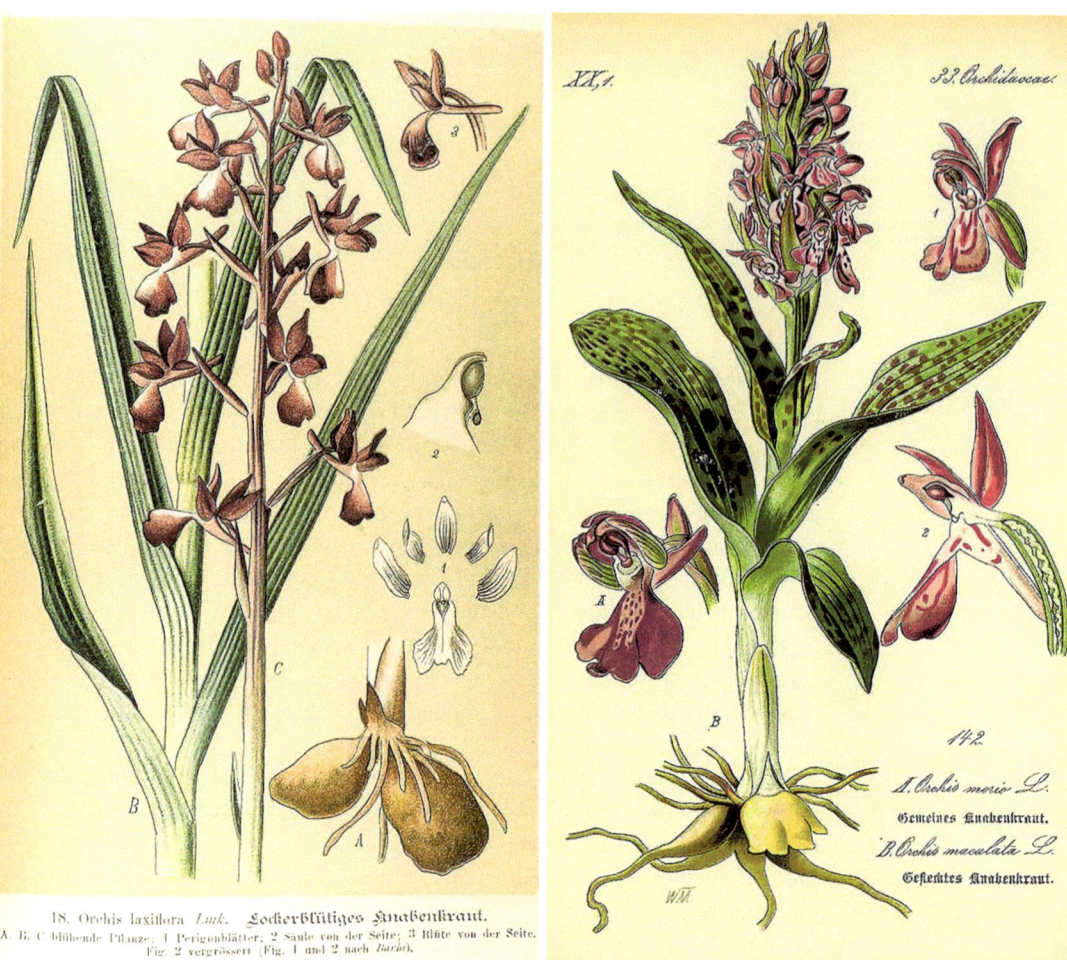

18. Orchis laxiflora *Lmk.* Lockerblütiges Knabenkraut.
A, B, C blühende Pflanze; 1 Perigonblätter; 2 Säule von der Seite; 3 Blüte von der Seite.
Fig. 2 vergrössert (Fig. 1 und 2 nach *Barón*),

Three Mediterranean terrestrial orchids with tubers that were touted as aphrodisiacs fpr nearly two millennia. Left: *Anacamptis laxiflora.* From: Schultze M, *Die Orchidaceen Deutschlands, Deutsch-Oesterreichs und der Schweiz*, t. 18 (1894). Right: *Anacamptis morio* [as *Orchis morio*, Fig. A]; and *Dactylorhiza maculata* [as *Orchis maculata*, Fig. B]. From: Thome OW, *Flora von Duetschland Osterreich und der Schweiz, Tafein*, vol. 1: t.142 (1885)

Five groups of herbal orchids that are of the greatest economic importance and with the longest recorded history of usage are individually described in separate chapters to demonstrate how studies should be conducted on the remaining 800 medicinal orchid species. These five items are *Tianma, Shihu, Baiji, Salep* and *Vanilla*.

Whereas an effort is made to provide an overview of medicinal orchids throughout the world, more comprehensive information about such usage in Meso and South America might only be available from Spanish and Portuguese sources which are not accessible to the author. Other than South Africa, tribal usage in many parts of that continent has not been properly studied and records of usage in much of Indonesia, Papua New Guinea and Australia are not as extensive as those of continental Asia. The approximately

200 medicinal orchid species with localized provincial usage in China are not discussed in this book. Interested readers may refer to my *Medicinal Orchids of Asia* (Springer 2016) where they are individually described in detail.

I made an exception with Xizang Province (Tibet) because *bcud len* is unique in its original application by hermits striving for spiritual advancement and survival while dwelling in remote caves. I am grateful to Charles Oliphant, PhD, for introducing me to this secret Tibetan practice of *bcud len* and the opportunity to read his thesis which revealed another unique aspect of medicinal orchid usage. This is discussed in Chap. 5.

Grateful thanks are due to Henry Oakeley, MD, FRCP, for valuable advice on my first few chapters and his photographs of European orchid species; Professor Ong Siew Chey, MD, Wu Dongyun, MD, Janet Loh and Sohjardto Wibowo, MD, for assistance with the translations of Chinese and Dutch texts; Joseph Arditti, PhD, Tan Wee Kiat, PhD, Tim Yam, PhD, Hew Choy Sin, PhD, and Chang Yoon Ching, PhD, and numerous research scientists whose papers we have quoted for making available resource materials; and also the library staff of the Singapore Botanic Gardens, in particular Christina Loh and Zakiah bte Agil. I also wish to thank Bhakta Bahadur Raskoti, PhD, and Professor Lokesh Shakya and Professor S.K. Ghimire for supplying me with photographs of some Nepali medicinal orchids; Professor Luo Yibo, PhD, for photos of cultivation and conservation of *Dendrobium* in China; Professor Huang Weichang for photographs of *Bletilla* species; the late Peter O'Byrne for rare southeast Asian species; Seol-Jong Kim for his photos of *chikanda*; Todd Boland for the photographs of Canadian *Cypripedium*; Nima Gyeltshen for photographs of Bhutanese *Cypripedium*s; Mak Chin On; the late Professor Rapee Sagarik, Apichart, Nantiya Vaddhanaphuti, PhD, Santi Watthana, PhD, Suyanee Vessabutr, PhD, and Peter Williams for showing me Thai native orchids; Irawati, PhD, for orchids at Bogor; Michael Ooi for the orchids at Gunong Jerai; Cheah Wah Sang, Tan Eng Khoon, Robert Ang, Tony Tan and many friends in Peninsular Malaysia; Rajendra Yonzone, PhD, for the orchids of Sikkim; Udai Pradhan and Ganesh Pradhan for orchids in Kalimpong; Ngawang Gyeltshen and Nima Gyeltshen for the orchids of Bhutan; and Teo Peng Seng, Phua Gik Song, Neo Tuan Hong, Christopher Teo, John Elliott, PhD, and other members of the Orchid Society of Southeast Asia for access to their flowering plants. Publications by researchers in Africa, the Americas and Australia are primary sources of my chapters on medicinal orchids from these regions, and I am thankful to their authors for the opportunity to study their material. In keeping with the historical perspective of this book, I made extensive use of classic botanical illustrations, and I am grateful to Plant Illustrations Organization and Missouri Botanic Gardens for access to their rich resources.

I hope this book will expand the interest of nature lovers in orchids and provide material for students of ethnobotany. It is not to be regarded as a text of complementary medicine. The great majority of reports on medicinal orchid usage are based on interviews with village elders, herbalists and traders and should therefore be regarded as hearsay or anecdotes: few researchers witnessed actual treatments and followed the results. There is a dearth of sound scientific and clinical studies or population surveys on the actual effects of orchids on humans.

In the absence of such evidence, I do not endorse the use of orchids as aphrodisiac or medicine.

Singapore   Eng Soon Teoh, MD
2019

Contents

About the Author

Eng Soon Teoh is a gynaecologist practising in Singapore and a past president of the Orchid Society of Southeast Asia. A long involvement in laboratory and clinical research and a 50-year experience in growing orchids provided him with the necessary knowledge and experience to review the topic of medicinal orchids. Dr. Teoh has published numerous papers in peer-reviewed medical journals and was a recipient of several local and international medical awards. He also wrote books on Pregnancy, Infertility and Menopause for the general public. Pregnancy went into eight reprints and Menopause received the top award for Non-Fiction in Singapore. Another best-selling book was *Orchids of Asia* which sold 25,000 copies. His latest book, *Medicinal Orchids of Asia* published by Springer in 2016 has been well received. Orchids as Aphrodisiac, Medicine or Food is a companion to the former, more technical book.

Orchids as Medicine: A Historical Overview

It was not so long ago that modern pharmacopoeia, such as the British Pharmacopoeia (BP), United States Pharmacopoeia (USP), German and Russian Pharmacopoeia and other European Pharmacopoeia, contained prescriptions which specified the use of orchids to treat various medical conditions, for instance, extract of *Cypripedium* for nerve disorders and salep (terrestrial orchid tubers) as nutrient for the infirm (Fig. 1.1).

Egyptian medical papyri are the oldest medical text still extant. They are from the early days of the Second Millennium BCE and predate the Exodus. Ancient Egyptians believed, as many cultures did, that diseases were caused by demons and rituals were required to obtain a cure. The taking or application of medicines was supplementary; nevertheless, hundreds of herbs and minerals were employed. The presence of an orchid among these drugs has not been suggested. The Bible also failed to mention an orchid.

Shen Nong Bencao Jing, the oldest Chinese herbal (or *Materia Medica*), is attributed to the Chinese father of agriculture and herbalist who promoted the cultivation of various cereals according to climate and soil and the necessity to include soya. It was alleged that he personally tasted every single herb before recommending its use. Of the herbs mentioned in *Shen Nong Bencao Jing*, there are four orchids, *Chih Jian*

(*Gastrodia elata*), *Baiji* (*Bletilla striata*) and *Shih Hu*, the last consisting of two orchids, *Dendrobium officinale* (syn. *Dendrobium catenatum*) and *Dendrobium moniliforme*. The earliest copy of this *Herbal* dates back to the first century CE (Han Dynasty), but it still influences the practice of Traditional Chinese Medicine (TCM). In his vastly expanded Chinese pharmacopoeia, *Bencao Gangmu*, compiled during the Ming Dynasty (1368–1644), Li Shizhen (1518–1593) included *Dendrobium nobile* and other *Dendrobium* species for use as *Shih Hu* (Figs. 1.2 and 1.3).

The wisdom of *Shen Nong Bencao Jing* is demonstrated in its description of the five divine crops, four cereals which are individually suited to separate geographic locations in China: rice which grows in the warm, wet south; wheat suited for a cooler climate; millet which requires very little water and barley which completes its life cycle in 3 months and is thus eminently suitable for places like Tibet; and finally, soya bean, a non-cereal. What was not known before the discovery by modern science was that of the eight essential amino acids that form the protein matrix in human and which humans cannot manufacture, rice contains only seven. But soya bean provides the missing essential amino acid. Whether similar wisdom exists in the medicinal herbs enumerated remains to be investigated. TCM claims that *Chih Jian* is neuro-protective, whereas *Baiji* stops bleeding and heals wounds. Among its several

© Springer Nature Switzerland AG 2019
E. S. Teoh, *Orchids as Aphrodisiac, Medicine or Food*, https://doi.org/10.1007/978-3-030-18255-7_1

Fig. 1.1 *Cypripedium parviflorum* (yellow moccasin orchid) was used by North American Indians to treat disorders of the nervous system. It was also employed by early European settlers in North America and eventually found its way into the official pharmacopoeia of the United States, Great Britain, several European countries and even India. (Photo: Todd Boland)

properties, *Shih Hu* restores kidney *yin*, which one could interpret as a euphemism for having aphrodisiac properties. Scientists in China are actively studying the four medicinal orchids.

Mankind has always been interested in aphrodisiacs. According to the *Theory of Signatures* propounded by ancient Greek Medicine, the appearance of a herb determines its properties. Tubers of many Mediterranean orchids resemble testicles. The word 'orchid' itself is derived from the Greek word *orchis* (testicle). Alluding to its aphrodisiacal property, another name for orchids was *Satyrion* (Latin *Satyr* and *ion*, resulting in the state of a Satyr). Greek, Roman and Arab herbalists and a historian of great repute reinforced this belief in their publications, sometimes with lurid anecdotes. Salep bars flourished in European towns and villages until the new trade routes brought in tea, coffee and chocolate from far-flung countries, and science disproved the value of salep (Figs. 1.4 and 1.5).

Meanwhile, cities along the ancient Eurasian trade routes, such as Samarkand, Constantinople, Genoa and Venice, grew rich by trading in spices, silk and other exotic luxuries and by taxing merchants during their passage. Venice was the entry point for spices to Europe: it controlled the trade from the eight to the fifteenth century.

The maritime route was faster and reduced the burden of taxes that traders had to pay at overland city stops. However, this route was initially controlled by Arabs and later by Ottoman Turks. It brought great wealth to the Abbasid Caliphate. The need to pay heavy duty to a Muslim nation was resented by Catholic Europe, but only Spain and Portugal made attempts to bypass the established trade route.

When Christopher Columbus proposed sailing west across the uncharted Atlantic Ocean, nobody in Portugal, Venice or Genoa agreed to support him. Spain's Isabella did. Spain was rewarded by the Pope with all the land in the west, but in a later treaty, Brazil was given to Portugal. This appears

Fig. 1.3 Shen Nong. The legendary herbalist tasted so many poisonous herbs daily that bumps appeared on his head

Fig. 1.2 Shen Nong as the Father of Agriculture. Chinese woodcut

to be a family matter because King Manuel I of Portugal was the son-in-law of Ferdinand and Isabella of Spain.

The Aztec, Mayan and Inca civilizations in the New World were annihilated by Spanish conquistadors who apparently became psychotic when they saw the abundance of gold: that and the introduction of old-world diseases to which new world natives were not immune. Moctezuma offered Cortez chocolate flavoured with vanilla, but he was told that the Spaniards 'suffered from a disease of the heart that could only be eliminated by gold'. Brought to Europe, vanilla was promoted as an aphrodisiac. Elizabeth I loved it. Her physician told her it could be added to any food. Today vanilla is ubiquitous in confectionery and other foods (Fig. 1.6).

Francisco Hernandez de Toledo, naturalist and physician to Philipp II of Spain, led a scientific expedition to investigate the medicinal plants of the New World from 1570 to 1577. He returned with vanilla, pineapple, cocoa, maize, passionfruit, hallucinogenic plants and seeds. A 1628 Latin redaction of his writings entitled *Rerum medicarum Novae Hispaniae thesaurus* published in Rome is regarded as the first botanical work from the Americas. It was originally published in Mexico in 1615. But, in fact, two native Mexican Catholic monks, Martinus de la Cruz and Johannus Badanius, had brought out an illustrated *Herbal* in Latin very much earlier in 1558. This Codex de la Cruz-Badanius was given to King Charles V (reigned as Holy Roman Emperor 1519–1556) and later to Cardinal Francesco Barberini when he visited the king. Bias against native efforts at that time probably caused the book to remain unknown and hidden in the Vatican Library until it was discovered by

Fig. 1.4 Page from Dioscorides, *Materia Medica* illustrating *Anacamptis morio*, one of many aphrodisiac orchids

Charles Upson Clark, a history professor from Columbia University in 1929. Similar works recording medicinal knowledge of Mayan and Inca civilizations do not exist. Most of the Mayan Codices were consecrated to fire by a bigoted Franciscan bishop in 1562, and fragments that remain do not contain any botanical information of value. In remote Central and South American villages today, people still believe that disease is caused by spirits and herbal cures usually involve magic and shamans. Around 65 to 70 species of orchids are recorded to have medicinal usage in Meso and South America.

North American Indians also employed orchids for healing. Various *Cypripedium* species were used to treat anxiety, hysteria, fits, spasms

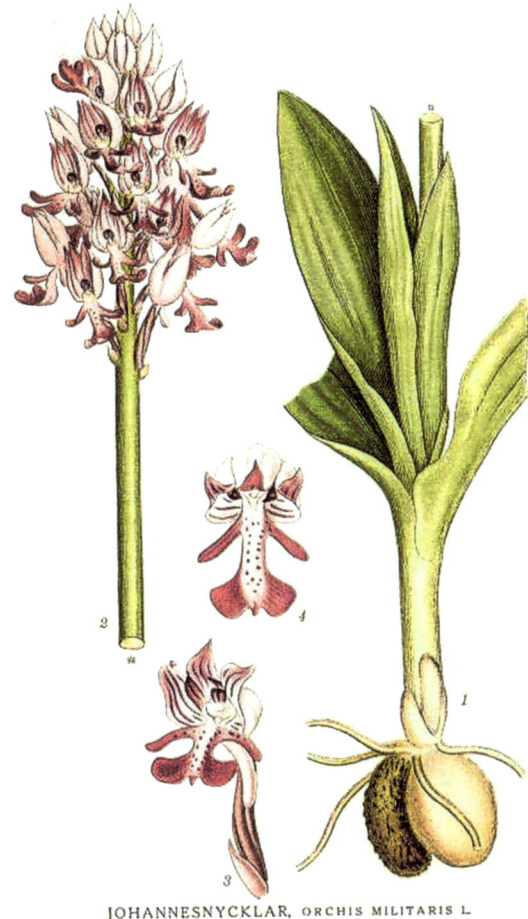

JOHANNESNYCKLAR, ORCHIS MILITARIS L.

Fig. 1.5 *Orchis militaris*. From: Lindman GAM, *Bilder ur Nordens Flora,* vol. 2, t. 402 (1922–1928)

and other disorders affecting the nerves. Early European settlers relied on such Indian remedies because they seldom had enough medicinal supplies from Europe. Orchid remedies were included in the official *Pharmacopoeia of the United States* until the twentieth century. *Cypripedium parviflorum* and *Cypripedium pubescens* were also included as a nerve medicine in British and several European pharmacopoeia. Depending on where they lived, North American Indians would be familiar with different orchid species, and accordingly, different plants were employed by widely separated tribes. Altogether, about two dozen North American orchids are employed in native tribal medicine (Fig. 1.7).

Portugal was also a sea-faring nation. King Manuel was the son-in-law of Spain's Ferdinand and Isabella. Not wanting to be outdone by his in-laws, he sponsored an effort to discover a new route to India by going round Africa. Vasco de Gama managed to round the Cape of Good Hope and reach Calicut in 1498. At Calicut, de Gama treated the natives with appalling cruelty, in one instance, setting fire to a pilgrim boat and watching women and children set ablaze or drown. In this manner, he managed take over control of the spice trade in this trading centre.

In 1581 the provinces of the Netherlands declared independence from Spain. The Dutch were a hardworking, innovative, entrepreneurial and sea-faring people. Following the formation of the Dutch East India Company (VOC) in 1602, the militant Dutch navy rapidly expelled the Portuguese from the Malay Archipelago and the Malabar Coast, leaving Portugal with small enclaves like Malacca and Goa. During her occupation of the East Indies and Malabar, VOC produced a doctor, a unique biologist with no formal training in science and a military administrator, all turned naturalists who made immortal contributions to botany and medicine.

A few years after being dispatched as midshipman and ensign to Amboin in 1654, Rumphius (Georg Eberhard Rumpf, 1627–1702) was given dispensation by the Governor-General in Batavia that enabled him to study the flora and fauna of the region. Despite going blind because of glaucoma, losing his wife and daughter in an earthquake, having part of his library destroyed by fire and losing the first version of his manuscript when the ship carrying it was sunk by the French navy, Rumphius managed to produce a second version of the *Herbarium Amboinense* which contained descriptions of 1200 species accompanied by line drawings of 350 plants. Rumphius described 35 orchids. (When Karel Heyne sent collectors to search for useful Indonesian plants during the 1900s, he only managed to add seven orchid species to the list.) Rumphius had help from many people, including his wife Susanna and his son Paul August who was an artist. For economic reasons, the VOC did not immediately set about to publish the work.

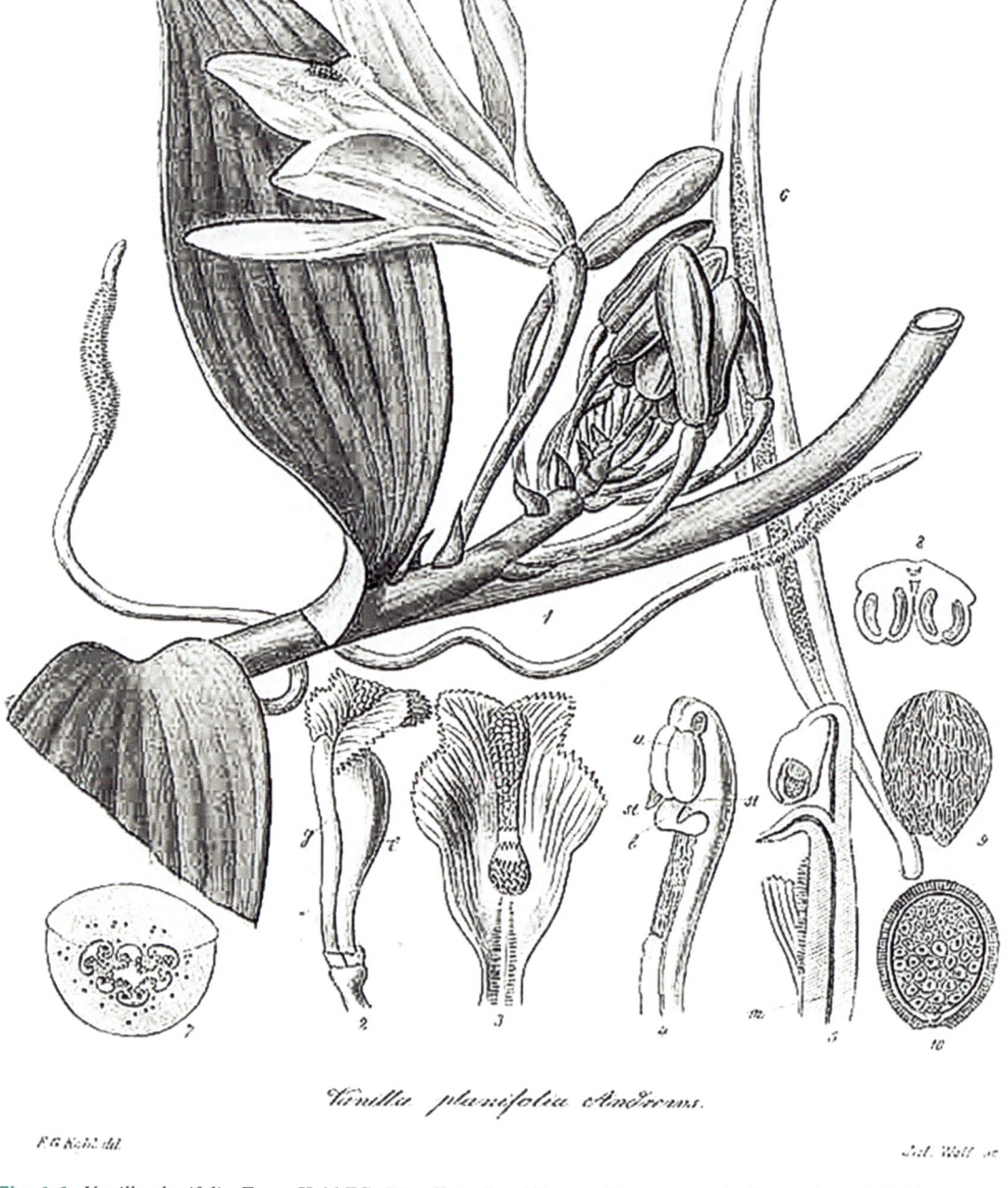

Vanilla planifolia Andrews.

Fig. 1.6 *Vanilla planifolia.* From: Kohl FG, *Die officinellen Pflanzen, Pharmacopoeia Germanica*, t. 25 (1891–1895) [artist: Kohl FG]. Courtesy of Universitats und Landesbibliothek, Dusseldorf, Germany

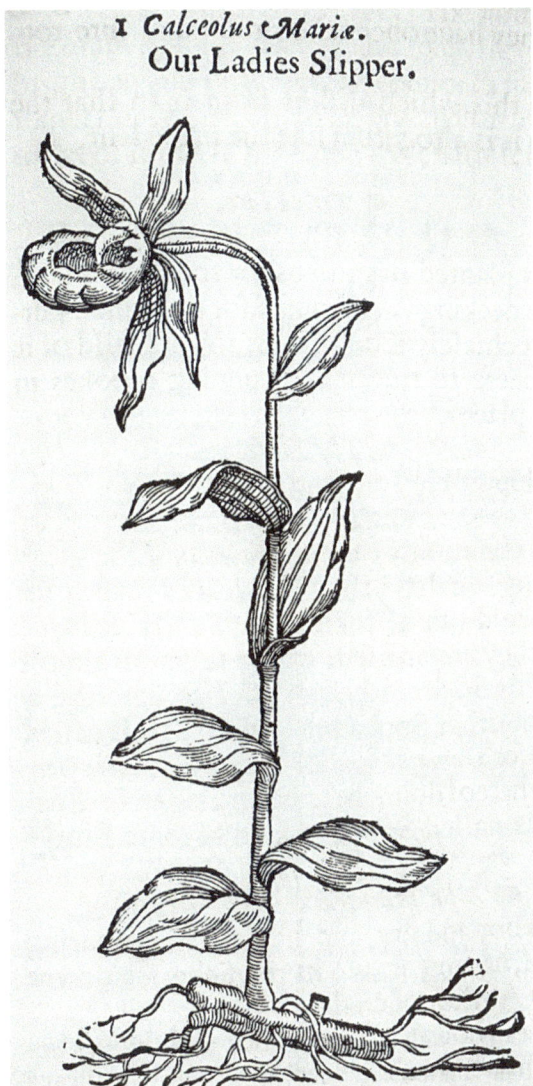

Fig. 1.7 *Cypripedium calceolus* (as *Calceolus Maria*. Our Ladies Slipper). From: Gerard J, *Herball*, (1597). Whereas many authors mentioned that American Indians employed *Cypripedium calceolus*, the orchid they used was actually *Cypripedium parvifolium* because *Cypripedium calceolus* is a Eurasian and not an American species

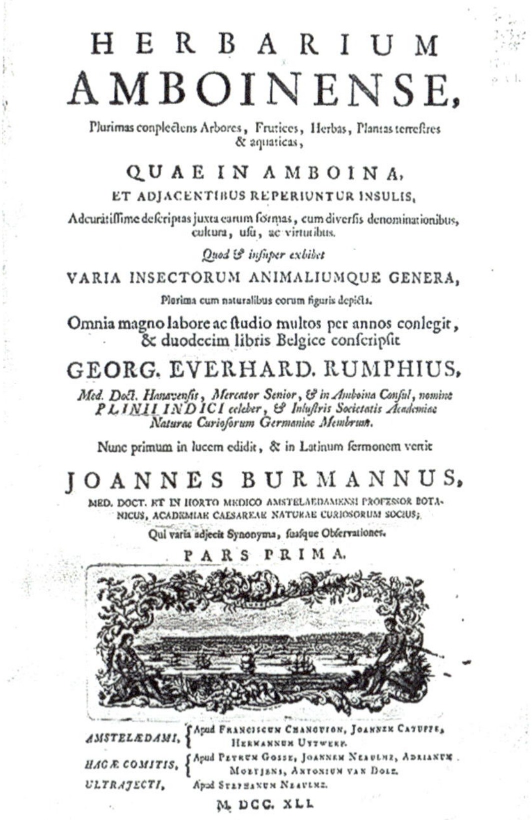

HERBARIUM AMBOINENSE,

Plurimas conplectens Arbores, Frutices, Herbas, Plantas terrestres & aquaticas,

QUAE IN AMBOINA, ET ADJACENTIBUS REPERIUNTUR INSULIS,

Adcuratissime descriptas juxta earum formas, cum diversis denominationibus, cultura, usu, ac virtutibus.

Quod & insuper exhibet

VARIA INSECTORUM ANIMALIUMQUE GENERA,

Plurima cum naturalibus eorum figuris depicta.

Omnia magno labore ac studio multos per annos conlegit, & duodecim libris Belgice conscripsit

GEORG. EVERHARD. RUMPHIUS,

Med. Doct. Hanavensis, Mercator Senior, & in Amboina Consul, nomine PLINII INDICI celeber, & Inlustris Societatis Academiae Naturae Curiosorum Germaniae Membrum.

Nunc primum in lucem edidit, & in Latinum sermonem vertit

JOANNES BURMANNUS,

MED. DOCT. ET IN HORTO MEDICO AMSTELAEDAMENSI PROFESSOR BOTANICUS, ACADEMIAE CAESAREAE NATURAE CURIOSORUM SOCIUS;

Qui varia adjecit Synonyma, suasque Observationes.

PARS PRIMA.

AMSTELÆDAMI, Apud FRANCISCUM CHANGUION, JOANNEM CATUFFE, HERMANNUM UYTWERF.
HAGÆ COMITIS, Apud PETRUM GOSSE, JOANNEM NEAULME, ADRIANUM MOETJENS, ANTONIUM VAN DOLE.
ULTRAJECTI, Apud STEPHANUM NEAULME.

M. DCC. XLI.

Fig. 1.8 *Herbarium Amboinense* vol. 5 by Georg Eberhard Rumphius (1741), title page

Herbarium Amboinense was finally published posthumously in 1741, in six folio volumes. It is a classic in botanical literature (Fig. 1.8).

Shortly after Hendrik van Rheede (1714–1773) arrived in southern India in 1669 to take over as Administrator of Dutch Malabar for the next 8 years, he assembled a team of nearly 100 scholars, botanists, physicians, native healers, professors of medicine, clergymen, translators, illustrators and engravers, both Indian and foreign, to work on the flora of the Malabar coast. The first draft of *Hortus Indicus Malabaricus* was completed in 1675, but it took 30 years to complete the publication. Originally rendered in Latin, it was subsequently translated into Sanskrit, Arabic, Malayalam and English. The 12 volumes describing 742 plants were accompanied by 794 beautiful copper-plate illustrations. Another classic of botanical literature, *Hortus Indicus Malabaricus*, was praised by Carl Linnaeus for its accuracy (Figs. 1.9 and 1.10).

van Rheede was probably inspired by the work of Jacobus Bontius (1592–1631), a Dutch physician of Jewish descent who had earlier studied the flora of Batavia. Bontius recruited the help of

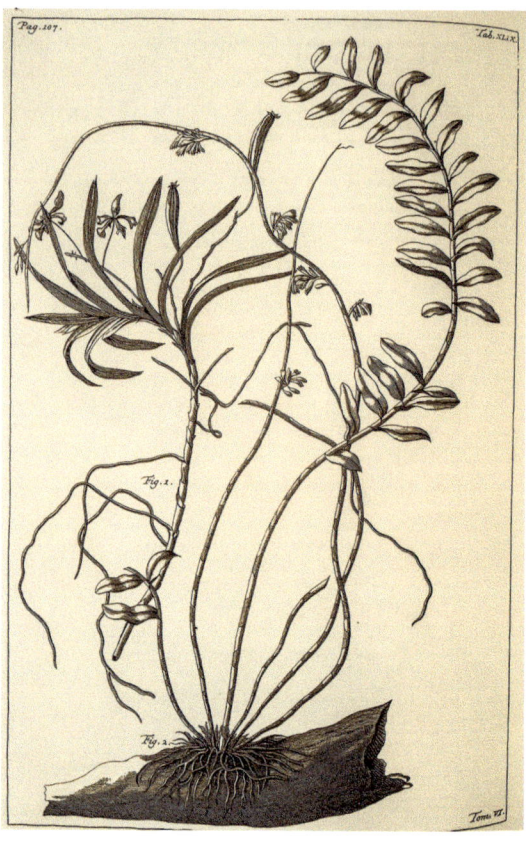

Fig. 1.9 Illustration of two orchids (*Dendrobium* and *Vanda*) in Rumphius GE, *Herbarium Amboinense* (1741)

Fig. 1.10 *Hortus Indicus Malabaricus* by van Rheede HV, tot Drakenstein (1678), title page

friendly natives to obtain his source materials because it was unsafe for him to wander more than 3 km beyond Batavia. He included descriptions of food plants, spices and other natural products with commercial possibilities in his four volume *De medicina Indorum*, a seminal text in tropical medicine that contained the first descriptions of beriberi, cholera, dysentery and the *Orang hutan*.

However, it was for different reasons that van Rheede also recruited the help of friendly natives. Being neither naturalist nor physician, and certainly not an artist himself, he resorted to his position of authority to recruit the best talents for his work. By so doing, van Rheede produced a work that matched the contributions of the observant Bontius and the indomitable Rumphius in the field of natural studies.

Although the British were late in extending their maritime influence to Asia, by 1765, the British East India Company (founded in 1600) managed to control the vast subcontinent of India, then wrested Batavia from the Dutch, established the Straits Settlements by 1819 and exerted political dominance over the sultanates of Malaya and British Borneo soon after. Meanwhile in London, Joseph Banks was elected president of the Royal Society in 1778, a position he held for 42 years during which time he promoted British interest in economic and beautiful plants. Shortly after graduating from Oxford, Banks accompanied James Cook on the first voyage of the *Endeavour*. Impressed by the varied flora of the lands he visited, Banks dispatched botanists to collect plants from many parts of the world and to have them established at Kew Botanic Gardens, thereby laying the foundation for Kew to be the leading Botanical Garden in the world. Half a century later, another botanist and explorer, Joseph Dalton Hooker (1817–1911), became president of the Royal Society. Hooker visited Indian Himalaya in 1847–1851, and subsequently, he published a seven-volume *Flora of India*. He is well known as a taxonomist. He was a close friend of Charles Darwin and was Director of Kew Botanic Gardens for 20 years. His name is attached to numerous orchids: best known in Singapore as *Papilionanthe hookeriana*, the parent of many hybrids that initiated the tropical orchid cut-flower industry. Some people regard William Roxburgh (1751–1815) as the father of Indian botany. Joining the Indian Medical Service as a surgeon in 1776, Roxburgh displayed such an interest and knowledge of plants that he was invited to take charge of the Calcutta Botanic Gardens in 1793. Roxburgh's descriptions of Indian orchids were usually accompanied by drawings prepared by Indian artists. His two-volume *Flora Indica: or Descriptions of Indian Plants* was published post-humously in Serampore in 1820 and 1824.

During the nineteenth century 'consumption of the exotic spread through the gardens and libraries of the wealthy and the well-to-do, amidst a growing vogue in natural history. . . . the Shows of London brought the exotic and fantastic to a still wider audience' (Millar 2011). Growing affluence and interest in beautiful orchids and exotica supported two large British nurseries, Veitch and Sanders, veritable institutions that were responsible for making the orchid-growing hobby what it is today. Veitch financed many plant collectors, William and Thomas Lobb being the two most famous. By 1914, Veitch Nurseries had introduced 1281 new plants to Europe. The first man-made orchid hybrid, *Calanthe* Dominii raised by John Dominy at Veitch nursery in Exeter flowered in 1856: presently, there are well over a hundred thousand registered hybrids. Veitch's competitor was Frederick Sander of St Albans who employed 23 plant collectors working in South America and Asia. During the 1880s and 1890s, Sanders

handled 2 million orchid plants. One might be appalled by the callousness of some collectors and by the disregard of their patron, but that was how the orchid industry originated (Fig. 1.11).

The classical studies of Bontius, Rumphius, van Rheede and Hooker were tough acts to follow. Later works from the region never matched the intensity of the early efforts, but nevertheless, they contained new and useful information. In 1933, Jesuit chemist J F Caius described 20 species of medicinal orchids in 14 genera employed in Malabar. He conducted tests on 300 herbs employed to treat snakebites. Two orchid species and all the remaining 298 herbs failed the tests. Caius reported that wealthy Orientals were

Fig. 1.11 *Cattleya labiata*, the orchid that launched the orchid mania of the Victorian era. From: Houtte L van, *Flore des serres et des jardins de l'Europe*, vol. 7: t. 660 (1852)

known to have paid handsome prices for pounded potatoes and gum because salep enjoyed an immense reputation as aphrodisiac, restorative and fattener. Substitution was as prevalent in the herb markets in 1938 as it is today.

After Indian independence, numerous studies were conducted on tribal medicine practiced by remote, isolated hill tribes. Generally the tribes employed few orchids for treatment. Many studies did not report orchids being employed medicinally. Nevertheless, India is a big country, and the total number of orchid species that have been employed medicinally total 112.

Following the establishment of the Straits Settlements between 1786 and 1819, many British botanists worked in the Far East. Charles Curtis (1853–1928) was one of the collectors sent to the Far East by Veitch to collect orchids and other interesting plants. Later appointed as superintendent of Penang Botanic Garden, Curtis managed to convert a disused granite quarry into the beautifully landscaped garden. He was a contemporary of Henry Ridley, and both men played a role in establishing the Malayan rubber industry (Reinikka 1972), but Ridley is better known for his contribution. Appointed director of the Gardens and Forests of the Straits Settlements in 1888, Ridley made a detailed study of the plants in the region and sent thousands of herbarium specimens to Kew. He wrote a five-volume *Flora of the Malay Peninsula* and commented on the medicinal usage of several orchid plants. In 1930, IH Burkhill and Mohamed Haniff described their observations on 17 species of medicinal orchids in their publication, *Malay Village Medicine*. There was much similarity with traditional Indian or Indonesian medicinal practices. This was not the case with Thai herbal medicine which employed 42 orchid species.

Tubers of many Australian orchids are eaten as food, but Australian aborigines employ only four species to treat skin disease or dysentery or to be used as contraceptive. *Chikanda*, a cake made with orchid tubers, is a popular delicacy in Central Africa, and there is a belief in Malawi that eating orchids protects one against illnesses. Transnational trade of orchids in Central Africa is worrying conservationists. The Royal Botanic

Gardens at Kew has launched a project sponsored by the Darwin Initiative to promote sustainable harvest in the region. In Central and South Africa, 46 orchid species are commonly used as protective or love charms; 65 medicinal species have been identified.

It was not scientific curiosity which led to the identification of hundreds of medicinal orchid species in China. Between the founding of the People's Republic of China in 1949 and Richard Nixon's visit in 1972, the 'Bamboo Curtain' isolated China from the rest of the world. The country was devastated by decades of war, foreign and warlord pillage and exploitation, and even if she had access, China was too poor to afford modern drugs. The country had to rely on its home-grown herbal remedies. Knowledge of all provincial remedies was soon collected and compiled in a new *Materia Medica* which vastly expanded the knowledge on the use of hundreds of medicinal orchid species.

The outward-looking policy responsible for China's spectacular economic advancement in the last 30 years also provided opportunities for great innovations in the technological and scientific arena. Medicinal plants are now examined at the molecular level. DNA studies permit accurate identification of species. Many phytochemicals have been isolated and their properties elucidated. New findings are announced almost on a weekly basis. Several difficulties have to be overcome before proper clinical studies can be performed on herbal remedies, but it is to be hoped that these can be resolved. Clinical studies on pure compounds should be more meaningful.

Some medicinal orchids are now cultivated on a massive scale in China, but still this does not prevent stripping of medicinal species from the wild. CITIES has drawn up rules to prevent collection of endangered plants from the wild and cross-border trade in wild orchids, but these need universal enforcement which is not forthcoming. In Indian Himalaya and Central Africa, underprivileged families living in rural communities depend on herb collection for their livelihood. The practice is not sustainable in the long run unless these people are taught to care for young plants, not over-collect, and vast expanse of land be constantly seeded with orchids. There is an on-going experiment to cultivate *Dendrobium officinale* on rocks in China because rock-grown *shihu* being preferred over the nursery grown herb can fetch a fourfold higher price.

Cosmetics are employed by women, and nowadays sometimes also by men, to improve their attractiveness and to reduce the ravages of time on skin. Since this boosts one's sense of well-being, it can be construed as a health benefit, so I take the liberty, only here, of briefly mentioning the use of orchids in cosmetics.

Over 20 cosmetic products in the market include orchid extracts among their constituents. The orchids employed are *Bletilla striata*, *Brassocattleya* Marcella, *Calanthe discolor*, *Cattleya*, *Cymbidium goeringii*, *Cymbidium* Great Flower Marie, *Cymbidium kanran*, *Cycnoches cooperi*, *Cypripedium pubescens*, *Dendrobium bigibbum* (syn. *Dendrobium phalaenopsis*), *Dendrobium chrysotoxum*, *Dendrobium moniliforme*, *Dendrobium nobile*, *Gastrodia elata*, *Orchis maculata*, *Orchis mascula*, *Orchis morio*, *Paphiopedilum* Maudiae, *Phalaenopsis amabilis*, *Phalaenopsis javanica*, *Phalaenopsis lobbii*, *Vanda coerulea*, *Vanda falcata* (syn. *Neofinetia falcata*) and *Vanda tessellata*. The orchid extract contributes to skin conditioning (cleanser, face mask, moisturizer, emollient), UV protection, whitening of skin and hair care. One product claims to promote hair growth. During the Tokugawa Period (1603–1867), Japanese nobility hung flowering *Vanda falcata* in their palanquins to enjoy the penetrating fragrance during their travels (Teoh 1982, 2011, 2016; Singh et al. 2016) (Figs. 1.12 and 1.13).

Orchids, indeed plants in general, were valued as food or medicine long before they became horticultural darlings. The fact that Confucius (551–479 BCE) could refer to a room pervaded with the fragrance of *Cymbidium* suggests that the orchid was cultivated in the sixth-century BCE.

Fig. 1.12 *Vanda coerulea*, a pink form (©Teoh Eng Soon 2019. All Rights Reserved)

Fig. 1.13 *Vanda falcata*
(syn. *Neofinetia falcata*)
(©Teoh Eng Soon 2019.
All Rights Reserved)

Orchid cultivation was popularized during the
Tang Dynasty (618–907). *Cymbidium ensifolium*
and *Dendrobium moniliforme* were described by
Ji Han in *Nan Fang Cao Mu Zhuang* (南方草木
状), a Chinese botanical work in three volumes
published during the Jin Dynasty (290–307) that
described the morphology of more than 80 plants
originating in southern China. The former species
was ornamental, the latter ornamental and medicinal. Chinese tradition maintains that orchids
were already included among the medicinal
herbs since the dawn of Chinese history. That
being said, a discussion of medicinal orchids
brings us right back to the beginning of human
interest in orchids.

References

Millar DP (2011) Introduction. In: Millar DP, Reid PH
    (eds) Visions of empire. Voyages, botany, and repre-
    sentations of nature. Cambridge University Press,
    Cambridge
Reinikka MA (1972) A history of the orchid. University of
    Miami Press, Miami, FL
Singh DR, Kishore R, Kumar R, Singh A (2016) Orchid
    preparations. Technical Bulletin No. -00. ICAR-
    National Research Centre for Orchids, Pakyong,
    Sikkim-737106
Teoh ES (1982, reprinted 2008) A joy forever. Vanda Miss
    Joaquim, Singapore's national flower. Times Media
    Press Pte Ltd/Marshall Cavendish, Singapore
Teoh ES (2011) Medicinal Asian orchids. Proceedings of
    the 20th World Orchid Conference, Singapore
Teoh ES (2016) Medicinal orchids of Asia. Springer
    (Nature), Cham

An Ancient Fantasy: *Salep* as Aphrodisiac

2

Once a fashionable drink touted to boost one's libido and performance, salep only fell into disrepute when physicians demanded proof before advocating any medicine. Salep, *saloop* or *salepi* is still being sold in Turkey and Greece by street vendors who push mobile carts holding pots of the liquid in busy city squares; but they are not as ubiquitous as they were a century or two ago. Business is not brisk because only old folk drink *salepi*, youngsters preferring to sip coffee in roadside cafes instead. Another reason is that the orchid species which provide the tubers for the drink have become so threatened with extinction that Turkey, their principal source, has banned their harvesting and export. Traders have now turned their attention to northern Iran and the orchids in that area are currently under threat. Over-collection was the reason the orchids became scarce, the belief that they were aphrodisiacs the culprit.

Although salep drinking boomed during the Ottoman Empire, its reputation dates from a much earlier period. Theophrastus (371–287 BC), Dioscorides (40–90 CE), Pliny the Elder (23–79 CE), Avicenna (980–1037), Maimonides (1135–1204) and botanists of the Renaissance period, these were the learned people who endowed salep with its undeserved reputation through the ages. Ancient Greek medicine subscribed to the *Doctrine of Signatures* which proclaimed that the usage of a medicinal plant may be gleaned by its resemblance to a body part. Since the paired tubers of many Mediterranean orchids resemble testes, their functions were linked to sex and procreation (Fig. 2.1). *Salab* (*sahlab*) is an Arabic word which means *fox's testicle*, whereas *orchis* in Greek means simply *testicle*. Middle Eastern herbals refer to the drug prepared from orchid tubers as *Khus yatu's salab* (fox's testicle) or *Khus yaty'l klab* (dog's testicle). It was stated that the odour of fresh tubers was similar to the smell of human semen and tubers could even induce 'an aphrodisiac effect if clasped in the hand' (Dymock et al. 1893). The association between orchid and sex was reaffirmed again and again in such old common names for the orchid species that constituted salep, for instance, *Satyrion,* goat's testicles, dog's testicles, hare's testicles. In Greek legend, these orchid plants arose from spilled semen of cavorting satyrs. The famous Flemish physician, Mathias de l'Obel (1538–1616) used the term *Testiculus vulpinus* (fox testicle) to describe an orchid species. During the fifteenth century, Jerome Bock (Hieronymus Tragus, 1489–1554), also alluding to the *Doctrine of Signatures,* concluded that since the flowers of some European terrestrial orchids resembled bees and other insects, they were begot by winged arthropods (Emboden 1974) (Figs. 2.2 and 2.3).

European fascination with *Orchis* and *salep* vastly exaggerated the potency of this group of orchids. It all started with the description by Theophrastus (271–287 BC) who stated in the

© Springer Nature Switzerland AG 2019
E. S. Teoh, *Orchids as Aphrodisiac, Medicine or Food*, https://doi.org/10.1007/978-3-030-18255-7_2

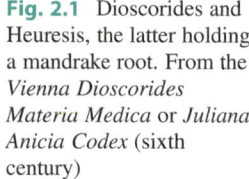

Fig. 2.1 Dioscorides and Heuresis, the latter holding a mandrake root. From the *Vienna Dioscorides Materia Medica* or *Juliana Anicia Codex* (sixth century)

seminal botanical work, *Historia Plantarum* or *Enquiry into Plants* that

> This is the so-called salep (Mediterranean terrestrial orchids) which has a double bulb, one large and one small. The larger, given in the milk of a mountain goat, produces more vigour in sexual intercourse: the smaller inhibits and forestalls. . . . . . . It is odd, certainly that both powers should be found in the same plant: but that a plant should have one or other power need not surprise us. We may remember Aristophilus, the druggist from Plataea, used to say that he had drugs with exactly these effects, one to improve sexual powers, one to inhibit: and that the impotence produced by the latter is general and lasts for a limited time, say 2–3 months, so that it can be used on slaves who require to be restrained and corrected (Dalby 2013).

Gaius Plinius Secundus (23–79 CE), or Pliny the Elder, described both aphrodisiacs and antiaphrodisiacs in *Naturalis Historiae* (*c. 77 C. E.*): to *Orchis* and *Serapias* (now classified under *Epipactis*), he attributed the former property; to the larger 'or some say, the harder bulb of *Orchis* when drunk in water'. The lesser or softer bulb taken in goat's milk repressed the sexual appetite. Furthermore, 'the root of the former *orchis* given to drink in the milk of an ewe bred up at home of a cade lamb, causeth a man's member to rise and stand; but the same taken in water, maketh it go down again and lie'. Thus it would seem that both the choice of orchid bulb and solvent had to be correct! Such orchids were reported to be equally effective when fed to goats, rams and stallions. Pliny offered mead or the juice of lettuce as an antidote when one became excessively lusty after consuming the orchids (Figs. 2.4, 2.5 and 2.6).

Fig. 2.2 Hieronymus Tragus Bock at 46. From: *New Kreutterbuch* (1546). Artist: David Kandel

Fig. 2.4 Pedanius Dioscorides. From: Jean Antoine Sarrasin, *Dioscorides* (1598), title page

Salep orchids had other medicinal uses. The Roman historian, Pliny the Elder, reported that the roots of *Orchis* healed mouth sores and it was used to clear phlegm from the chest (Turner 1962). When 'bruised and applied to the place', they healed the king's evil (scrofula, swollen lymph nodes in the neck caused by spread of tuberculosis) (Grieve 1971).

Pedanius Dioscorides (40–90 CE) alleged that the consumption of *Satyrion* not only stirred the fleshy lust but, additionally, 'if men ate the fat tubers they would beget male children, whereas if women ate the lesser, dry or barren root which was withered and shriveled, they would bring forth girls' (quoted by Leonhard Fuchs in 1542).

The *Materia Medica of Dioscorides* became the authoritative herbal text almost immediately after its composition around 50–70 C.E. For the next 1500 years, it enjoyed wide circulation in Greek, Latin and Arabic, supplemented by commentaries by Persian and Arab physicians. In his *al-Qanun fi al-Tibb* (Canon of Healing),

Fig. 2.3 Theophrastus *Historia Plantarum*. Title page of a 1644 edition

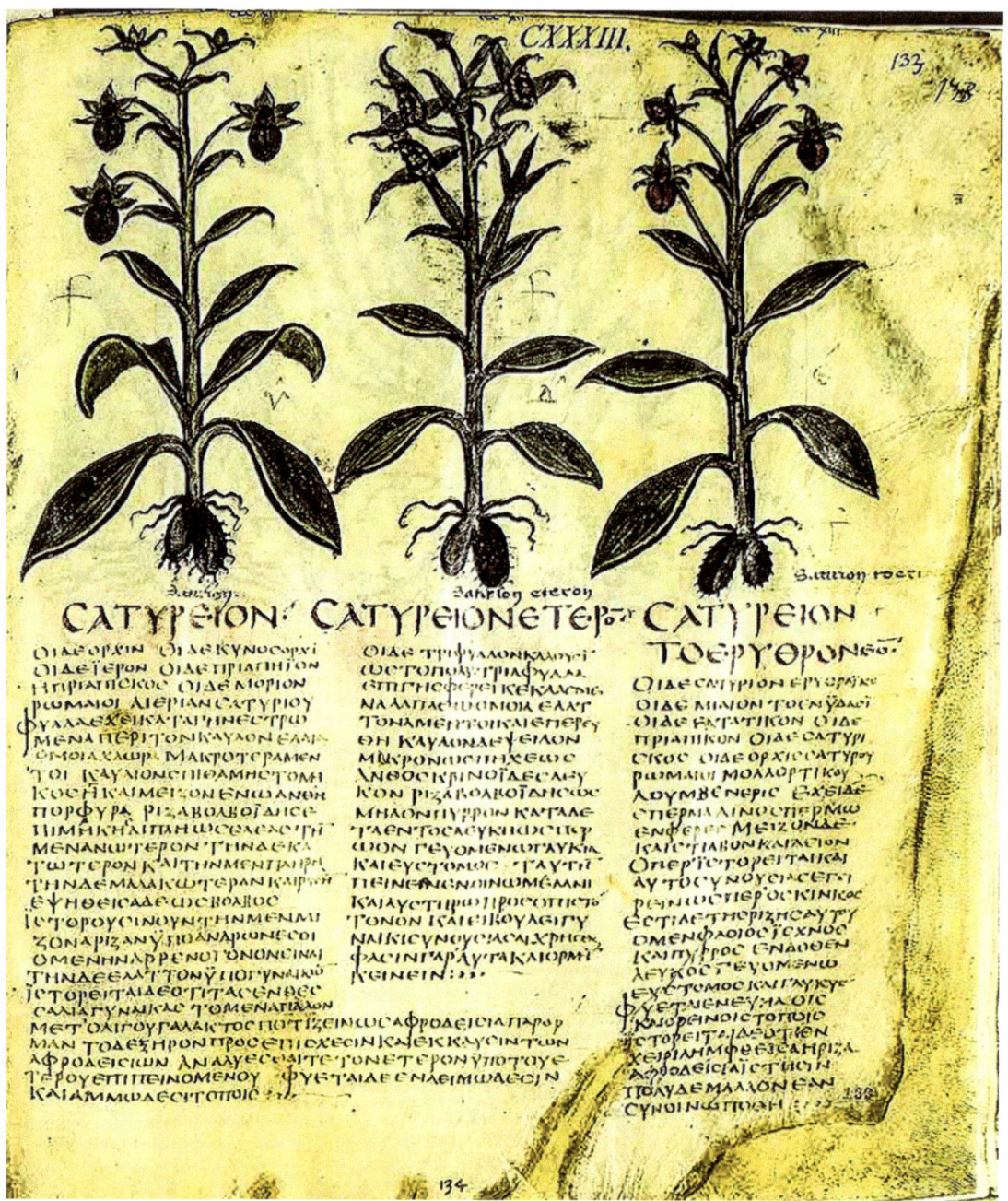

Fig. 2.5 *Ophrys* species. From Dioscorides *Erbario Greco* (*Materia Medica*), t. 133, Fig. 2 (487–580)

Avicenna (Ibn Sina) (980–1037) stated that orchid tubers were employed as aphrodisiac or appetite stimulant, but in addition it encouraged mucus production and promoted recovery from stroke. He mentioned an orchid species *Alisma* sive (or) *Damasonia* which served to relive cough and asthma. Early European pharmacopoeia were largely based on Dioscorides and secondarily on Avicenna. *Historia Plantarum* or *Enquiry into Plants* by Theophrastus was not available in

Fig. 2.6 Gaius Plinius Secundus (Pliny the Elder) (23–79 C.E.)

Latin until 1483 when it was translated into that language from Greek by Theodorus Gaza (c. 1398–1475) (Figs. 2.7 and 2.8).

Botanical Renaissance

The attitude of inquiry promoted by the Renaissance (fourteenth to seventeenth century) saw the emergence of Natural Sciences which initially focused on the restoration of the knowledge of the ancients, much of which focused on plants and herbal remedies. European botanists or naturalists during the fifteenth to seventeenth century were usually physicians, later, joined by several apothecaries, whose curiosity of herbal remedies led them to study the plants themselves, eventually leading to botany for its own sake. The development of printing from the mid-fifteenth century witnessed the authorship of several

Fig. 2.7 Dioscorides as featured in an Arabic Materia Medica

Fig. 2.8 Avicenna, portrait on a silver vase. Museum at BuAli Mausoleum, Hamadan, Iran

classical botanical compendiums that sought to promote wider knowledge of medicinal herbs and especially the *Materia Medica of Dioscorides*, with some enlarging their *Herbals* by including the knowledge of local plants and those introduced from distant lands (Figs. 2.9, 2.10 and 2.11).

Jean Ruel (1474–1537), the French physician who taught at the University of Paris, showed special interest in Dioscorides. He published a Latin translation of Dioscorides' *Materia Medica* in 1516. His own work, *De Natura Stirpium* published in 1536 was devoid of illustrations. He was the first European after Theophrastus to compose a general treatise on botany. His translation was used by Pietro Matthiolus (1501–1577) to produce numerous editions of the translation heavily illustrated with accurate woodcuts, with associated commentaries that made Matthiolus the foremost publisher of Dioscorides' work.

Afterwards Germany took the lead in fostering this interest with illustrious physicians like Otto Brunfels (1488–1534), Hieronymus Bock

(1498–1554) and Leonhart Fuchs (1501–1566) (Table 2.1) (Figs. 2.12, 2.13, 2.14 and 2.15).

In the mid-fifteenth century, the German goldsmith, Johannes Gutenberg invented the casting of individual letters that enabled these letters to be fitted into words and to be reused repeatedly. This simplified printing in Europe and provided the opportunity for physicians to publish. The work of the four illustrious German physicians, published in Latin, consisted of descriptions of hundreds of plants and their usage, accompanied by woodblock illustrations that were occasionally hand-coloured. Leonhart Fuchs's *De historia stirpium commentarii insignes maximis impensis et vigiliis elaborati adiectis earundem vivis plusquam quingentis imaginibus, nunquam antea ad naturae imitationem artificiosius effictis & expressis* (Notable commentaries on the history of plants prepared with great expense and care adorned with more than 500 lifelike pictures in imitation of nature, never hitherto drawn and printed with greater care) was illustrated with 512 woodcuts that accompanied the description of various plants.

Otto Brunfels' *Herbarum Vivae Eicones* (Living Portraits of Plants) published between 1530 and 1536 featured more than 120 outstanding woodcuts by Hans Weiditz (1495–1537) that established a standard for botanical illustration which required plants to be represented realistically as they appear seasonally in nature. The *Kreutterbuch* (plant book) by Hieronymus Bock (1539) was illustrated in a later edition of 1546 by David Kandel (1520–1592). The *Materia Medica of Dioscorides* being the standard reference at this time, it was natural that several German botanical publications attributed aphrodisiac properties to many terrestrial orchids.

In *De Historia Stirpium*, Fuchs described and illustrated three species of orchids that had the property of *ad venereum excitare* (to excite lust): (1) *Satyrium basilicum mas* (the 'male' *Satyrium basilicum*, *Gymnadenia conopsea*, known to Avicenna as *digiti citrine*, a reference to the finger-shaped roots that resembled the fruit of *Citrus medica* var. *sarcodactylis*); in German as *Kreutzbluomen*; and in Latin as *Satyrium regium* and *Palma Christi* (the Hand of Christ);

Fig. 2.9 Ruel J, Dioscorides Materia Medica. Title page of a 1516 translation

PEDANII DIOSCO=
RIDIS ANAZARBEI DE MEDICI=
NALI MATERIA LIBRI SEX, IOANNE RVELLIO SVESSIO=
nenfi interprete. Singulis cùm ftirpium, tum animantium hiftorijs,
ad naturæ æmulationem expreffis imaginibus, feu uiuis picturis,
ultra millenarium numerum adiectis: non fine multiplici peregrinatione, fum-
ptu maximo, ftudio atque diligentia fingulari, ex diuerfis regionibus con-
quifitis. Additis etiam Annotationibus fiue Scholijs bre-
uiffimis quidem, quæ tamen de medicinali materia omnem
controuerfiam facile tollant. Per GVALTHE
RVM H. RYFF, *Argentinum,*
Medicum, & Chirurgum,
Omnia ex doctiffimorum uirorum lucubrationibus iamprimùm
concinnata, & in lucem ædita.

Cum Indice quintuplici, copiofiffimo: quorum primus omnium ferè
fimplicium, quibus paffim utuntur Medici, nomenclaturas Græcas:
alter Latinas: tertius Officinis, Herbarijs, & Arabum fami-
liæ uulgares: quartus Germanicas: quintus Galli-
cas, miro ordine complectitur.

ACCESSERE IN EVNDEM AVTOREM
Scholia noua, cum nomenclaturis Græcis, Latinis, Hebraicis & Ger-
manicis, IOANNE LONICERO *autore.*

anno *1543*

Cum Gratia & Priuilegio Imperiali.

FRANC. *Apud Chr. Egenolphum.*

(2) *Satyrium basilicum femina* (the 'female' *Satyrium basilicum*, still *Gymnadenia conopsea*) which shared similar foreign names and properties as the first but in appearance was more robust, with larger, wider leaves and a fuller inflorescence; and (3) *Satyrium trifolium* (now known as *Platanthera bifolia* but then in Latin as *Testiculus vulpis aut sacerdotis*—Wolf or priest's testicles—saying it was named for the satyrs who played with the wood nymphs and found that the plant increased sexual activity (Fuchs 1542. Of *Orchis femina major*, Fuchs relying on the authority of Dioscorides stated that if a man ate its tubers, he would father a

Fig. 2.10 *Testiculus orchis; Cynosorchis*. From: Ruel J, *Dioscorides Materia Medica* (1516)

male child and if a woman ate the bulbs of *Orchis femina minor*, she would beget a girl. However, the uses of orchids were not entirely confined to sex and procreation. Tubers of *Triorchis serapias mas* (which Henry Oakeley identified as probably *Orchis morio*) and *Triorchis femina* (probably *Ophrys holosericea* according to Oakeley) removed swellings, checked herpes, abolished fistulae, soothed inflammation, healed sores, abscesses and mouth ulcers (Fuchs 1542). As the names of orchids have changed over time, to avoid confusion, the correct scientific names for relevant species are provided in Table 2.2.

Detailed woodcut illustrations for *De Historia Stirpium* were achieved through the combined effort of Albrecht Meyer who drew the plant portraits, Heinrich Fullmaurer who transferred the drawings to the woodblocks and Veit

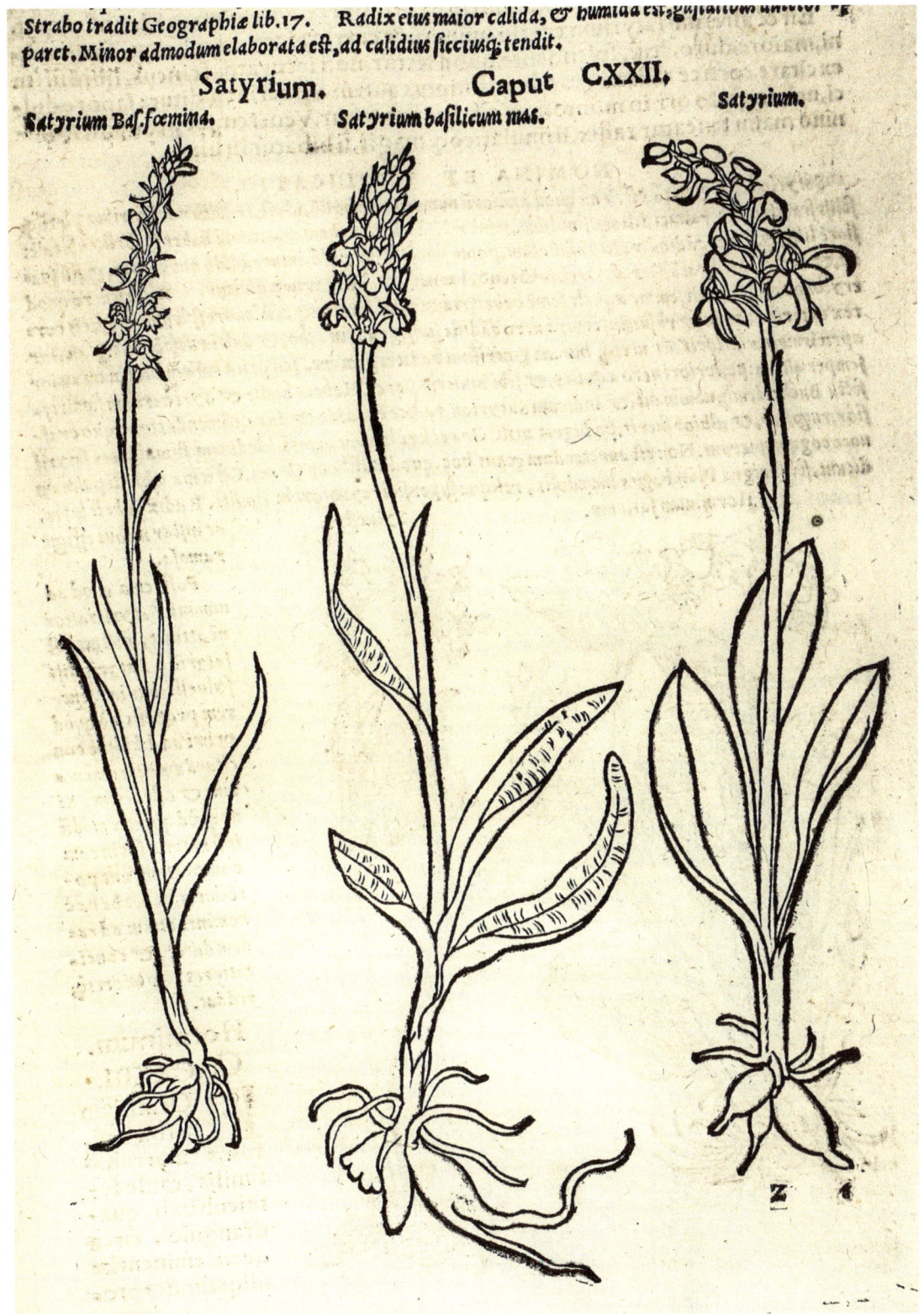

Strabo tradit Geographiæ lib.17. Radix eius maior calida, & humida est, gustatum
paret. Minor admodum elaborata est, ad calidius sicciusq; tendit.

Satyrium. Caput CXXII.

Satyrium Basilicum mas. Satyrium.

Fig. 2.11 *Dactylorhiza.* From: Ruel J, *Dioscorides Materia Medica* (1516)

Table 2.1 Principal contributors to the botanical renaissance

Botanist	Selected publications	Year
Jean Ruel (1474–1537)	*Materia Medica*	1516
	De Natura Stirpium	1536
Otto Brunfels (1488–1534)	*Herbarum Vivae Eicones* (Woodcuts by Hans Weiditz)	1530
Hieronymus Bock (1498–1530)	*Kreutterbuch* (plant book) (illustrator David Kandel)	1539
Leonhart Fuchs (1501–1566)	*De historia stirpium, commentarii insignes*	1542
Pietro Andrea Mattioli (Matthiolus) (1501–1577)	*Di Pedacio Dioscoride*	1544
	Commentarii	1554
	Compendium de Plantis Omnibus Una cum *Erum Iconibus*	1571
Mathias de l'Obel (1538–1616)	*Stirpium adversaria nova*	1570–1571
	Plantarum seu stirpium historia	1576
	Kruydtboeck	1581
Rembert Dodoens (1516–1585)	*Cruydeboeck*	1554
	Stirpium historiae pemptades sex	1574
Carolus Clusius (1526–1609)	*Rariorum aliquot stirpium per Hispanias observatarum historia: libris duobus expressas*	1575
	Dodoens R: *Histoire de plantes* (French translation from Dutch)	1557
	de Orta: *Aromatum et simplicium aliquot medicamentorum apud Indios nascentium historia* (Latin translation from Portuguese)	1567
	Monardes: *De simplicibus medicamentis ex occidental India delatis quorum in medicina usus est* (Latin translation from Spanish)	1574
John Gerard (1545–1612)	*Herball or Generall Historie of Plantes*	1597
John Parkinson (1567–1650)	*Theatrum Botanicum*	1640

Fig. 2.12 Otto Brunfels (1488–1534)

Rudolph Speckle who created the woodblocks. The printed images were then hand-coloured (Figs. 2.16, 2.17, 2.18, 2.19, 2.20 and 2.21).

During the sixteenth century, no physician exerted more influence than Pietro Andrea Matthiolus (1501–1577) who was the personal physician to the Holy Roman Emperor in Vienna, Maximilian II (r. 1564–1576). Matthiolus did not tolerate rivals: the Inquisition investigated people who disagreed with him. Thus, when Matthiolus' *Discorsi* (Commentaries on the *Materia Medica* of Dioscorides, in Italian) was published in 1544, Dioscorides' information on the aphrodisiac and antiaphrodisiac properties of salep reached a wide audience. In 1554 he produced a Latin edition with the same information on the plants described by Dioscorides, with his commentary and extra plants unknown to Dioscorides, augmented by 562 lifelike woodcut illustrations (Figs. 2.22 and 2.23).

Discoveries from the New World and Dutch exploration of India and the East Indies spurred

Fig. 2.14 Leonhart Fuchs (1501–1566)

Fig. 2.13 *Cynosorchis.* From: Brunfels O, *Herbarum Vivae Eicones*, 1530–1536. [Artist: Hans Weiditz]

the growth of intense botanical interest in the Netherlands. The movement was led by Mathias de l'Obel (1538–1616), Rembert Dodoens (1516–1585) and Carolus Clusius (1526–1609).

After graduating in medicine from Montpellier Mathias de l'Obel practised in Holland and England. Impressed with l'Obel's competency as superintendent of his medicinal garden in Hackney, Lord Zouch got him appointed as royal botanist. L'Obel also became a personal physician to James I. L'Obel's first work, co-authored with Pierre Pena, *Stirpium adversaria nova* written and published in England in 1571, was dedicated to Elizabeth I. It

described over 1200 plants collected in England, France and the Netherlands. The next work, *Plantarum seu stirpium historia* published 5 years later featured 1486 woodblock prints. Ten years later, his Dutch edition entitled *Kruydtboeck* was illustrated with 2187 woodcuts that included 46 orchids employed as salep. L'Obel befriended many English botanists including John Gerard who joined him and Clusius to study English plants in their natural habitats; John Gerard and John Parkinson used his materials in their publications (Figs. 2.24, 2.25, 2.26, 2.27, 2.28, 2.29, 2.30, 2.31 and 2.32).

Rembert Dodoens was also a physician to Maximilian II, but he turned down an appointment as physician to Philip II of Spain. He taught at the University of Leiden. His major contribution is the extensive herbal, *Cruydeboek* (1554) which emphasized the pharmacological effects of plants. Dodoens classified plants into groups according to their properties. Illustrated with

Fig. 2.15 Fuchs L, *De Historia Stirpium* (1551). Title page

715 woodcuts, it was translated into French by Clusius in 1557 and then into English in 1578. Henry Lyte's English translation entitled *A Niewe Herball* or *Historie of Plantes* in 1578 never achieved the fame of John Gerard's celebrated *Herball* of 1597 which was based on a translation from Dodoens' Latin edition—the *Stirpium Historia Pemptades Sex*—that had been made by a Dr. Priest of the London College of Physicians (Oakeley, personal communication).

Carolus Clusius (Charles de l'Ecluse) (1526–1609) also studied medicine at

Table 2.2 Orchid species providing tubers for preparation of salep

Species name in the references	Accepted name
Anacamptis laxiflora	*Anacamptis laxiflora* (Lam.) R.M. Bateman, Pridgeon & M.W. Chase
Anacamptis morio	*Anacampti morio* (L.) R.M. Batemn, Pridgeon & M.W. Chase
Anacamptis morio var. *longicornu*	*Anacamptis longicornu* subsp. *longicornu* (K. Koch) H. Kretzschmar, Eccarius & H. Dietr
Cymbidium aloifolium	*Cymbidium aloifolium* (L.) Sw.
Eulophia campestris Suresh	*Eulophia nuda* Lindl.
Eulophia herbacea	*Eulophia herbacea* Lindl.
Eulophia spectabilis	*Eulophia dabia* (D. Don) Hochr
Eulophia virens	*Eulophia* epidendraea (J. Koenig ex Retz.) C.E.C. Fisch
Habenaria commelinifolia	*Habenaria commelinifolia* (Roxb.) Wall. ex Lindl.
H. pectinata	*Habenaria pectinata* D. Don
Himantoglossum sp.	*Himantoglossum* sp. (e.g. *H. hircinum*)
Ophrys fuciflora	*Ophrys fuciflora* (F.W. Schmidt) Moench,
Orchis anatolica Boiss	*Orchis anatolica* Boiss
Orchis coriophora	*Anacamptis coriophora* (L.) R.M. Bateman, Pridgeon & M.W. Chase
Orchis hircine	*Himantoglossum hircinum* (L.) Spreng.
Orchis italic	*Orchis italica* Poir.
Orchis latifolia	*Dactylorhiza hatagirea* (D.Don) Soo
O. maculaae	*Dactylorhiza maculata* (L.) Soó
O. mascula	*Orchis mascula* (L.) L.
O. militaris	*O. militaris* L
Orchis morio	*Anacamptis morio* (L.) R.M. Bateman, Pridgeon & M.W. Chase
Orchis romana	*Dactylorhiza romana* (Sebast.) Soó
O. sambucina	*Dactylorhiza sambucina* (L.) Soó
O. ustulata	*Neotinea ustulata* (L.) R.M. Bateman, Pridgeon & M.W. Chase
Platanthera bifolia	*Platanthera bifolia* (L.) Rich
Satyrium hircimum	*Himantoglossum hircinum* (L.) Spreng.
Serapias	*Serapias* (L.)
Zeuxine strateumatica	*Zeuxine strateumatica* (L.) Schltr.

Note that there is considerable confusion in the naming of these orchids. Names employed in the articles referenced are placed in the left column, and the presently accepted names are in the right column

Montpellier between 1551 and 1554 under Guillaume Rondelet, but he probably did not graduate. He neither practised nor styled himself as a physician. He was more interested in plants. Clusius had the opportunity to study plants collected by Spain from the New World, and when he was director of the Imperial Medicinal Garden in Vienna in 1573–1576, he received exotic bulbs from Constantinople (Figs. 2.33 and 2.34).

Appointed professor at the University of Leiden in 1593, Clusius was given the responsibility of establishing the city's botanical garden, the *Hortus Academie*. There he planted potato from the Americas, chestnut from Asia Minor and tulips from Turkey. His studies on the tulip laid the foundation for the Dutch tulip industry. Clusius published many original works and translated the publications of other European botanists. In particular he translated into Latin the groundbreaking *Colóquios dos simples e drogas da India* (Discussion of the medicinal herbs and drugs of India), published in 1563. This with his translation of Nicolás Monardes' publication on the medicinal plants of the Americas and their uses, as *De simplicibus medicamentis ex occidentali India delatis quorum in medicina usus est* (Of herbal medicines from the West Indies set out with their uses) in 1574 made him one of the great physician botanists of his age. His major contribution came from the role that he

Fig. 2.16 Goatstones. *Orchis latifolia, Orchis mascula* [as *Orchis mas latifolia, Orchis mas angulifolia*]. From: Fuchs H, *De Historia Stirpium* (1551), p. 538

Fig. 2.17 *Orchis morio, Ophrys tenthredinifera* (?) [as *Triorchis serapias mas, Triorchis femina*]. From: Fuchs H, *De Historia Stirpium* (1551), p. 457

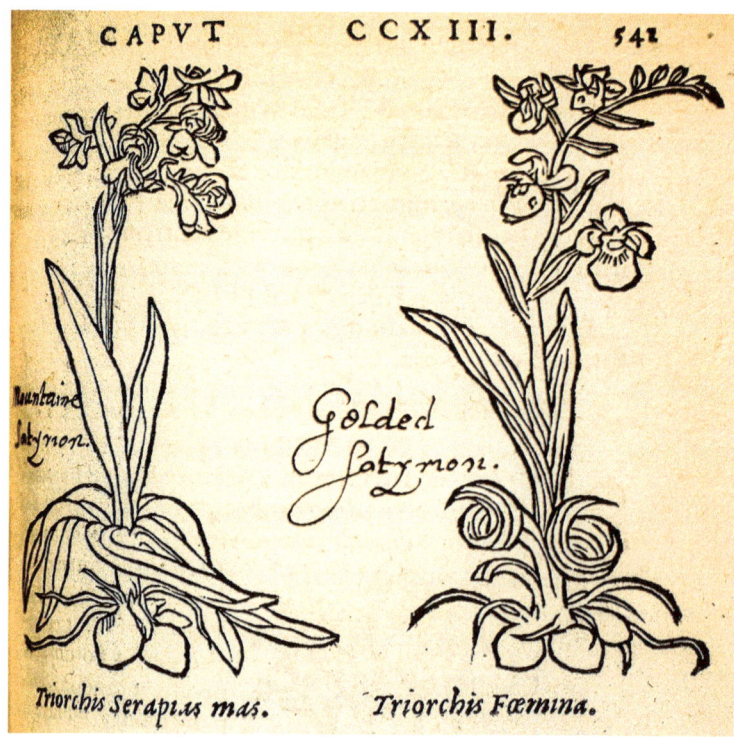

Fig. 2.18 *Gymnadenia conopsea* [as *Satyrium basilicum mas* and *Satyrium basilicum femina*]. From: Fuchs H, *De Historia Stirpium* (1551), p. 677

Fig. 2.19 *Orchis femina major, Orchis femina minor*. From: Fuchs H, *De Historia Stirpium* (1551), p. 539

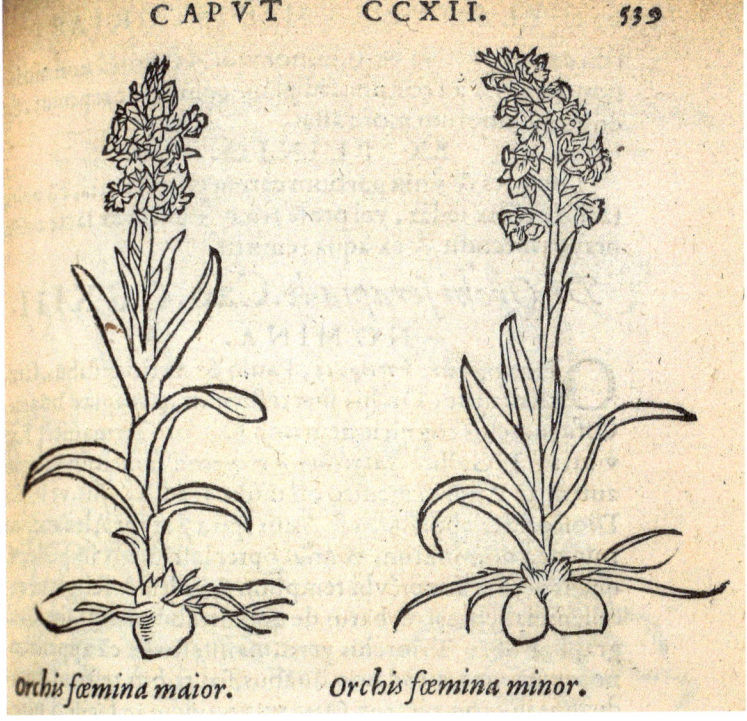

Fig. 2.20 Fuchs H, *De Historia Stirpium* (1551), title page

played in advising the Dutch East India Company (VOC) that botanical research must be a major undertaking of the Company, given that their principal interest in the India and the East Indies was to secure control of the spice trade.

This motivation lent support for the work of Rumphius in Ambon and van Rheede in India. Rumphius investigated Indonesian terrestrial orchids as potential sources for salep without success. Van Rheede showed no interest in orchids with alleged aphrodisiac properties, notwithstanding that Theophrastus stated that the Indian had an ointment that on one occasion enabled a big strong fellow to 'go with 70 women'. The startling medication was not taken internally. It was not salep (Dalby 2000).

Knowledge of the alleged 'uses and virtues' of orchids became public in England when William Turner (1509–1568) published his famous *Herbal* in 1551. In the preface to his book, Turner

Fig. 2.22 Matthiolus (Pietro Andrea Matthiolus) (1501–1577) [painting by Alessandro Bonvicino]

Fig. 2.21 *Platanthera bifolia* [as *Satyrium trifolium*]. From: Fuchs H, *De Historia Stirpium* (1551), p. 675

admitted that some people would accuse him of divulging to the general public what should have been reserved for a professional audience. However, Turner was a liberal who was ahead of his time and a person who openly spoke his mind. For instance, he voiced objection to Henry VIII's proclamation that reading of the Bible should be reserved for men of good social standing. His interest in botany inspired by his studies in Cambridge led him to study plants in their natural habitat, and in 1538 he published *Libellus de re herbaria novus*. He studied medicine at Ferrara

and Bologna and then returned to England to practise. In the *Herbal*, Turner described the virtues of Adders grasse (*Orchis mascula*)—that first, the root was edible; secondly, it was told that the women of Thessaly gave it with goat's milk to provoke the pleasure of the body, while it is tender, but they give the dry one to hinder and stop the pleasure of the body. He also described a second species which had the ability to 'reduce swellings, heal sores, fistulae and grievous sores of the mouth.' Turner was merely quoting from Fuchs (Figs. 2.35 and 2.36).

Friendship and botanical expeditions in England with Clusius and L'Obel may have inspired John Gerard (1545–1612) in 1597 to publish an English *Herball* that became the most popular botanical work in English during the seventeenth century. It is still widely quoted today. Gerard borrowed extensively from l'Obel and Dodoens, but he did not acknowledge their work, possibly because he had no academic training. He was trained by apprenticeship and practised as barber surgeon. Nevertheless, he was a well-respected gardener who had his own garden in Holborn, and he acted as superintendent of

Fig. 2.23 *Orchis, Anacamptis* [as *Testiculus*]. From: Matthiolus, *Sinensis Medici* (1569)

Lord Burghley's garden for over 20 years. Gerard's *Herball* emphasized the alleged properties of orchids and further popularized salep in London. The printing of books in spoken languages, such as Dutch, French, German and English allowed a wider readership. It encouraged

Fig. 2.24 *Dactylorhiza.* [as *Palma Christi*]. From: Matthiolus, *Sinensis Medici* (1569)

PALMA CHRISTI MAIOR.

Fig. 2.25 Mathias de l'Obel (1538–1616)

an interest in the medicinal properties of plants as originally described by Dioscorides and promoted the belief throughout Europe that salep was a potent aphrodisiac. Before the introduction of experimentation, serious authors by relying solely on authority had unintentionally become the spin doctors for salep (Fig. 2.37 and 2.38).

Meanwhile in Italy, a new botanical work by Giovanni Battista della Porta also stated the pro-creative properties of orchids. Coming from a sceptical, highly educated Italian, this was puzzling. Della Porta was a scholar, playwright, philosopher, naturalist, architect and inventor with publications that attested to an extraordinarily wide range of interests: he scoffed at the ancient belief that garlic could disable magnets when experiments failed to prove the belief. He collected a group of friends to form a scientific

Fig. 2.26 Rembert Dodoens (1516–1585)

society, the Academia Secretorum Naturae, to investigate the secrets of nature, and he headed the local chapter of the Accademia dei Lincei when Federico Cesi founded this scientific body in Rome in 1603. Thus it is a paradox that in his *Phytognomonica* published in 1588, della Porta chose to ignore experimentation and instead he revived the *Doctrine of Signatures* to explain why orchids were aphrodisiacs. This belief system was based on the 'deduction' that 'God gave a form and colour to plants so that man would know their properties.' Perhaps this was to satisfy the Inquisition which had called him up in the mid-1570s, disbanded his scientific society and prohibited publication of his philosophical works for a time. A devout Catholic, della Porta supported the Jesuits (Fig. 2.39).

He illustrated yet another orchid with testicular-shaped tubers and gave it the name *Tragorchis* (orchid of the River Seine) and subsequently bestowed the synonyms *Testiculus hircinus* (goat's testicle) and *Satyrium hircinum* (Satyr's testicle) (Heursel-De Meester and Delmotte, 1912); this is *Himantoglossum hircinum*, the lizard orchid. Persian salep was derived from *Dactylorhiza hatagirea*, whereas in

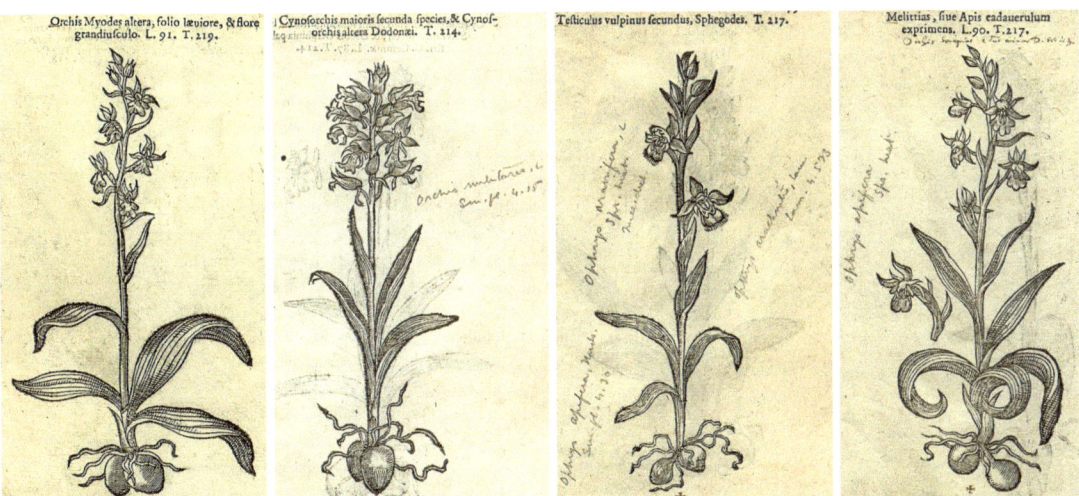

Fig. 2.27 (Left to right): 1. *Ophrys insectifera* [as *Orchis myodes altera*]; 2. *Orchis militaris* [as *Cynosorchis majoris secunda*]; 3. *Ophrys apifera* [as *Testiculus vulpinus secundus*]; 4. *Ophrys apifera* [as *Melittias, sine Apis cadauerulum exprimens*]. From: l'Obel M, *Plantarum seu stirpium icones,* vol. 1: (1581), pp. 180, 175, 179, 180

Fig. 2.28 (Left to right): *Anacamptis morio* [as *Cynosorchis Morio femina*]; 2. *Anacamptis papilionacea* [as *Orchis ornithophora, vel Ornites foliolaeni*]; 3. *Anacamptis coriophora* [as *Tragorchis minor & verior sive Coriosmites vel Coriophora flore instar simicum*]; 4. *Orchis militaris* [as *Orchis strateumatica, sive stratiotes major*]. From: l'Obel, op. cit

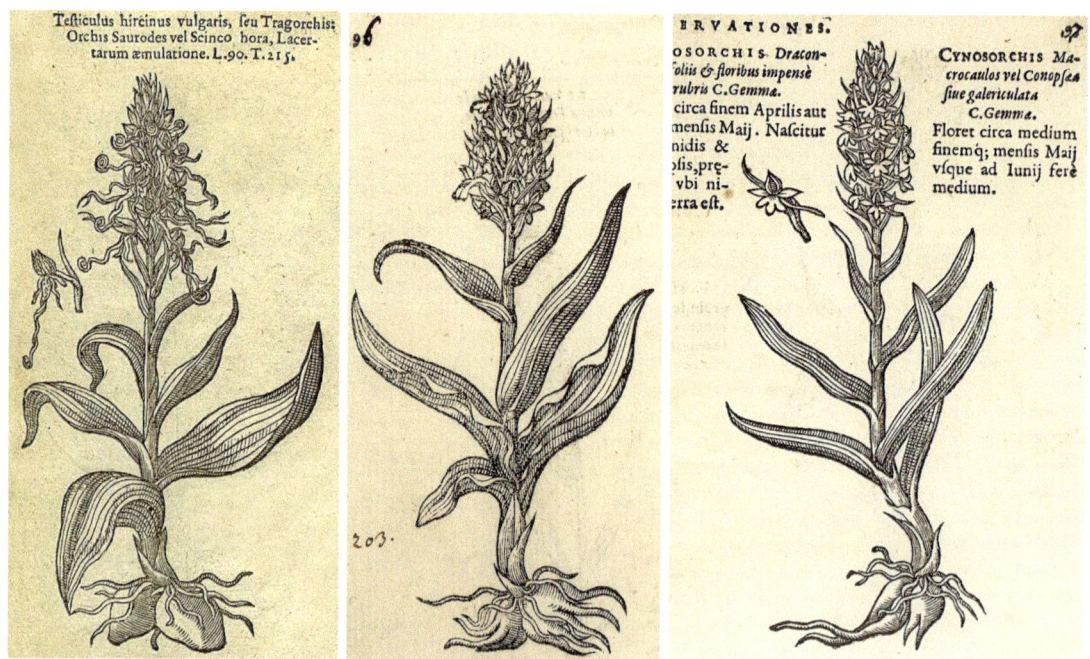

Fig. 2.29 (Left to right): 1. *Himantoglossum hircinum* [as *Testiculus hircinus vulgaris seu tragorchis*]; 2. *Dactylorhiza praetermissa* [as *Serapias palustris altera leptophylla*] 3. *Gymnadenia conopsea* [as *Cynosorchis macrocaulus vel Conopse sine galericulata C. Gemme*]. From l'Obel, op. cit

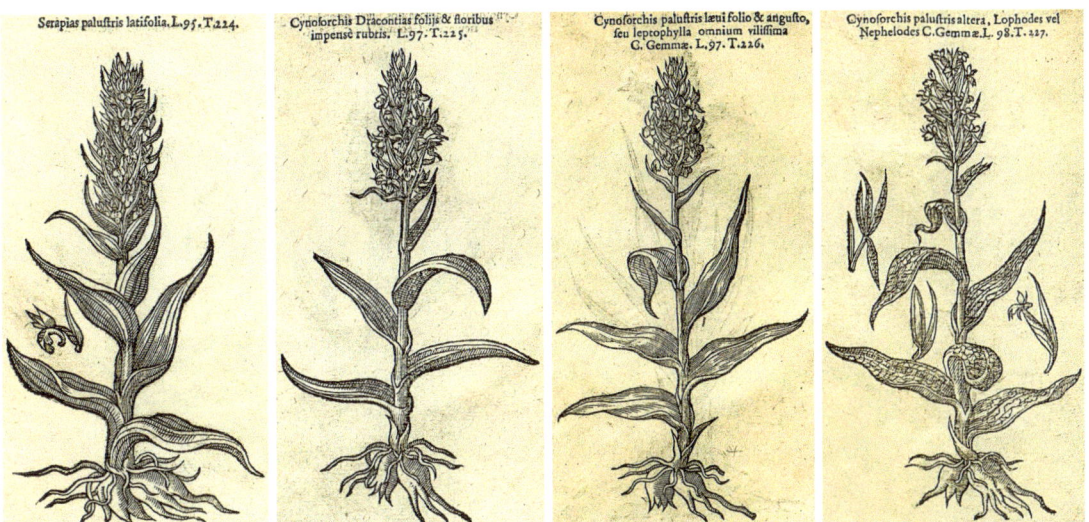

Fig. 2.30 (Left to right): 1–4. *Dactylorhiza incarnata*. [as 1. *Serapias palustris latifolia*; 2. *Cynosorchis draconitias folis et floribus impense rubis*; 3. *Cynosorchis palustris laevi folia et augusto seu leptophylla omnium vilissima*; 4. *Cynosorchis palustris altera Lophodes vel Nephelodes*]. From l'Obel, op. cit

the Himalayas, *Eulophia nuda* (syn. *Eulophia spectabilis*) and *Dactylorhiza hatagirea* were employed (Dymock et al. 1893; Teoh 2016). Interest in naming plants with similar properties led the Spanish physician, Joanne Fragoso (1530–1597) to mention in *De Succedaneis Medicamentis* (1575) that if one could not obtain *Satyrium* or *Testiculi canis*, the roots of garden rocket (*Eruca sativa*) was an alternative. This belief enjoyed a history as old as salep's. Virgil's *Moretum*, a poem in hexameter from the first century BCE that described the life or a farmer, carries the line, *et Venerem revocans eruca morantem* ('and rocket which revives a man's flagging potency') (Fairclough and Gould 2001) (Figs. 2.40 and 2.41).

Another prominent English herbalist of that period was John Parkinson (1567–1650) who was apothecary to James I. Parkinson was a founding member of the Worshipful Society of Apothecaries, a guild that could award a license to practise medicine in the United Kingdom and in many parts of the British Commonwealth up to the mid-twentieth century. He participated in the publication of the *London Pharmacopoeia* (*Pharmacopoeia Londinensis*) in 1618. He was a friend of Mathias de L'Obel whose papers he edited and

published in *Theatrum Botanicum* in 1640, a beautifully illustrated work on 3800 plants that included the first description and illustration of *Cypripedium calceolus*. In the final section of the publication, Parkinson described and illustrated what he considered to be 77 orchid species (albeit the actual number is far fewer by present reckoning) and detailed their properties and usage. In recognition of Parkinson's contribution to botany, he was conferred the honorary title of *Botanicus Regius Primarius* (Royal Botanist of the First Rank) by Charles 1 (Figs. 2.42, 2.43, 2.44 and 2.45).

In England, Nicholas Culpeper (1616–1654) continued to promote the ancient reputation of *Orchis* (Culpeper 1652) in his *The English Physician* published in 1652 and subsequently republished with revisions and additions to the text. With Cambridge education and upbringing by his grandfather, an English priest much interested in astrology, this apothecary believed in the astrological influence of herbs, a common belief among educated men of his era. On *Orchis* he stated:

'They are hot and moist in operation, under the dominion of Dame Venus, and provoke lust exceedingly; which, it is said, the dry and

Fig. 2.31 Left: *Orchis purpurea* [as *Testiculus IIII*] in Matthiolus *Sinensis Medici* (1569). Right: *Orchis purpurea* subspecies *purpurea*, colour print by Sowerby (1791) from *English Botany* (1840)

withered roots restrain again. They are held to kill worms in children; also being bruised and applied to the place, to help the king's evil'. (Text quoted from Culpepper's *English Physician and Complete Herbal*, 1652)

England became a focal point for botanical science in the mid-eighteenth century when Joseph Banks returned from a voyage with Captain Cook that circumnavigated the world. The Royal Society elected him as their president, a position he retained for 41 years. Banks advised the setting up of the Royal Botanical Gardens at Kew. Prior to 1796 details in plant drawings were

limited by what could be achieved by woodcarvers who produced the blocks for printing. With the invention of lithography in that year, it became possible to reproduce in print fine drawings of botanical artists. This inspired James Sowerby to decide on a career in botanical illustration after he graduated from the Royal Academy. He was responsible for many of the early colour illustrations in *The Botanical Magazine*. From 1790 to 1813 Sowerby worked on the 36-volume *English Botany* (popularly known as *Sowerby's Botany*) which contained 2592 of his hand-coloured engravings that

Fig. 2.32 Left: *Ophrys sphegodes* [as *Testiculus vulpinus fecundus* or fox testicles] from l'Obel, *Plantarum seu stirpium icones* vol. 1 (1891). Right: *Ophrys sphegodes* [as *Ophrys aranifera*] in Sowerby's *English Botany* (1840) Fuchs *De Historia Stirpium* (1551) and l'Obels *Plantarum* *seu stirpium icones* (1581) established very high standards for woodblock illustration of orchids. These woodblocks were borrowed for reproduction by several later authors of *Herbals*

included numerous illustrations of terrestrial orchids. Hand painting was employed to introduce colour for limited editions before technological advancement made it possible to print drawings in multiple colours for a wider readership. This chapter showcases some of his works with photographs generously supplied by

Dr. Henry Oakeley (Figs. 2.46, 2.47, 2.48, 2.49 and 2.50).

Belief in the aphrodisiacal properties of salep persisted into the nineteenth century (Parkins 1809). In 1829, it was still included in the *Edinburgh Pharmacopoeia* on account of its mucilaginous content. However in 1792,

Fig. 2.33 *Anacamptis pyramidalis* [as *Orchis Batavica*]; *Dactylorhiza sambucina* [as *Orchis Pannoniva*] in Clusius C, Atrebatis. Impp. Caess Angg., MaximilianiII, Rudolphi II, arlee quondamfamilaris. Rariorum plantarum historia: quae accesserini proxima pagina docebit (1601)

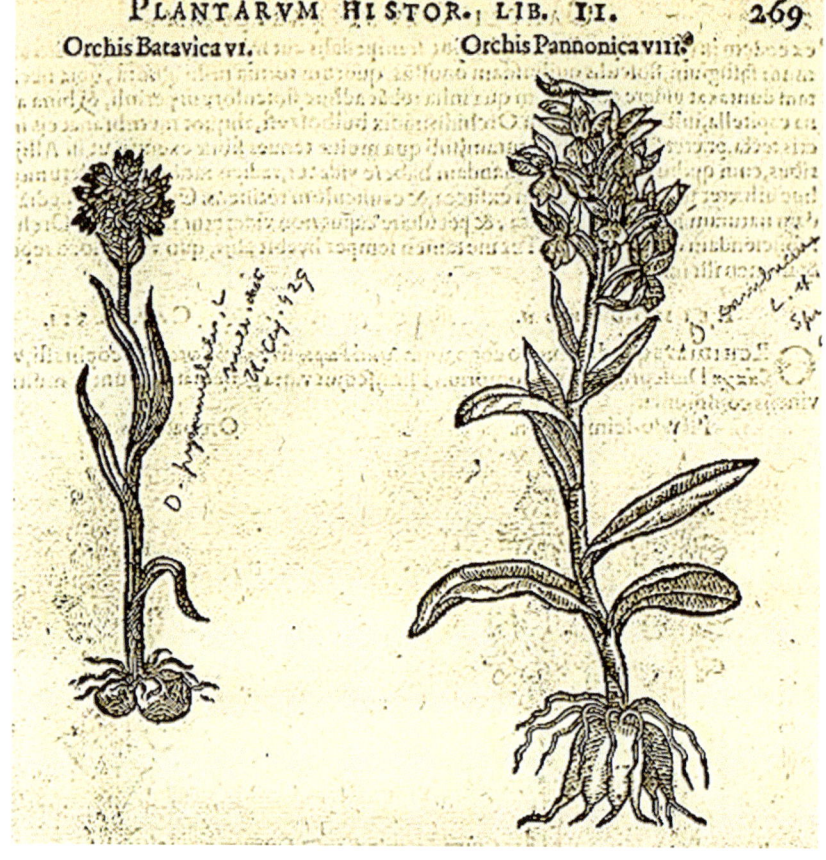

Fig. 2.34 Carolus Clusius (Charles de l'Ecluse) (15226–1609)

Fig. 2.35 *Anacamptis laxiflora*. Painting by J. Bourdichon in the beautifully illustrated prayer book for Anne of Brittany, Queen of France: *Grandes Heures Anne de Bretagne*, (1503–1508), p. 235. Flowers are sometimes included in religious paintings, e.g. in Bottichelli's *Adoration of the Magi*, Uffizo Gallery, Florence, Italy

Fig. 2.36 Orchid painting (*Gymnadenia conopsea*?) from *Grandes Heures Anne de Bretagne* (1503–1508). [artist: J. Bourdichon]

Dr. William Woodville and others already published their doubts its legendary properties (Woodville 1792; Hooper and Akerly 1829). Salep was not much appreciated in the United States where its main use was limited to the preparation of 'Castillon powders, a nutritive and bland article of diet for invalids'. For aphrodisiac, salep was replaced by *Vanilla* beans (Griffith (1847).

Salep Orchids in Literature

In many cultures, stories abound on the origin of unique plants. Greek legend mentioned that Orchis was the son of nymph sired by a satyr. Like father like son, Orchis was given to lust. Overcome by wine during a Bacchanalia, Orchis committed an unpardonable crime: he seduced a priestess. When they discovered the act, the enraged Bacchanals attacked Orchis and tore him to pieces, scattering bits of his body all over

Fig. 2.37 *Orchis mascula*. Photo: Henry Oakeley

Fig. 2.38 Gerard's *Herball* (1597), cover

the ground. His satyr father implored the gods to restore Orchis back to life, and in a partial granting of this request, the gods caused *Orchis* plants to arise from the fragments of his remains. This kindness had consequences for the plant assumed the shape of paired testicles and transmitted the trait of Orchis to whomsoever ate its tubers (Figs. 2.51, 2.52, 2.53 and 2.54).

In the Roman play *Satyricon*, Petronius (c. 27–66) featured the use of orchids by prostitutes: ' . . . she brought me a goblet full of satyrion and, with jests and quips and a host of marvelous tales, induced me to drink up nearly all the liquor'. Elsewhere, 'We saw in the chambers persons of both sexes acting in such a way that I concluded they must all have been drinking satyrion' (Wedeck 1961).

Later writers revived this stigmatization of orchids. Bernard Shaw (1856–1950) described the character of a courtesan as 'orchidaceous'.

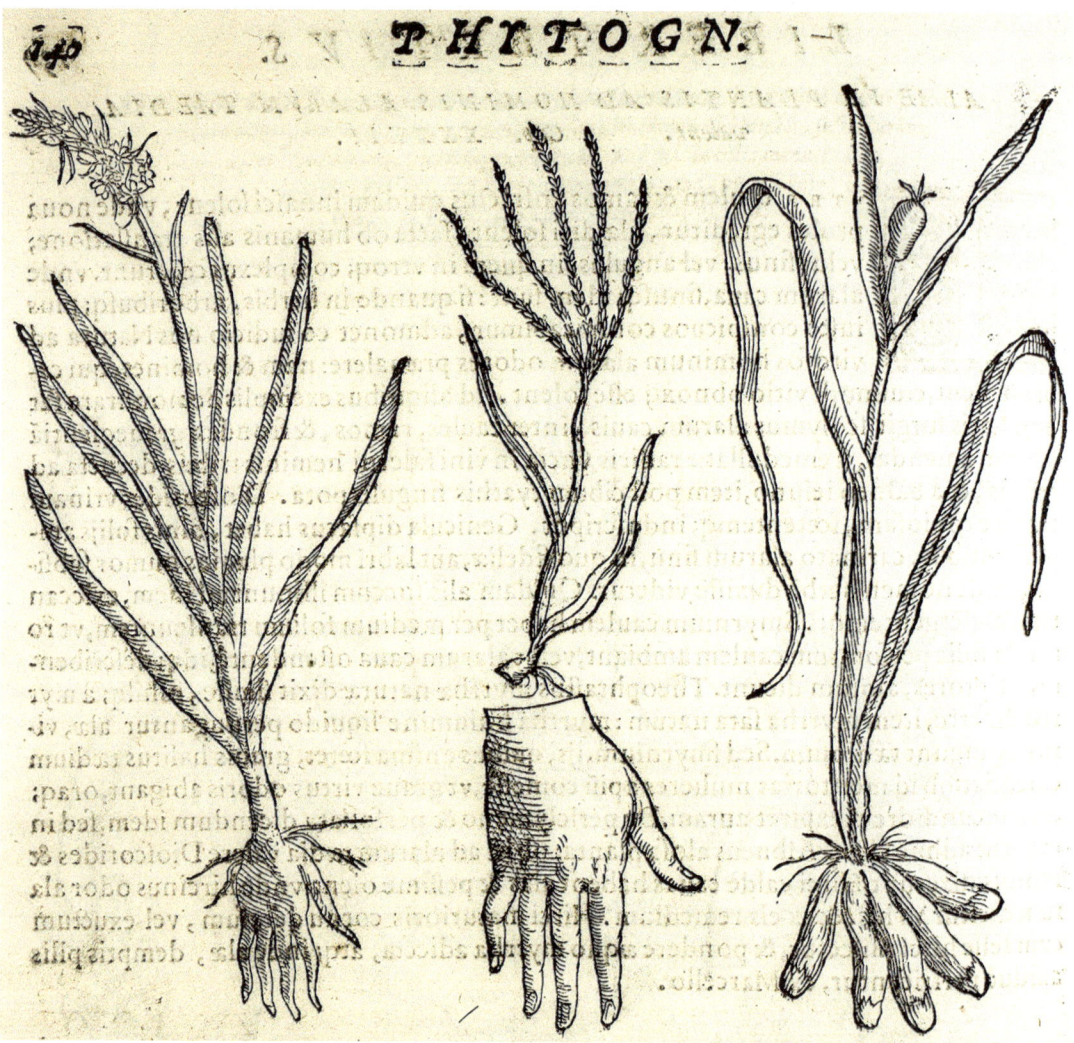

Fig. 2.39 *Orchis; Orchis; Spiranthes* spp. from della Porta, *Phytognomonica* (1588) with semi-schematic portrayal of the tubers in accord with the *Doctrine of Signatures*

Arnold Bennett (1867–1931) introduced a new verb, *orchidize* for the passionate behaviour of a courtesan (Lewis 2009).

During the Dark Ages, it was alleged that witches employed tubers of the early purple orchid (*Orchis mascula*) to produce philtres (aphrodisiac drinks), the fresh tuber ensuring true love and the withered one to check improper passions (Grieve 1971). Centuries later, and in a different continent, around 1670 to be more precise, a physician by the name of John Gent reported that he had watched 'a wanton woman' gathering

the American slipper orchid, *Cypripedium parviflorum* (syn. *Cypripedium pubescens*) to use as dogs' stones (presumably meaning aphrodisiac). Roots of the orchid, mixed with wine, gave an 'amorous cup which wrought the desire effect' (Emboden 1974).

Witchcraft and love potions were much loved by storytellers. Shakespeare referred to the *salep* orchid in *Hamlet*:

'There with fantastic garlands did she come

Of crow flowers, nettles, daisies, and long purples

Fig. 2.40 *Himantoglossum* sp. [as *Orchis anthropopor-phyra*]. From: Colonna F, *Ekpharesis Altera* (1616)

That liberal shepherds give a grosser name,
But our cold maids do dead men's fingers call them;'
(Shakespeare (1600), *Hamlet* Act. 4, scene 7)
The long purples and dead men's fingers referred to *Orchis mascula, O. maculata* or *Dactylorhiza* species (Lawler 1982; Bulpitt 2005). The orchid forming part of the fairy garland, together with stitchwort, harebell and wild thyme, was also known as the cuckoo flower, so called because it bloomed when the first cuckoo was heard in spring (Emboden 1974) (Fig. 2.55).

Fig. 2.41 *Orchis italica* [as *Orchis anthropoporphyra*] with flower depicted in fanciful humanoid form. From: Colonna F, *Minus cognitarum stirpum Pars altera* (1616)

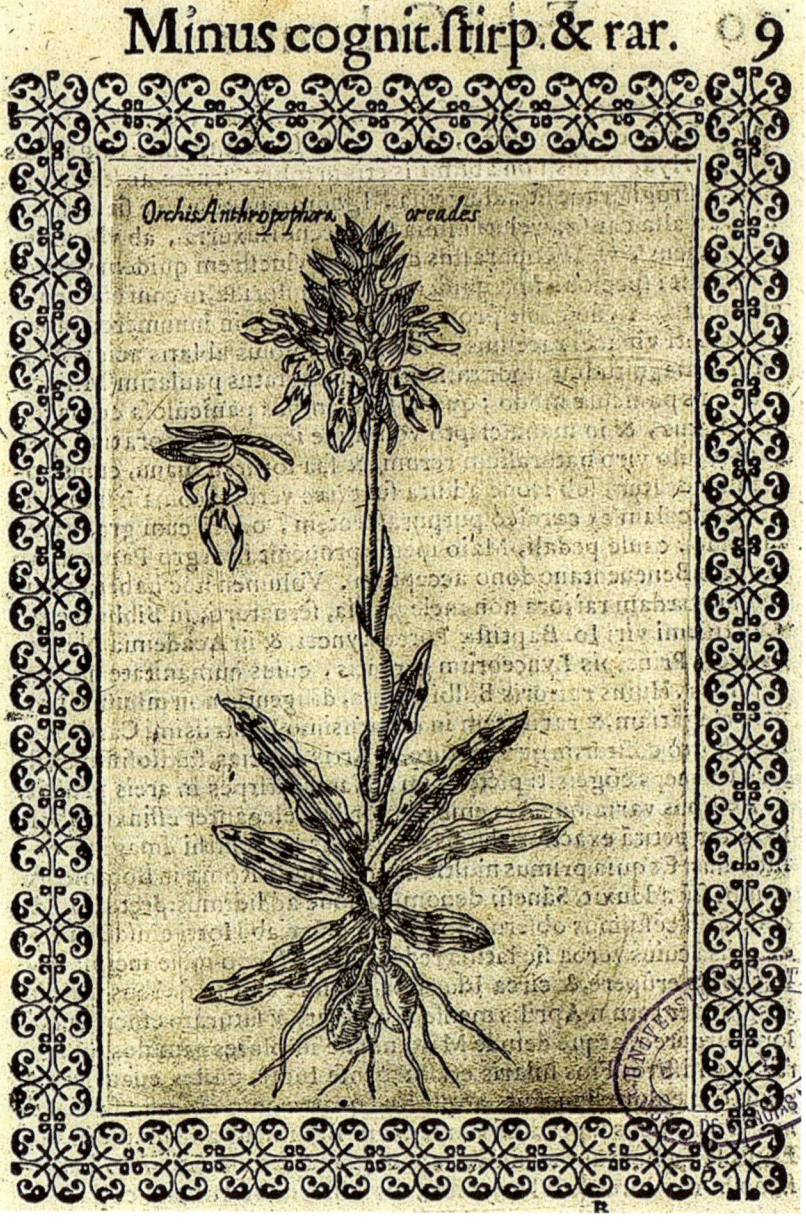

Old English texts make for fascinating reading. 'The roots are round and something long/ two together small as an olive/the one is higher up which is the fuller/and the one which groweth lower/and is the softer and fuller of wrinkles' is how the tubers were described (Turner 1551).

The Beverage

Salep is a drink prepared from the starchy substance in the tuberous roots of many species of *Orchis* and other bulbous terrestrial orchids that are distributed in the Mediterranean, Middle East

Fig. 2.42 John Parkinson (1567–1650)

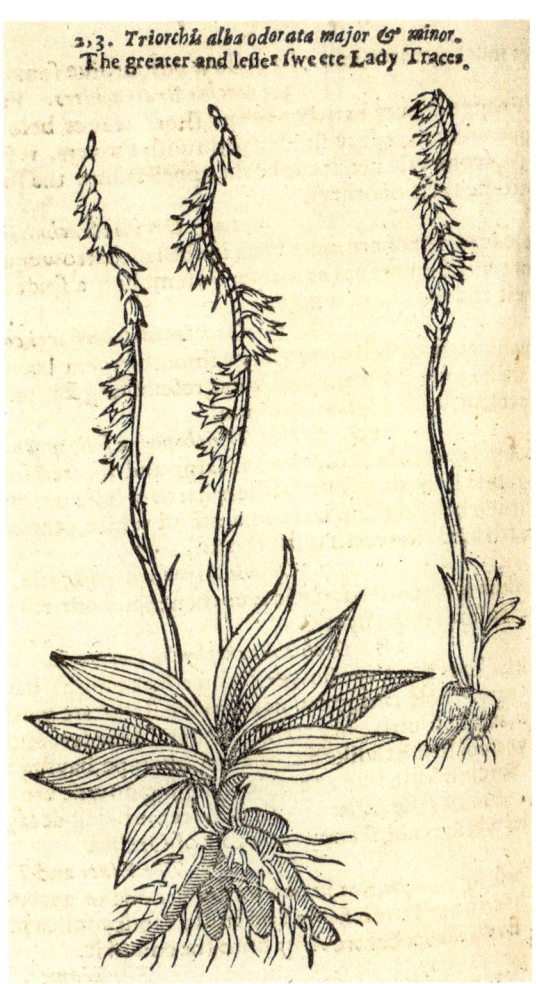

Fig. 2.43 *Spiranthes autumnalis* [as *Triorchis alba odorata major* and *minor*]. From: Parkinson J, *Theatrum Botanicum* (1640)

and Northern India. The term *salep* is an English corruption of the Arabic *sahlab*. It was sometimes spelt as *saloop* or *sahlop*. The drink originated in the Middle East, in the region of Asia Minor (Turkey) from whence it spread in all directions but no further East than India and Bangladesh.

Salep is sweet and has a faint, somewhat unpleasant smell. Sassafras chips (aromatic bark from a North American tree) were sometimes added to give the drink a flavour. Alternatively amber, cloves, cinnamon and ginger were added to salep (Fernie 1914). In any case, if one desired an aphrodisiac, one would be prepared to ignore the smell. Before the introduction of coffee, from the Middle Ages to the nineteenth century, many European cities sported salep bars. London had its share of salep bars. The English writer, Charles Lamb (1775–1834) mentioned a 'Salopian shop' in Fleet Street: he commented that to many, the taste of salep had 'a delicacy beyond the China luxury'. Furthermore, 'a basin of salep at three-halfpence, with a slice of bread was ideal breakfast for a chimney-sweep' (Grieve 1971).

In Turkey where salep is still widely enjoyed and often made into ice cream, 30 species of terrestrial orchids belonging to the genera *Orchis, Anacamptis, Himantoglossum, Ophrys, Serapias* and *Dactylorhiza* (syn. *Aceras*) are employed to prepare salep (Sezik 1967, 1990; Tekinsen and Guner 2009). Mucilage (glucomannan) and starch are the major components. Composition of the tubers varies from species to species. The starch content of old tubers collected in autumn is low. Mucilage content ranges from a low of 9.6% in *Ophrys fuciflora* to a high of 61.05% in *Orchis romana*, whereas starch content is high in *Ophrys fuciflora* (18.7%) and low in *Orchis romana*

Fig. 2.44 *Spiranthes autumnalis* [as *Tetrorchis* or *Triorchis alba spiralis* or *autumnalis*]. From: l'Obel (1576) op. cit

to identify 150 individual salep tubers from 31 batches purchased in 12 Iranian cities, Ghorbani and colleagues discovered that species in the genera *Orchis* (34%), *Anacamptis* (27%) and *Dactylorhiza* (19%) were the most common (Ghorbani et al. 2017).

To meet the enormous appetite for salep, the various species of terrestrial orchids were stripped from their habitats by the millions, and today they are endangered plants in Turkey. It is illegal to export terrestrial orchid tubers from Turkey. Indeed CITES regulations prohibit the importation of these terrestrial orchid tubers into any country in the world without a CITES export and import permit. Salep exporters generally deal in flour that has been artificially flavoured. Turkish traders have now induced villagers in northern Iran to collect salep orchids for them, a move that would seriously threaten the survival of Iranian orchids (Ghorbani et al. 2014).

Lawler (1984) gave us some idea of what happened in the past. In 1892, Istanbul exported 19 tons of salep and retained another 10 tons as reserve, or for its own use. In 1879, Izmir exported 6.4 tons and, from 1905 to 1908, 10.5 tons. It was estimated that in the last quarter of the nineteenth century, 125 tons of such orchid bulbs were dug up annually. Further afield in Nepal, about 5 tons of *Dactylorhiza hatagirea* (syn. *Orchis latifolia*) tubers are available for export each year. It is claimed that these orchids are replanted, but experts think that the following year's harvest probably comes from the smaller bulbs which were originally ignored. *Anacamptis laxiflora* (syn. *Orchis laxiflora*) and *A. morio* var. *longicornu* (syn. *O. longicornu*) are two additional species employed in Western Asia, but the latter species is found predominantly in the Mediterranean countries and Africa. It is probably imported into Turkey (Lawler 1984).

Indian salep is not the same as the Mediterranean variety. *Salep misri* sold in the Indian bazaars are derived from various species of *Eulophia*, in particular *Eulophia nuda* (syn. *E. spectabilis*), *Eulophia dabia* (syn. *E. campestris*) and *Eulophia herbacea* (Chopra 1933). Other orchids that constitute Indian salep are *Eulophia virens*, *Habenaria commelinifolia*,

(0.48%) (Sezik 1990). The best salep are said to come from *Orchis coriophora*, *O. longicruris*, *O. mascula*, *O. militaris* and *Platanthera bifolia* (Sezik 1967).

Orchid species constituting salep vary from region to region, being dependent on which species are more prevalent in any one location. *Orchis anatolica* is found throughout Turkey and is present in most salep. In northwestern Turkey, *Orchis* and *Anacamptis* species are common, while in eastern Turkey *Orchis morio* thrives on the chalky soil. Using DNA barcoding

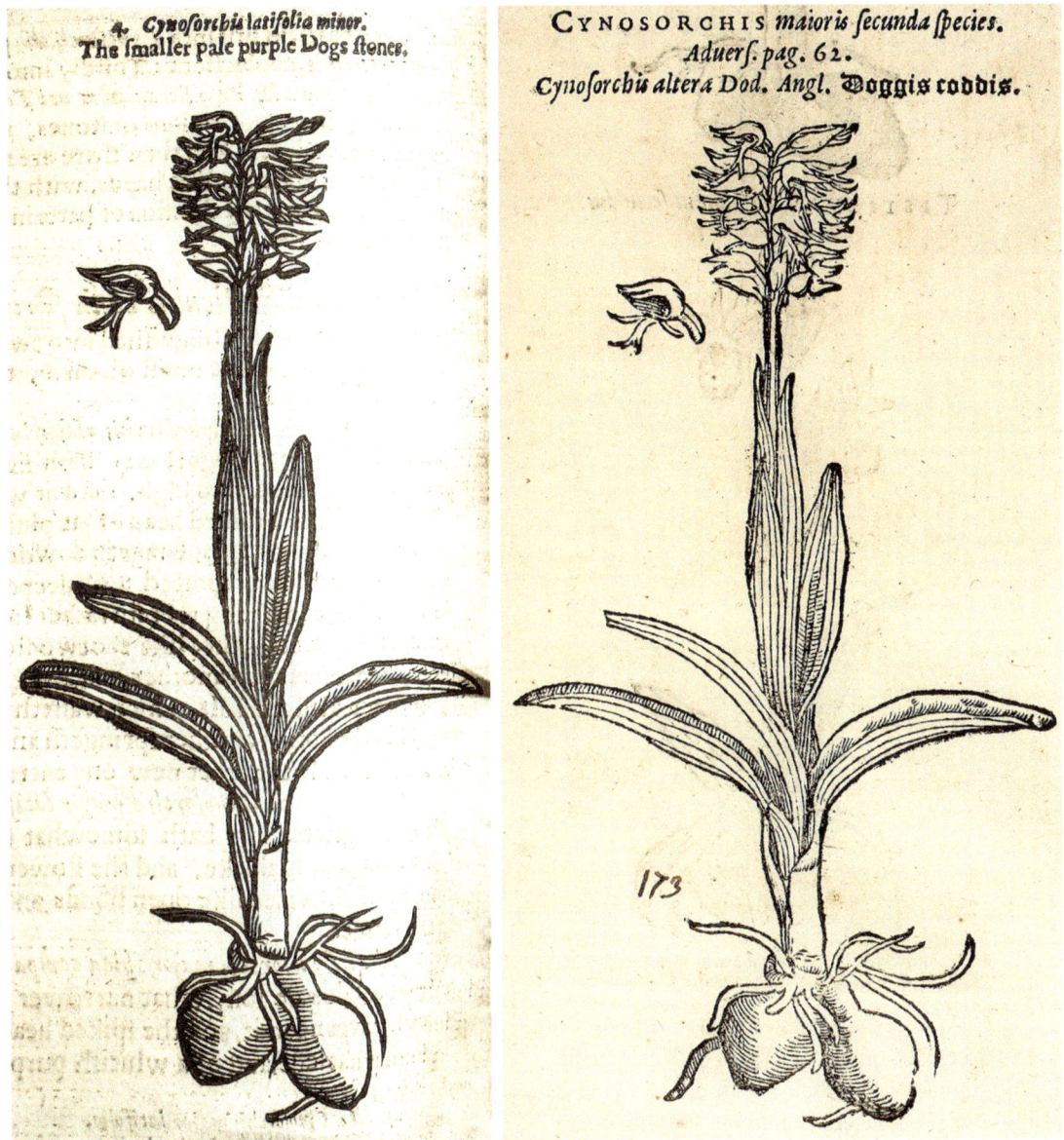

Fig. 2.45 *Orchis italica*. LEFT: in Parkinson (1640) [as *Cynosorchis latifolia minor*, smaller pale purple dog stones]. RIGHT: in l'Obel (1576) [as *Cynosorchis Altera Dog Angl*, Doggie cobbies]

H. pectinata, Zeuxine strateumatica and even the epiphytic *Cymbidium aloifolium* (Lawler 1982).

Preparation of Salep

In *Medical Botany* (1792), William Woodville quoted a Dr. Percival that the proper way to prepare salep was as follows:

> The best way to use it is to wash the new root in water; separate it from the brown skin which covers it by using a brush and dipping it in hot water. When a sufficient number of roots have been thus cleaned, they are to be spread out on a tin plate, and placed in an oven heated to the usual degree, where they are to remain for 6 or 10 min, in which time they will have lost their milky whiteness, and acquired a transparency like horn without any diminution in bulk. When arrived at this state, they are to be removed in order to be dried and hardened in the air, which will require several days to effect; or by using a very gentle heat, they may be finished in a few hours.

Fig. 2.46 *Anacamptis pyramidalis* [as *Orchis pyramidalis*]. From Sowerby J, *English Botany* (1840)

In Turkey, the orchid tubers are collected at the end of summer when their starch content is at a maximum. If collected in autumn, the tubers are old and their starch content is low. Washing and boiling for a short period removes their bitter taste. After manual removal of the outer covering, the tubers are dried in the sun or in the low heat of an oven until they turn from milky white to translucent. A few days of air-drying hardens the tubers. In this state, they can be stored for long periods. Dried tubers are grounded before use (Ercisli and Esitken 2002). Grounded powder has a slight yellow tinge like *baiji* (*Bletilla striata*). Salep orchids are present in pills sold as natural 'health' products that claim to promote male fertility and correct erectile dysfunction.

Fig. 2.47 *Dactylorhiza incarnata* [as *Orchis latifolia*]. From Sowerby J, *English Botany* (1840)

Medicinal Usage

Although it is patently untrue, it was once believed that salep contained the greatest amount of nourishment in the smallest bulk and was thus useful in times of privation or famine (Culpeper 1652). A small amount of salep in a large volume of warm water converted into a jelly-like substance which was believed to be superior to rice. It masked the taste of salt water. To protect against famine at sea, it was proposed that salep should constitute part of a ship's provision at all

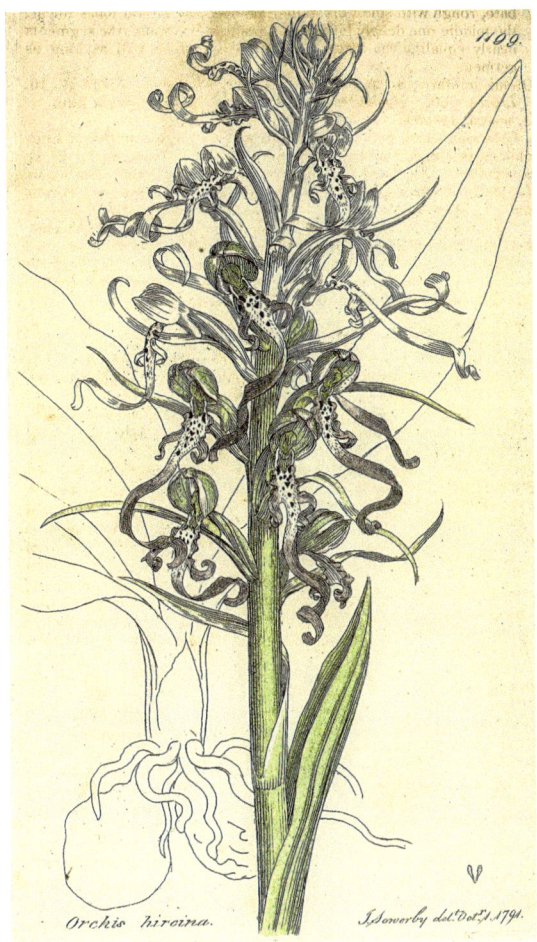

Fig. 2.48 *Himantoglossum hircinum* [as *Orchis hircina*]. From Sowerby J, *English Botany* (1840)

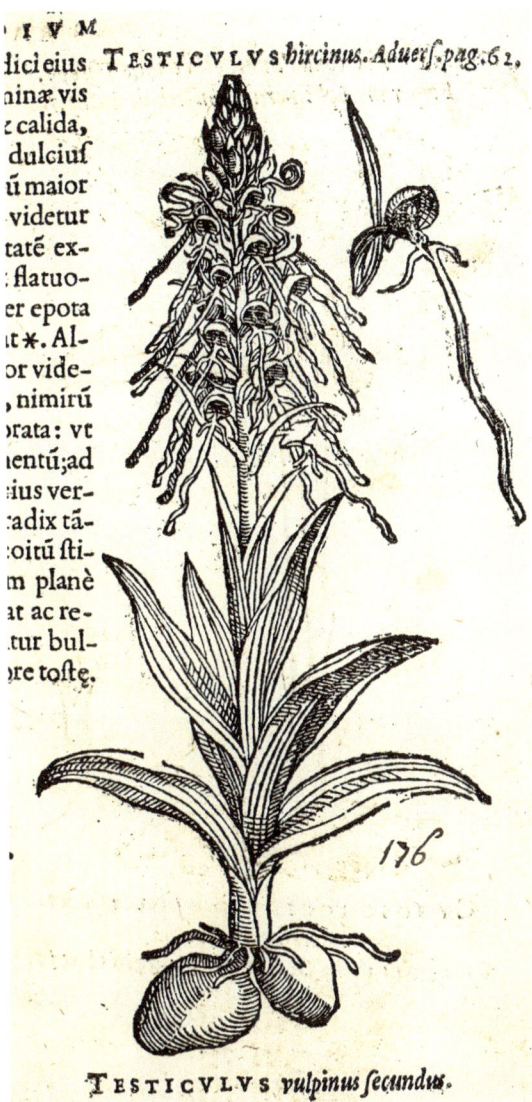

Fig. 2.49 *Himantoglossum hircinum* [as *Testiculus vulpinus fecundus*]. From: l'Obel, (1576) op cit

times. It was also considered to have a role in treating scurvy, diarrhoea, dysentery and fever (Woodville 1792; Hooper and Akerly 1829). It was seldom employed in the United States, except in the composition of Castillon powders. Numerous tuberous orchids apart from *Orchis* have also been used as a nutrient in other parts of the world, generally serving an alternative food for tribals when they were deprived of other nutrients (Hedley 1888; Low 1987).

Powdered salep is not readily miscible with water and Fernie (1914) recommended that it should first be stirred with wine. When the powder is well dissolved, water is added and the mixture brought to boil. Fernie's recipe called for 1 drachm (approximately 4 g) of salep powder to be dissolved in 1.5 fluid drachm (6 ml) of spirit, followed by 0.5 pint (284 ml) of water. Fernie (1914) claimed that 'salep is a most useful article of diet for those who suffer from chronic diarrhoea'.

Salep, not *khus yaty'l klab,* was the official name when it entered the standard pharmacopoeias of Europe, the United States, Japan and some South American nations in the nineteenth century. It was employed as a restorative or tonic, an emollient and to treat

Fig. 2.51 *Ophrys apifera*. From Sowerby J, *English Botany* (1840)

Fig. 2.50 *Anacamptis morio* [as *Orchis morio*]. From Sowerby J, *English Botany* (1840)

stomachache, heartburn, bilious colic, diarrhoea, dysentery and other intestinal disorders. It was used for coughs, colds, tuberculosis and other respiratory disorders. It was used to treat infections of the bladder and kidneys, strangury (painful urination in drips) and renal stones or to treat acute and chronic fevers, whatever their cause. It was employed to stop bleeding and haemoptysis. It was employed to correct infertility, prevent abortion, facilitate childbirth, expel the afterbirth and cure venereal disease and sexual misconduct. It was recommended for diabetes, scurvy, arthritis, scrofula, paralysis, nervous exhaustion, hoarseness and poisoning and to assist recovery from a prolonged illness. *Saloop* was a sovereign cure for drunkenness (Lawler 1984), almost a panacea. But that was all in the past. Salep is not employed in modern medicine.

Fig. 2.52 *Spiranthes autumnalis*. From Sowerby J, *English Botany* (1840)

Composition of Salep

Despite earlier claims of its properties, the constituents of salep are unremarkable. Main constituents are mucilage, starch, cellulose, sugar (glucose, mannose, glucomannan), some proteins, a bit of fat, a trace of acetic acid, water and ash, the last consisting predominantly of chlorides and phosphates of potassium and calcium and sometimes calcium oxalate. Different orchids and their source determine the actual

Fig. 2.53 *Gymnadenia conopsea*. From Sowerby J, *English Botany* (1840)

Fig. 2.54 *Orchis anthropoporphyra* [as *Aceras anthropoporphyra*] From Sowerby J, *English Botany* (1840)

amounts of the various components. Coumarin is present in its volatile oil.

In modern Turkey, salep is employed as a stabilizer, particularly in Kahramanmaras-type ice cream. It is said to improve the taste and slows the melting of the ice cream. This property is attributable to glucomannan and to a lesser extent to starch. Tekinsen and Guner (2009) proposed that species which contain the highest amount of glucomannan and starch, like *Orchis italica*, *O. morio* and *O. anatolica*, should be more valuable and effort should be made to turn them into commercial crops through tissue

culture and cultivation. Instead, it is reported that despite a 1995 prohibition of export of salep in either tuber or powder form, 120 million wild orchid plants are damaged annually in Turkey alone (Tekinsen and Guner 2009). The reason: it takes 1000–4000 dried tubers of the terrestrial orchids to make 1 kg of salep and annual salep production in Turkey is around 45,000 kg (Krentz 2002). Effort should be made to cultivate the salep orchids, but the choices need to be based on further research and factors like ease of propagation and field tests. In vitro, asymbiotic

Fig. 2.55 *Orchis mascula*. From Sowerby J, *English Botany* (1840)

Fig. 2.56 *Orchis simia*. Photo: Henry Oakeley

germination of *Orchis mascula* has been achieved using max medium supplemented with benzyl adenine and activated charcoal (Valetta et al. 2008).

Two phytoalexins, orchinol and p-hydroxybenzyl alcohol, have been isolated from *Orchis mascula*, *O. militaris*, *O. sambucina* and *Anacamptis morio* (syn. *O. morio*) and in small amounts in *Dactylorhiza hatagirea* (syn. *Orchis latifolia*), but the two compounds were not found to be present in *Orchis maculata* and *Neotinea ustulata* (syn. *Orchis ustulata*). Orchinol was the first phyto-alexin to be characterized (Nuesch 1963). The phenolic glycoside, loroglossin is also present in *Orchis mascula* (Veitch and Grayer 2001). They may play an import role in the ecology of *Orchis*, but currently no medical application has been discovered. Nevertheless, a herbal capsule containing ten herbs among which is *Orchis latifolia* (=*Dactylorhiza hatagirea*) claims to promote health and vitality for men and to 'strengthen muscles, vitalize and counteract premature aging by reducing the formation of free radicals'. Even in first world countries, such preparations may occasionally be found in health supplement market which is not stringently regulated (Fig. 2.56).

References

Chopra RN (1933) The indigenous drugs of India. The Art Press, Calcutta. Republished as Chopra's Indigenous Plants of India, 2nd ed. Kolkata: Academic Publishers (1986)

Culpeper N (1652) Culpepper's English physician and complete herbal. British Directory Office (stated as: printed for the author), London

Dalby A (2000) The name of the rose again; or, what happened to Theophrastus on Aphrodisiacs? Petits Propos Culinaires 64:9–15

Dymock W, Warden CJH, Hooper D (1893) A history of the principal drugs of vegetable origin met with in British India. Education Soc. Press, Bombay

Emboden WA (1974) Bizarre plants. Magical, monstrous, mythical. Macmillan, New York

Ercisli S, Esitken A, (2002) Orchids (salep) growing in Turkey. Proc. 12 WOC, Shah Alam, Malaysia, pp 242–244

Fairclough HR (tran.), Gould (ed) (2001) Virgil Aeneid 7–12. Vergiliana. Cambridge, Loeb Classic Library

Fernie WT (1914) Herbal simples approved for modern uses of cure, 3rd edn. John Wright & Sons, Bristol

Fragoso L (1575) De Succedaneis Medicamentis. P. Cosin, Madrid

Fuchs L (Fuchsius Leonhardus) (1542) De historia stirpium

Ghorbani A, Gravendeel B, Naghibi F, de Booer H (2014) Wild orchid tuber collection in Iran: a wake-up call for conservation. Biodivers Conserv 23:2749–2760

Ghorbani A et al (2017) DNA barcoding of tuberous Orchidoideae: a resource for identification of orchids used in Salep. Mol Ecol Resour 17(2):342–352

Grieve M (1971) A modern herbal, vol II. Hafner Publishing Co, New York

Griffith RE (1847) Medical botany: descriptions of the more important plants used in medicine, with their history, properties and mode of administration. Lea and Blanchard, Philadelphia

Hedley C (1888) Uses of Queensland plants. Proc Royal Society of Queensland 5:10–13

Hooper R, Akerly S (1829) Lexicon medicum or medical dictionary, 4th American edn. Collins and Hannay, New York, p 137

Krentz CAJ (2002) Turkiye'nin orkideleri salep, dondurma ve katliam. Yesil Atlas Degisi 5:99–109. (quoted by Tekinsen, 2009)

Lawler LJ (1984) Ethnobotany of the orchidaceae. In: Arditti J (ed) Orchid biology reviews & perspectives, vol 3. Cornell University Press, Ithaca

Lewis MWH (2009) Power and passion: the orchid in literature. In: Arditti J (ed) Orchid biology reviews and perspectives, vol V. Timber Press, Portland

Low T (1987) Australian wild foods. Ground orchids – salute to Saloop. Australian Natural History 22 (5):202–203

Nuesch J (1963) Defense reactions in orchid bulbs. Symp Soc Gen Microbiol 13:335–343

Parkins (1809) The English Physician, enlarged with 369 medicines made of English herbs not in any former impression of Culpeper's British Herbal. Crosby & Co., London

Sezik E (1967) Turkiye'nin Salepgilleri Ticari Salep Cesitleri ve Ozellikle Mugla Salebi Uzerinde Arastirmalar. Doctoral Thesis. Istanbul Universitesi Eczacihk Fakultesinde (In Turkish. Summary in English)

Sezik E (1990) Turkiye'nin orkideleri. Bilim ve Teknik 269:5–8. (quoted by Ericisli, Esitken, 2002)

Shakespeare W (1600) Hamlet

Tekinsen KK, Guner A (2009) Chemical composition and physicochemical properties of tubera salep produced from some Orchidaceae species. Food Chem 121:468–471

Teoh ES (2016) Medicinal orchids of Asia. Springer, Cham

Turner P (ed) (1962) Selections from The History of the World commonly called The Natural History of C. Plinus Secundus. Translated into English by P. Holland. Centaur Press, London

Turner W (1551) A new herbal. S. Mierdman, London. Parts II and III edited by GTL Chapman, F McCombie and A Wesencraft. Cambridge: Cambridge University Press, 1995

Valetta A, Attorre F, Bruno F, Pasqua G (2008) In-vitro asymbiotic germination of *Orchis mascula* L. Plant Biosyst 142(3):653–655

Veitch NC, Grayer B (2001) Phytochemistry of Habenaria and Orchis. In: Pridgeon AM, Cribb PJ, Chase MW, Rasmussen FN (eds) Genera Orchidacearum, Vol. 2. Orchidoideae (Part One) Orchidoideae. Oxford University Press, Oxford

Wedeck HE (1961) Dictionary of aphrodisiacs. Philosophical Library, New York, p 216

Woodville W (1792) Medical botany. James Phillips, London

No orchid species surpasses the reputation of *Tianma* (*Gastrodia elata*, R. Br.) in Traditional Chinese Medicine (TCM). Possibly the oldest orchid to be employed medicinally, *Tiama* was said to have been bestowed on the Chinese people by a divine person some 5000 years ago to tackle 'diseases of wind'. Its original name was *Chijian* (red arrow), alluding to the appearance of the upright, brick-red inflorescences. Since this arrow apparently appeared from nowhere, the Japanese call it *Oni No Yagara* or 'Orge's Arrow'. During the Song Dynasty, tubers of medicinal *Gastrodia elata* was conferred the name *Tianma* which alluded to its supposedly divine origin.

Another popular medicinal name for *Gastrodia elata* is *Ding Feng Cao* (wind-calming herb). Other names describe the appearance of tubers after they are processed: *Bai Long Pi* (white dragon skin) and *Bailongcao* (white dragon herb). The scientific name is derived from ancient Greek, *gastro* (stomach) or *gastrodes* (thick bellied) in reference to the sepals which bulge laterally (Mayr 1998) (Fig. 3.1).

Three Legends

Anyone familiar with TV commercials should know that legends provide the spin in selling a product. Since ancient times stories have been crafted to teach morals (*Aesop's Fables, Jataka Tales*), or to transmit critical observations about human behaviour (*Romance of the Three Kingdoms, Journey to the West*), and, in the case of medicine, to promote the use of certain herbs. Once the story is heard (or watched), most people find it difficult to shake off the subliminal suggestion that it is, in fact, the truth. This is a working hypothesis of spin doctors.

Thus it is not surprising that numerous legends are attached to medicinal orchids and other medicinal herbs. Cures are initially promoted by word of mouth usually based on a single or a handful of anecdotes. If such anecdotes chanced to be recorded by writers of some repute, such as scholars or officials, they became the stuff of legend. Such was the basis of traditional treatment.

1. A very long time ago, pestilence descended on an ancient village. One by one, people suffered from headaches or dizziness and became unconscious, and afterwards they could not move their limbs in a proper fashion. Village doctors did not understand the illness. They were unable to offer a cure. Help was sought from nearby towns and villages, but no healer was forthcoming with a remedy. One day, a stranger arrived at the village. When he observed that many people were afflicted with the disease, he told the young innkeeper that an old man in a remote village had cured people suffering from a similar malady. He did not know the name of the village. He could only point to its direction. On hearing this, the

© Springer Nature Switzerland AG 2019
E. S. Teoh, *Orchids as Aphrodisiac, Medicine or Food*, https://doi.org/10.1007/978-3-030-18255-7_3

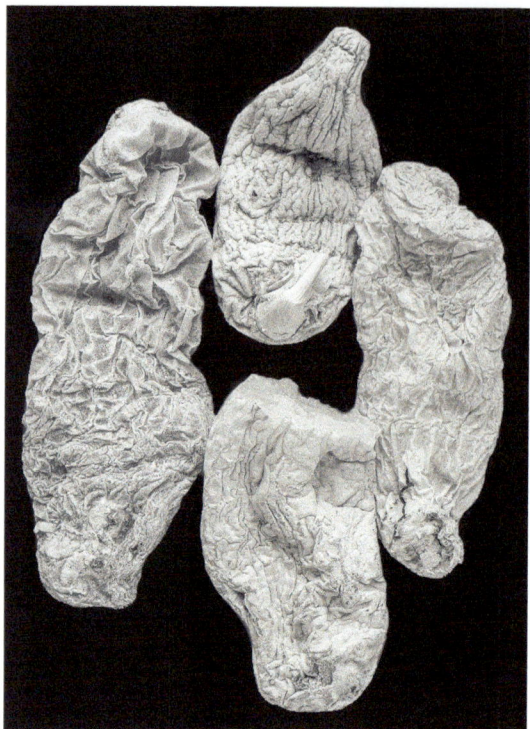

innkeeper closed his premises and went in search of the old man. Along the way, he asked about the healer, but no one seemed to know. After travelling for many days, he reached the end of the beaten track where he met an old man who told him that if he journeyed on into the forest he would presently meet the healer that he sought. The path was difficult and hazardous; nevertheless, the indomitable innkeeper journeyed on. Finally, hungry, exhausted and despite his youth, the innkeeper could not walk any further. He collapsed. When he regained consciousness he found himself in a hut with the same elderly man that he had met on the road hovering over him. He tried to get up but he could hardly move his limbs. The old man fed him a brew made with some tubers which looked like wrinkled sweet potato and advised him to rest. He was administered the same brew over the next few days. Gradually he regained

his strength and eventually, full control of his limbs. The old man then showed him how to locate the tubers. Amidst fallen trees and decaying wood in the forest, there were strange-looking plants which had the appearance of red arrows plunged into the ground. Apparently these were flowers produced by the tubers. The plants were devoid of leaves.

The two men managed to collect a bundle of tubers by digging into the ground below the arrows. The old man then presented them to the innkeeper. He advised the young man to use the tubers to treat the sick people in the village. He should keep several to plant in the forest beneath oak trees, and the following year, he might return to the spot to collect more tubers—this despite the fact that he would not see plants growing there. Thanking the old man, the innkeeper went into the hut to collect his few belongings. When he emerged a minute later, he found the bundle of orchid tubers on the doorstep, but the old man had vanished.

People say that *Tianma* was a gift from the divine.

2. The second legend explains the first name of the orchid and its usage.

When a shepherd suffered from a stroke, he had no choice but to send his son to tend their goats. On the mountainside, the young lad noticed that doves had made a nest on a tree. Hoping to find some eggs in the nest, or perhaps even capture a bird for dinner, the boy climbed up the tree. When he reached the higher branches, a sudden gust of wind shook the entire tree. The boy looked down to see how far he would fall if he could not hold on. Serendipitously, he saw red arrows embedded vertically in the ground below some old oak trees. When it was safe to continue his climb, the lad managed to find and collect some eggs from the nest. Back on the ground, he discovered that the 'red arrows' were really inflorescences bearing numerous buds. But strangely the plants were devoid of leaves. Curious, he dug into the ground whereupon he uncovered succulent tubers that looked very much like potatoes. He surmised that these might make an excellent gruel for his father when cooked with eggs. Indeed, fed on this mix for many days, the ailing father recovered completely. When news of his miraculous recovery spread, people came to collect *Chijian* (red arrow) tubers to use for treating strokes and other 'diseases of wind' (Fig. 3.2).

3. Herb collectors ensure their livelihood by not over-collecting the plants regardless of price. Conservationists today encourage such practice, referring to it as 'sustainable exploitation'.

A long time ago, there was a collector who so dominated the local market with *Gastrodia elata* tubers that he became known as Gastrodia Man. Then came a time when he could not find any tuber in the forest because he had over-collected the previous year. He resorted to seed sowing, but the following year, still no tuber could be found. Thinking that someone had beaten him in harvesting the crop, he decided the next year to live by the field where he had sown the seeds. The result was still the same: no *Tianma*. The third year was also barren. Gastrodia Man concluded that man could not cultivate the plant which was actually a gift from heaven. It deserved the name *Tianma* which translates as 'heaven's fibre' (Figs. 3.3 and 3.4).

The term *Tianma* was first recorded during the Song Dynasty (960–1279) in *Kai Bao Bencao* published in 973 (Chen and Tang 1982) indicating that the last legend is more recent than the second. Since *Tianma* alludes to the herb's alleged divine origin, it immediately took precedence over other names in popular usage.

麻 天 箭 赤

Fig. 3.2 *Tianma*. Diagrammatic representation of the tuberous (but leafless) orchid plant in Li Shizhen's *Bencao Gangmu* (1593). Published posthumously, this drawing of a plant bearing leaves would not have been approved by the author

Fig. 3.3 *Gastrodia elata*, inflorescence (Photo: Qwert 1234, *Wikimedia Commons*)

Fig. 3.4 *Gastrodia elata*, seedpods on old inflorescence (Photo: Qwert 1234, *Wikimedia Commons*)

Symbiosis

Indeed the orchid was not successfully cultivated until the 1960s when the Chinese scientist, Xu Jintan discovered that not only did *Gastrodia elata* require fungal assistance for germination, it needed a second mycorrhiza for continued survival and growth. This discovery paved the way for commercial cultivation of *Gastrodia elata* in China to meet the overwhelming demand that was fast depleting natural sources. Today, marketed *Tianma* is mostly farmed, not collected from the wild. It is estimated that up to 80% of *Tianma* consumed in China comes from farms.

Gastrodia elata is widely distributed in China occurring in Sichuan, Yunnan, Guizhou, Guangxi and Hainan, but it is under threat because of over-collection for herbal usage. During the 1960s over a hundred tons of tubers were harvested annually. The Chinese government now bans export of *Tianma* collected from the wild, and *Tianma*

requires government approval to be exported or imported. This restricts trade to cultivated *Tianma*.

Meanwhile, Korean scientists have recently established an in vitro production system of *Gastrodia elata* using symbiotic seed germination (Park et al. 2008). In a WHO publication of the medicinal plants of Korea, *Gastrodia elata* is the only orchid listed among the 150 medicinal plants (Han et al. 1998). Its Korean name is *Cheon ma*.

Parasitic and leafless, totally devoid of chlorophyll, *Gastrodia elata* lives underground and only becomes visible when it sends out its yellowish-red flower stalk or 'crimson arrow' (*Chi Jian*). The tuber is irregular, somewhat ovoid-shaped, 8–12 cm long and 3–7 cm in thickness (Chen et al. 1999). It resembles a sweet potato tuber. Inflorescence arises from the growing pole of a mature tuber. It is 5–30 cm tall and carries 30–50, white to orange flowers that are loosely arranged, almost in opposite facing pairs, over the distal third of the raceme. Unlike most orchid flowers, its lip is at the top and faces downwards. Flowering season is May to July all over China (Chen et al. 1999; Perner and Luo 2007; Jin et al. 2009).

From as early as the fourth century, several Chinese writers already commented that *Tianma* was similar to a mushroom or fungus. The Daoist master and pharmacologist, Tao Hongjing (425–536) stated in his famous *Tujing yanyi bencao* that *chijian* was a fungus (*zhi*) which swayed in the wind and even when the air was still. In the *Tang Bencao* which was edited by a team of scholars headed by Su Jing and published in 659, it was mentioned that 'in the *Baopuzi,* authored by Ge Hong (283–343), a formula for immortality included *duyaozhi*' (a spontaneously waving fungus plant). The famous Ming Dynasty pharmacologist, Li Shizhen (1518–1593) noted that *Tianma* was a parasitic organism whose origin was unknown. All rhizomes had a dense coverage of minute hair-like roots; like *fuling* (*Poria cocos*), another fungus that was employed medicinally. When devoid of this *fuling,* spiking did not succeed (Fig. 3.5).

Orchid-fungal symbiosis became common knowledge after Bernard in 1909 and Burgeff in

Fig. 3.5 *Gastrodia elata*
tuber covered with mycelia
of its symbiotic fungus.
Armillaria mellea. From:
Kusano (1911): *Gastrodia
elata and its symbiotic
association with Armillaria
mellea.* J. Coll agri Tokyo
4: 1–33

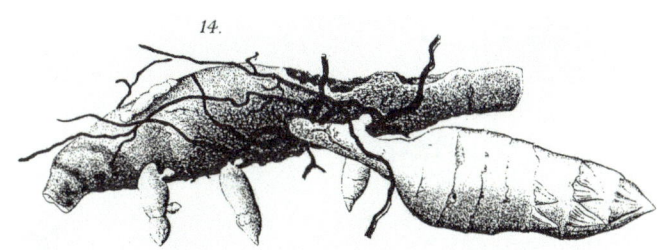

1909 established that mycorrhiza was essential for orchid seed germination (Arditti 1992; Bernard 1908, 1909; Burgeff 1909). Kusano (1911) at the University of Tokyo then discovered that *Mycena osmundicola* supported germination of *Gastrodia elata.* However, this knowledge did not lead to laboratory or open cultivation of the commercially important *Tianma* for another 50 years because the orchid needed to establish a second symbiotic relationship with a different mycorrhiza for its tubers to thrive. During the 1960s, Xu and Guo achieved limited clonal propagation of *Gastrodia elata* by culturing budding divisions of the tubers and then transferring them to timber infected with *Armillaria mellea* (honey mushroom) for growth (Xu and Guo 2000). Unfortunately serial passage led to degeneration of the tubers and reduction in yield, so this method involving vegetative propagation had limited commercial application (Fig. 3.6).

Sexual reproduction of *Gastrodia elata* eluded Chinese scientists until someone noticed that the orchid germinated on fallen oak leaves (*Quercus* spp., Fagaceae) in the absence of *Armillaria mellea.* The search for mycorrhiza responsible for germination led to the isolation of 12 species from the *G. elata* protocorms, and one of them, *Mycena osmundicola,* supports *Gastrodia elata* seed germination to an 80% level. This was essentially a reconfirmation of the Tokyo findings of 1911 (Kusano 1911; Xu 1989, 1990a, b). Subsequently, it was discovered that mycorrhiza isolated from other orchid species also supported *Gastrodia elata* seed germination. Among these were *Mycena orchidicola* isolated from *Cymbidium sinense* (Fan et al. 1996), *Mycena*

anoectochila from *Anoectochilus formosanus* (Guo et al. 1997), *Epulorhiza albertaensis* from *Eria szetchuanica* (Fan and Guo 1998) and *Mycena dendrobii* from *Dendrobium officinale* (=*Dendrobium catenatum*) (Guo and Fan 1999). On the other hand, the *Armillaria mellea* on which adult *Gastrodia elata* is so dependent inhibited seed germination of *Gastrodia elata* (Xu 1989; Xu and Guo 2000).

It was noted that *Gastrodia* protocorms (seedlings) failed to thrive unless they established relationship with a second fungus, *Armillaria mellea* (the edible honey mushroom). Scientists from the Medicines Institute of the Chinese Academy of Medical Sciences led by Xu Jintan concluded that *Gastrodia elata* needed *Mycena osmundicola* for seed germination; following which, it needed to switch to another mycorrhiza that would supply it with nutrients, hormones and other materials essential for its protection. Armed with this knowledge several Chinese teams set out to cultivate *Tianma* on a commercial scale. Today the best *Tianma* comes from Guizhou Province, and it is almost entirely cultivated.

In nature, the relationship of *Gastrodia elata* to mycorrhiza is far more complex. Sixty-two strains of endophytic fungi have been isolated from *Gastrodia elata* collected from different areas of Guizhou, Shanxi, Sichuan and Yunnan. They belong to 13 genera, 9 families, 7 orders, 5 classes and 3 phyla. Fungal species differ in different geographic regions and parts of the orchid plant (Mo et al. 2009). In Hokkaido, Japan, *Armillaria nabsnona* is present in the tubers of *Gastrodia elata* (Sekizaki et al. 2008).

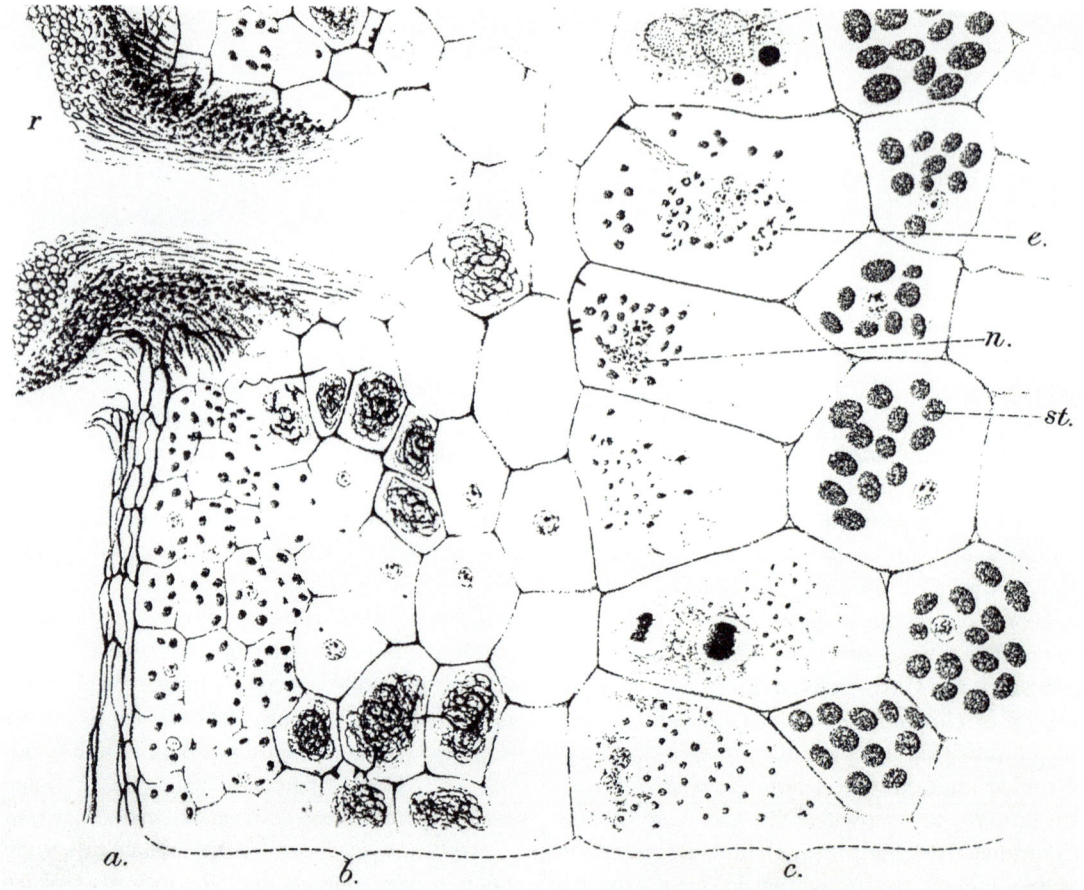

Fig. 3.6 Pelotons (mycorrhizal bodies within host cells) in *Gastrodia elata* tuber, the fungus and host coexisting with mutual benefit. From: Kusano (1911), Ibid

Grades of *Tianma*: Adulterants and Substitutes

Protocorms become small, white *Gastrodia elata* within a year. After winter they bud, blossom and bear fruit, taking 3 years to complete the life cycle (Xu 1989). Commercial *Gastrodia elata* is most commonly harvested in winter. Sometimes, the orchid is also harvested in spring. Wild herbs are known, respectively, as *Winter Gastrodia* (*Dong Tianma* or *Dong Ma*) and *Spring Gastrodia* (*Chun Tianma*), with the former considered superior. *Tianma* from Sichuan is known as *Chuan Tianma*. Harvesting of cultivated *Tianma* is timed for November, at the start of winter (Figs. 3.7, 3.8, 3.9 and 3.10).

The strain of cocultured *Armillaria mellea* determines the yield of cultivated *Gastrodia elata*, strain Av-4 providing the greatest yield in both wild cultivation and outdoor-box planting (Ji and Li 2009). Traditionally, wood of *Quercus fabri* (Chinese oak, native to China and Korea) is employed as a substrate for culture of *Gastrodia elata*. Nevertheless, recently, it was discovered that satisfactory yield of *Tianma* tubers can also be achieved through cultivation in wood of *Betula luminifera* (birch), *Populus adenopoda* (aspen) and *Juglans regia* (walnut). In this study, the highest yield was achieved by using wood from *Betula luminifera* (Rong and Cai 2010).

Excellent *Tianma* should have big, stout, compact rhizomes without holes at its core; is creamy white, devoid of dark spots; translucent and shiny

Fig. 3.7 Small *Tianma* tubers on sale in Yunnan. (©Teoh Eng Soon 2019. All Rights Reserved.)

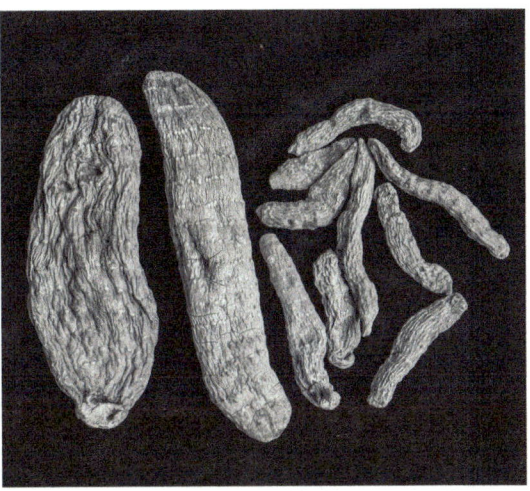

Fig. 3.9 *Tianma* tubers bought in Guihou. (©Teoh Eng Soon 2019. All Rights Reserved.)

when sectioned. *Gastrodia elata* put up for sale is sometimes adulterated by various tubers of several ornamental garden plants such as *Mirabilis jalapa* (common name: marvel of Peru), *Dahlia pinnata*, *Canna edulis*, *Cacalia tangutica* and even potato (*Solanum tuberosum*) (Bensky et al. 2004). Besides *Gastrodia elata* there are over 20 species of *Gastrodia* distributed throughout the world, in Southern and Tropical Africa (Cribb et al. 2010; Hsu and Kuo 2010), India and eastwards through the rest of Asia to Australia, New Zealand and the Pacific Islands. They are not cultivated and, as far as I am aware, they have not been used as substitutes for *Gastrodia elata*. Tubers of *Gastrodia sesamoides*

Fig. 3.8 Additional varieties of *Tianma* tubers offered in Yunnan. (©Teoh Eng Soon 2019. All Rights Reserved.)

which is native to Australia and New Zealand are eaten by their respective natives, the Woiworung in Queensland and Maoris in New Zealand.

Maoris believe that the plant is supernatural. When one goes digging for *Hupari* (the Maori name for *Gastrodia sesamoides*), one must not call it by its real name, or else it would disappear. If one needs to mention it, it should be referred to as *maukuuku*. So goes the Maori admonition.

Processing the Tubers and Usage

After harvesting, stems and roots are removed from the rhizomes. Mud is washed off and the rhizomes are soaked in clean water. The rough outer coat is stripped off, and the rhizomes are steamed or boiled until they become totally translucent and devoid of white spots. Boiling time varies according to tuber size: those weighing more than 150 g require 10–15 min, whereas small tubers weighing 70–100 g are boiled for only 5–8 min. Tubers are then are placed on a heated brick stove until 70–80% dry at which stage they are flattened by hand before they become totally dry.

Tianma is marketed in the form of hard, dry rhizomes. When needed the rhizome is soaked in water until it is about 70% wet. It is then ready to

Fig. 3.10 *Tianma, Shihu* (processed *Dendrobium* pseudobulbs) and turmeric on sale in a wholesale market in Guangzhou, China. (©Teoh Eng Soon 2019. All Rights Reserved.)

be cut into slices which are dried again. Sliced *Tianma* is fried in low to medium heat until both sides turn a light yellow. After transfer to paper, the slices are sprayed with water, placed in a pot and cooked in low heat until the surface turns to dark yellow. As a decoction, the usual dosage is 3–9 g daily; as a powder 1–1.5 g per dose. *Tianma* is described as sweet in taste, neutral in property. According to TCM, it nourishes the liver meridian (Yang et al. 1999).

Sometimes the herb is cooked with meat and served as soup. That is because it is 'sweet, neutral and moist, with a nature that is slightly moistening and possessing a tonic effect'. It 'nourishes the yin fluid, calms the liver and extinguishes wind'. It is used to treat all forms of *internal wind* (disorders of the nervous system), such as 'wind' stroke, seizures, aphasia, slurred speech,

blurred vision, numbness, tingling of the extremities, headache, vertigo and dizziness. In his *Grand Materia Medica* published in 1596, the celebrated Ming Dynasty herbalist, Li Shizhen (1518–1593), quoted past masters who said that without *Tianma* it was impossible to treat 'darkening of the vision with dizziness, a wind disorder'. Thus, *Tianma* is also known as the wind-calming herb (*Ding Feng Cao*).

During a visit to China in September 1971, 6 months before the visit of President Nixon, Victor W Sidel, a public health professor at Albert Einstein College of Medicine at the Bronx in New York City, was told that China was relying on a barefoot doctor's system to cope with the health needs of its massive rural population. In addition to being farmers, these health workers were trained to deliver both preventive and treatment services in the village. For the latter purpose, they were provided with a manual, one of which was prepared by the Revolutionary Health Care Committee of Hunan Province in 1970. The English translation of the *Barefoot Doctors Manual* provides a glimpse of the symptoms and syndromes which required the use of *Tianma* in rural China in the 1960s and 1970s. The *Manual* states that *Tianma* is advocated for 'lateral and frontal headaches, dizziness and fainting, epileptic fits, muddled speech, numbness, joint pains, infantile diarrhea, apprehension and convulsive spasms' (Anonymous, trans., 1974).

A number of prescriptions employing *Tianma* are reproduced in *Zhongyao Da Cidian* (*Chinese Medical Encyclopedia*). There are provincial variations in its usage and the *Encyclopedia* provides a wide range of indications. Generally *Tianma* is used to treat lesions affecting ectodermal tissues (i.e. brain, nerves and skin), but there is also a claim that it aids in tissue repair and boosts immunity (Hu 2005).

Much is lost in the process of translation because concepts of Traditional Chinese Medicine (TCM) differ considerably from concepts of Western medicine. In TCM, joint pains may be attributed to 'wind', but this would be only applied to reflected pain caused by nipping of a spinal nerve or cold injury to a nerve: it would hardly apply to pain caused by septic arthritis. In *Barefoot Doctor's Journal* two recipes for treating dizziness contain *Tianma*. However, dizziness is qualified according to whether it is due to (1) sputum and moisture stagnation, (2) abnormal *liver yang dominance* or (3) energy and blood deficiencies: prescriptions are individualized accordingly. 'Dizziness due to sputum and moisture stagnation' is characterized by heavy headedness, chest discomfort, nausea, vomiting, an urge to spit, coated tongue and a slippery pulse. In addition to acupuncture and massage, the following herbal remedy is provided:

Decoction prepared with:
Processed *pan hsia*
Toasted *pai shu*

Tienma
Fu-ling
Chen-pi (dried orange peel) 2 qian (10 g)

'Dizziness arising from *liver yang dominance*' is characterized by restlessness; anger; insomnia; a dry, bitter taste in the mouth; white to pale yellow, furry tongue; and a bounding pulse. Quelling *liver yang dominance* is best treated with a decoction prepared with:

Tianma
Guoteng (*Uncaria sinensis*)
Mother of pearl (precooked)
Magnetite (precooked)
Hsia ku tsao
Lung tan tsao (*Gentiana*)
Yeh chiao teng

This prescription is a variation from the famous *Tianma Gouteng Yin* (*Gastrodia and Uncaria Decoction*) which was formulated in *Zabing Zhengzhi Xinyi* (*New Significance of Patterns and Treatments in Miscellaneous Diseases*) during the Qing Dynasty (1644–1911, but exact publication date and authorship are unknown). The latter contains *Tianma, Uncaria, Haliotis, Achyranthes, Gardenia, Scute, Eucommia, Leonurus, Loranthus, Polygonium* and *Hoelen* (Dharmananda 2009).

Some publications reported that *Tianma* was used to treat internal haemorrhage. However, the *New Compilation of Materia Medica* warns that 'It definitely should not be used lightly in those with *qi* and blood deficiency'. Nor should it be used 'whenever patients feel a lack of saliva with dry tongue and mouth, dry sore throat, constipation with dry stools, blazing fire causing dizziness, blood deficiency headaches without pathogenic wind' (Bensky et al. 2004).

The inflorescence of *Gastrodia elata* was considered to be a tonic with aphrodisiac properties, and it has its own distinctive medicinal name (*huan-t'ung-tzu*) (Li 1596; Stuart 1911).

Tianma has a very low toxicity, and side effects are uncommon when it is consumed in the normal manner. However, it is also available in injectable form and here there are some dangers. There are sporadic reports of allergic reactions, vertigo, anaphylactic shock and acute renal failure. Administration of a very high dosage by whichever route (80 g in 3 h) resulted in poisoning, manifested by flushing, headache, dizziness, visual disturbance, general weakness, loss of muscle coordination and loss of consciousness. *Tianma* is present in varying amounts (2.8–27%) in 9% of patented herbal products manufactured in China. Three products are intended for use in infants, but *Tianma* is contraindicated during pregnancy (Fratkin 1997).

Tianma has been extensively investigated in China and Japan. A few tertiary medical centres in the USA and elsewhere are exploring the possibility of using *Tianma* to treat Parkinsonism and strokes. Presently, there is no suitable medication to treat these conditions. Three compounds from *Tianma* demonstrate an anticonvulsant effect, and one of them, hydroxybenzaldehyde, is more effective than the proven anticonvulsant, valproic acid. In rodents, memory loss and chemically induced abnormal behaviour were suppressed by purified compounds derived from *Gastrodia elata.*

Tianma is used in the Traditional Chinese Medicine (TCM) to treat hypertension and strokes (both to prevent and to treat strokes). In 2003, a Korean team of scientists decided to see what it will do for brain cells when these are denied their oxygen supply for a brief period. In their experiments, gerbils (rodents) were given oral doses of the ether fraction of methanol extracts of *Gastrodia elata* at the rates of 200 mg/kg or 500 mg/kg per day for 14 days. The animals were then subjected to transient global ischaemia, i.e. blood supply to the brain was temporarily interrupted. At 200 mg/kg, the GE extracts failed to attenuate nerve cell damage at the hippocampus, an area of the brain that is involved in memory, learning, emotion and spatial navigation. However, the higher dosage of 500 mg/kg/day reduced hippocampal neuronal damage (Kim et al. 2003). Stroke reduction was also evident when rodents were pretreated with *Tianma* for 2 weeks before being subjected to

transient middle cerebral artery occlusion for 1 h, i.e. the main blood supply to the brain was stopped for an hour (Yu et al. 2005). Similar experiments in China showed that gastrodin was also effective in preventing strokes in rats (Zeng et al. 2006; Bei et al. 2007). Gastrodin protected nerve cells in tissue culture from programmed cell death (apoptosis) when they were deprived of oxygen (Zeng et al. 2007). A Singapore study involving tissue culture found that crude *Tianma* extract promoted nerve cell and dendritic proliferation (Nah et al. 2010).

Constituents of Tianma

Tianma contains a wide range of compounds, viz. (1) phenolic compounds, vanillin, vanillyl alcohol, *p*-hydroxybenzyl alcohol, *p*-hydroxybenzaldehyde, 3,4-dihydroxybenzaldehyde, 4-4 dihydroxydibenzyl methane, *p*-hydroxybenzyl ethyl ether, 4-4 dihydroxydibenzyl ether, 4-ethyloxytolyl-4-hydroxybenzyl ether and 4-hydroxybenzyl methyl ether; (2) glycosides, gastrodin, *p*-hydroxymethyl phenyl-beta-D-glucopyranoside, *bis*(4-hydroxybenzyl) ether mono-beta-D-glucopyranoside, 4(beta-D-glucopyranosyloxy) benzyl alcohol and parishin; and (3) other constituents such as organic acids, *p*-hydroxy benzylaldehye, beta-sitosterol and daucosterol.

The principal active compound is gastrodin which is also found in the mycelia of *Armillaria mellea*. Gastrodin can be produced by *Armillaria mellea* in the absence of the orchid. Vanillin has similarly been prepared using mould and fungal fermentation of wood pulp. For practical purposes, gastrodin obtained from batch fermentation of *Armillaria* is indistinguishable from the gastrodin present in the orchid.

Gastrodin is a simple compound whose formula is known. Medicinal preparations both oral and intravenous employ synthetic gastrodin. Chemical synthesis is a complicated procedure invariably associated with serious environmental pollution. Hopefully, this may be avoided by using the system for enzymatic synthesis which employs gastrodin biosynthesis enzyme (GBE) to biotransform *p*-hydroxy-benzaldehyde into gastrodin. The enzyme itself, GBE, is obtained from the bread mould, *Rhizopus chinensis* SAITO AS3.1165. The method was developed by a team of scientists at the Northwestern University in Xi'an (Zhu et al. 2010).

Both *Tianma* and pure gastrodin are being used by TCM practitioners to treat hypertension, strokes and Parkinsonism. It is postulated that in the human body, gastrodin is converted into methyldopa, L-dopa or a similar substance. For many decades around the mid-twentieth century, methyldopa was widely used to treat hypertension, particularly in pregnant women. However, it is a slow-acting drug with rather low potency. It cannot be relied upon in an emergency, and it is inadequate for the treatment of severe hypertension. Methyldopa has been retained in the modern scientific pharmacopoeia, but like the Indian reserpine, it has generally been superseded by far more effective drugs even in third world countries. L-dopa is used to treat Parkinsonism. Work is underway to modify gastrodin in an effort to find more effective therapies.

A report published in a Chinese journal reported that when 1000 mg of gastrodin was administered intravenously to 63 patients whose hypertension had not responded to other forms of treatment, it brought down the systolic blood pressure after 4 weeks of treatment (Zhang et al. 2008). Current antihypertensive drugs effectively lower diastolic blood pressure, but they are not very effective in lowering systolic blood pressure. It is now recognized that even isolated raised systolic blood pressure can be as damaging as, if not more damaging than, a raised diastolic blood pressure. Therefore a drug that selectively and effectively lowers systolic blood pressure would be a valuable addition to the therapeutic armamentarium. Nevertheless, a good antihypertensive agent must be effective when administered orally, and it should not be necessary to administer it intravenously. Moreover, 1000 mg is a massive dose. A lot of work remains to be done before gastrodin, or its analogue can become a useful therapeutic agent for treating hypertension.

Gastrodin, vanillin and a few other compounds present in *Gastrodia* are present in other parasitic or terrestrial orchids that enjoy a close relationship with fungi, for instance, *Galeola faberi*, *Cremastra appendiculata* (syn. *Cremastra wallichii*) (Liu et al. 2008) and *Coeloglossum viride*. The last orchid, *Coeloglossum viride*, is the famous Tibetan drug *wangla* (Huang et al. 2002) which is purportedly used to treat dementia, but *Galeola faberi* and *Cremastra appendiculata* do not have a similar usage.

Gastrol and ten known phenolic compounds isolated by methanol extraction from *Tianma* were shown to possess smooth muscle relaxant activity of isolated guinea pig ileum (Hayashi et al. 2002). Sulphated derivatives of two glucans, WGEW and AGEW, from *Gastrodia elata* were found to possess strong anti-dengue virus bioactivities, and potency of antiviral activity was directly correlated with the degree of substitution (Qiu et al. 2007). The development of an effective anti-dengue medication will be an important contribution to medicine because dengue is a common, serious illness in tropical Asia.

Anticoagulant Effects

Anticoagulant effects of Chinese herbs can be a boon or a bane depending on how the herb is used and an understanding that it has this effect of increasing the bleeding tendency. A novel phenolic compound from *Tianma*, 4,4'-dihydroxybenzyl sulfone, suppressed platelet aggregation by U466 19 (Pyo et al. 2004). Gastrodin binds to fibrinogen causing fibrinogen depletion and prolongation of coagulation time without affecting the kaolin partial thromboplastin time (KPTT) or prothrombin time (PT) in rats. It inhibits the formation of clots and the risk of thrombosis (Liu et al. 2006). The effect appears to be also attributable to polysaccharide 2–1 from *Gastrodia elata* which prolongs both clotting and bleeding time and prevents platelet aggregation (Ding et al. 2007). There is interest in studying the use of *Tianma* as a possible antithrombotic agent. It may have an edge over aspirin because it does not cause gastric bleeding. However, this implies that one should stop taking *Tianma* for several weeks prior to any surgery or invasive intervention such as angioplasty. Also it should not be used together with aspirin, heparin and other anticoagulants.

Comment

The numerous publications of well-designed and careful research mentioned in the foregoing discussion provide evidence in support of a neuroprotective effect of *Tianma*. Nevertheless, in a recent review on the efficacy of TCM for stroke published in the refereed journal *Stroke*, the authors commented that despite finding 11,234 articles on TCM stroke therapies, the review team with participants from Italy and China were unable to conduct a meta-analysis. They found 34 randomized controlled trials, and all but one reported results in favour of TCM, which to the reviewers suggests a strong publication bias. Wide variations in the studies prevented them from being pooled for analysis. In other words, from the perspective of modern medicine, proof that the *Tianma* is effective in preventing memory loss and that it promotes recovery from stroke in human subjects is still lacking. Such proof has to be provided by well-designed, randomized controlled clinical trials, the more the better. Granted, there is a plethora of health supplements in the billion-dollar market with even weaker evidence than *Tianma*, but that is not a valid excuse for a potentially valuable medicine to go untested. Randomized clinical trials should be performed for the advancement of TCM, perhaps beginning with *Tianma* and going on to other promising herbal remedies.

References

Anonymous (trans) (1974) Barefoot doctors manual. NIH, Bathesda

Arditti J (1992) Fundamentals of orchid biology. Wiley, New York

Bensky D, Clavey S, Stoger E, Gambie A (2004) Herbal medicine materia medica, 3rd edn. Eastland Press, Seattle, WA

Bernard N (1908) La culture des orchids dans ses rapports avec la symbiose. J Soc Nat Hort France 4th Ser 24: 180–185

Bernard N (1909) L'Evolution dans la symbiose, les orchidees et leur champignons commensaux. Ann Sci Nat Bot Ser 9 9:1–196

Bie X, Chen Y, Han J et al (2007) Effects of gastrodin on amino acids after cerebral ischemia-reperfusion injury in rat striatum. Asia Pac J Clin Nutr 16(Suppl 1): 305–308

Burgeff H (1909) Die Wurzelpilze der orchideen ihre kultur und ihr leben in der Pflanze. G Fisher, Jena, p 220

Chen SC, Tang T (1982) A general review of the orchid flora of China. In: Arditti J (ed) Orchid biology. Reviews and perspectives II. Cornell University Press, Ithaca

Chen SC, Tsi ZH, Luo YB (1999) Native orchids of China in colour. Science Press, Beijing

Cribb P, Fischer E, Killmann D (2010) A revision of *Gastrodia* (Orchidaceae: Epidendroideae, Gastrodieae) in tropical Africa. Kew Bull 65(2):315–321

Dharmananda S (2009) Modernizing Chinese medicine. The case of Amillaria as Gastrodia substitute. Google 2009

Ding CS, Shen YS, Li G et al (2007) Study of a glycoprotein from *Gastrodia elata*: its effects of anticoagulation and antithrombosis. Zhongguo Zhong Yao Za Zhi 32(11):1060–1064

Fan L, Guo SX (1998) The development of orchid mycorrhizal fungi research. Microbiology 25:227–230

Fan L, Guo SX, Cao WQ et al (1996) Isolation, culture, identification and biological activity of *Mycena orchidicola* sp. nov in *Cymbidium sinense* (Orchidaceae). Acta Mycol Sin 15(4):252–255

Fratkin J (1997) Chinese herbal patent formulas: a practical guide. Pelanduk Publications, Selangor, Darul Ehsan

Guo SX, Fan L (1999) *Mycena dendrobii*, a new mycorrhizal fungus. Mycosystema 18:141–144

Guo SX, Fan L, Cao WQ et al (1997) *Mycena anoectochila* sp nov isolated from mycorrhizal roots of *Anoectochilus roxburghii* from Xishuangbanna, China. Mycologia 89:952–954

Han ST et al (1998) Medicinal plants in the Republic of Korea: information on 150 commonly used medicinal plants. WHO Regional Office of the Western Pacific, Manila

Hayashi J, Sekine T, Deguchi S et al (2002) Phenolic compounds from *Gastrodia* rhizome and relaxant effects of related compounds on isolated smooth muscle preparation. Phytochemistry 59(5): 513–519

Hsu TC, Kuo CM (2010) Supplements to the orchid flora of Taiwan (IV): four additions to the genus gastrodia. Taiwania 55(3):243–248

Hu SY (2005) Food plants of China. The Chinese University Press, Hong Kong

Huang SY, Shi JG, Yang YC, Hu SL (2002) Studies on chemical constituents from Tibetan medicine wangle (rhizome of *Coeloglossum viride* var. bracteatum). Zhongguo Zhong Yao Za Zhi 27(2):118–120

Ji N, Li Y (2009) Effect of different strains of *Armillaria mellea* on the yield of *Gastrodia elata* f. glauca. J Fungal Res 6(4):231–233

Jin XH, Zhao XD, Shi XC (2009) Native orchids from Gaoligongshan Mountains, China. Science Press, Beijing

Kim HJ, Lee SR, Moon KD (2003) Ether fraction of methanol extracts of *Gastrodia elata*, medicinal herb protects against neuronal cell damage after transient global ischemia in gerbils. Phytother Res 17: 909–912

Kusano S (1911) *Gastrodia elata* and its symbiotic association with *Armillaria mellea*, vol IV(I), vol 4016. College of Agriculture, Imperial University of Tokyo, Tokyo, pp 1–73

Li S (1596) Bencao Gangmu

Liu Y, Tang X, Pei J et al (2006) Gastrodin interaction with human fibrinogen: anticoagulant effects and 4124 binding sites. Chemistry 12:7807–7815

Liu J, Yu ZB, Ye YH, Zhou YW (2008) Chemical constituents from the tubers of *Cremastra appendiculata*. Yao Xue Xue Bao 43(2):181–184

Mayr H (1998) Orchid names and their meanings. Gantner-Verlag K.G., Vaduz

Mo L, Kang JC, He J et al (2009) A preliminary study of the composition of endophytic fungi from Gastrodia elata. J Fungal Res 6(4):211–215

Nah ELQ, Ng XW, Chen KS (2010) An in-vitro study of the effect of gastrodin on neuronal cells. Recent development in Chinese herbal medicine. Programs and Abstracts. Nanyang Technological University, p 67

Park EJ, Ahn JK, Lee WY, Kim ST (2008) Establishment of in-vitro production system of *Gastrodia elata* immature tubers followed by symbiotic seed germination. Plant Biol (Rockville) 2008:172–173

Perner H, Luo Y (2007) Orchids of Huanglong. Sichuan Fine Arts Publishing House, China

Pyo MK, Jin JL, Koo YK, Yun-Choi HS (2004) Phenolic and furan type compounds isolated from *Gastrodia elata* and their anti-platelet effects. Arch Pharm Res 27:381–385

Qiu H, Tang W, Tong X et al (2007) Structure elucidation and sulfated derivatives preparation of two alpha-D-glucans from *Gastrodia elata* Bl. and their anti-dengue virus bioactivities. Carbohydr Res 342:2230–2236

Rong LH, Cai CT (2010) Effect of different woods on the yield of *Gastrodia elata*. Wuhan Zhiwuxue Yanjiu 28(6):761–766

Sekizaki H, Kuninaga S, Yamamoto M (2008) Identification of *Armillaria nabsnona* in *Gastrodia* tubers. Biol Pharm Bull 31:1410–1414

Stuart GA (1911) Chinese materia medica: vegetable kingdom (A revision of a work by F. Porter Smith.). American Presbyterian Mission Press, Shanghai

Xu JT (1989) Studies on the life cycle of *Gastrodia elata*. Zhongguo Yi Xi Xue Ke Xue Yuan Xue Bao 11: 237–241

Xu J (1990a) Cytological observation on hyphae invading *Mycena osmundicola* in the process of germination of *Gastrodia elata* Bl. Zhongguo Yi Xue Ke Xue Yuan Xue Bao 12:313–317

Xu J (1990b) Studies on nutrition source of seeds germination of *Gastrodia elata* Bl. Zhongguo Yi Xue Ke Xue Yuan Xue Bao 12:431–434

Xu J, Guo S (2000) Retrospect on the research of the cultivation of *Gastrodia elata* Bl, a rare traditional Chinese medicine. Chin Med J 113:686–692

Yang YL, Wu KJ, Lu G (eds) (1999) Traditional Chinese materia medica. Wuhan University Press, Wuhan

Yu SJ, Kim JR, Lee CK et al (2005) *Gastrodia elata* Blume and an active component, p-hydroxybenzyl alcohol reduce focal ischemic brain injury through antioxidant related gene expression. Biol Pharm Bull 28:1016–1020

Zeng X, Zhang S, Zhang L et al (2006) A study of the neuroprotective effect of the phenolic glucoside gastrodin during cerebral ischemia in vivo and in vitro. Planta Med 72L:1359–1365

Zeng XH, Zhang Y, Zhang SM, Zheng XX (2007) A microdialysis study of effects of gastrodin on neuro-chemical changes in the ischemic/reperfused rat cerebral hippocampus. Biol Pharm Bull 30:801–804

Zhang Q, Yang YM, Yu YG (2008) Effects of gastrodin injection on blood pressure and vasoactive substances in treatment of old patients with refractory hypertension: a randomized controlled trial. Zhong Xi Yi Jie He Xue Bao 6:695–699

Zhu HL, Dai PG, Zhang W et al (2010) Enzymic synthesis of gastrodin through microbial transformation and purification of gastrodin biosynthesis enzyme. Biol Pharm Bull 33(10):1680–1684

Dwelling on Rocks (*Shihu*)

4

Dendrobium flowers play a prominent role in the orchid cut-flower industry initiated by Singapore in the 1960s and vastly expanded by Thailand after the 1980s. So commonplace is *Dendrobium* that even very young children call it by its botanical name. What is not widely known is the fact that *Dendrobium* is a massive family with approximately 900–1000 member species that may differ vastly in plant form and the appearance of the flowers (Fig. 4.1).

When Olof Swartz coined the name for this genus, he was impressed by the epiphytic nature of its members. *Dendrobium* is derived from two Greek words, *Dendron* (tree) and *bios* (life), i.e. the members live on trees. The ancient and current Chinese name for *Dendrobium* is *shihu*, *shi* (rock) and *hu* (living) referring to the fact that the two most ancient medicinal species thrive on rocks. In Traditional Chinese Medicine (TCM), *shihu* is used to boost a person's vitality and immunity, improve eyesight and moisten and clear the throat (Figs. 4.2, 4.3 and 4.4).

The earliest descriptions of *Dendrobium* are contained in the seminal Chinese pharmacopoeia, *Shennong Ben Cao Jing* (Materia Medica of the Divine Younghusband), earliest copies of which date to the first century (Han Dynasty). *Dendrobium officinale* and *D. moniliforme* or *shihu* were described as herbs that occur on rocks in Central China. During the Tang Dynasty (618–907), other saxicolous *Dendrobium* species distributed along China's two great rivers, the Huang He and Yangtze, were also accepted as *shihu*. This included *Dendrobium nobile*, which is both lithophytic and epiphytic, and since the flowers of this species occur in numerous colour forms ranging from light to dark pink, white, yellow, golden yellow and orange, other look-alike *Dendrobium* species came to be accepted as *shihu* during the Ming Dynasty (1368–1644). In 1960, when China was still very much isolated from the rest of the world and had to depend on its own natural resources to provide universal health care, the People's Health Publishing Agent in Beijing issued a new *Chinese Materia Medica*. Volume III of this publication devoted 16 pages to a discussion of *shihu* limited to 14 species of *Dendrobium* and one species of *Flickingeria*. Hu Shih Ying of the Arnold Laboratory at Harvard published an English translation of this material in 1968. However, a survey by Xu and his colleagues published in 2006 discovered that of the 74 species and 2 varieties of *Dendrobium* native to China, 32 were sold as *Huangcao shihu* (Golden Herb *Dendrobium*). The term *Huangcao* alludes to the yellow or golden appearance of dried *Dendrobium* pseudobulbs, not to the colour of the flowers, albeit several species of *Shihu* bore striking yellow flowers. Collectors of medicinal plants occasionally added orchids belonging to other genera into their collection of *shihu*. *Pholidota* is commonly accepted by dealers as a proper substitute (Bao and Shun 1999).

Fig. 4.1 *Dendrobium officinale* (syn. *Dendrobium catenatum*) (©Teoh Eng Soon 2019. All Rights Reserved.)

The genus *Dendrobium* is widely distributed from Central China southwards through Southeast Asia to Papua New Guinea, Australia and the Pacific Islands, with a southwesterly spread to the southern Himalayas, South India and Sri Lanka.

Members of the genus *Dendrobium* are divided into sections, grouped according to their physical characteristics, with some adjustment on placement now recommended on the basis of DNA fingerprinting. Broadly speaking, two groups are noteworthy. There is a southern clade (the *Spathulata* or *Ceratobium* section) centred in New Guinea characterized by heat tolerance, horn-shaped forms, perennial flowering and with long-lasting flowers that forms the backbone of cut-flower *Dendrobium*. The northern clade is made up of deciduous species that require a distinct dry season and low temperature to initiate flowering, and many members have a distinct flowering season. Species in this group are the ones employed in Traditional Chinese Medicine (TCM) under the term *shihu*.

Members of the northern and southern clades are not similar in their chemical constitution. In the early 1970s, Swedish scientist Bjorn Luning and his Australian counterpart, Frank Loffler

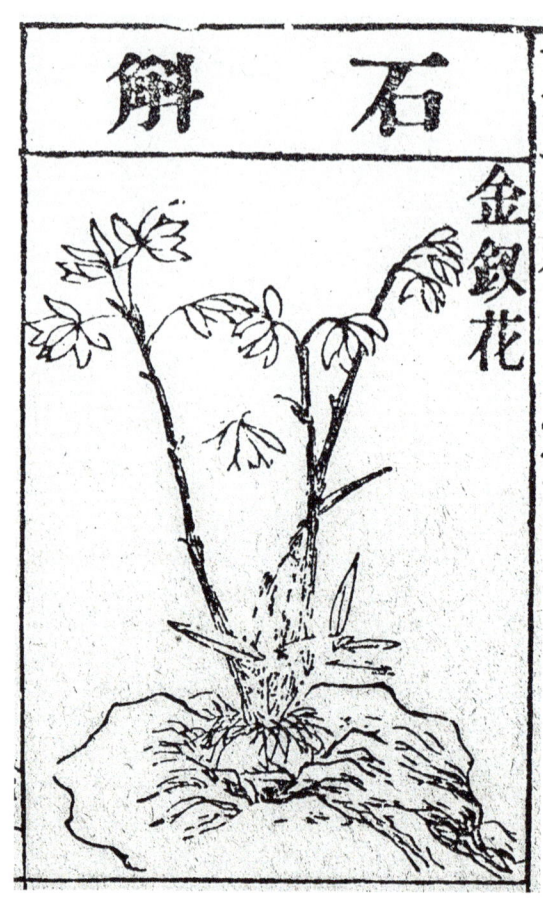

Fig. 4.2 *Shihu*. From: Li Shizhen, *Bencao Gangmu* (1593). Flowering *Dendrobium* plant growing on rocks

found that whereas many deciduous species produce alkaloids, the *Spathulata* species do not. In the past there has been no attempt to employ *Spathulata Dendrobium* as *shihu*, but now some sources warn that they may be present as cheap substitutes or contaminants. This would be serious because orchids grown for decorative purposes are constantly treated with pesticides which are harmful to human health (Fig. 4.5).

Shihu or *Shicao* (stone herb) was originally collected from Sichuan Province which is located in Central China. Plant was described as lithophytic and characterized by sympodial stems that bend downwards at their tips. Leaves are small, lanceolate; flowers large, white but tipped with purple on the sepal and petals. Since the stems are as strong as a hairpin, from ancient

Fig. 4.3 *Dendrobium nobile* (©Teoh Eng Soon 2019. All Rights Reserved.)

Fig. 4.4 *Dendrobium moniliforme* (©Teoh Eng Soon 2019. All Rights Reserved.)

times they have been referred to as *Jinchashihu* (golden hairpin *shihu*).

In the discussion of *shihu* in his monumental *Bencao Gangmu*, the famous Ming Dynasty physician, Li Shizhen (1518–1593) quoted numerous published sources. Li reported that *shihu* grew predominantly in the Valley of Liu An, on rocks located near streams and rivers. Plant carried green leaves that turned yellow when dried, and white roots which became soft when dry. Roots appeared at the internodes. Plantlets developed from cuttings placed on small stones, and this was employed as the mode of propagation. Containers should be small and left outdoors. If watered regularly, the plant would not perish: for

this reason, it had earned the name *Qiannianrun* (1000 years nourished and wet) (Figs. 4.6, 4.7, 4.8, 4.9, 4.10 and 4.11).

Some *shihu* plants were reported to bear red flowers. This was the first report of the inclusion of *Dendrobium nobile*. Another form of *shihu* had short stems; plant was dense and epiphytic. *Dendrobium jenkinsii* fitted this description. *Wuhu* was described as possessing long, thin stems. This referred to *Flickingeria* species.

As with many Chinese herbs, or even other products like coffee and vanilla, the quality of *shihu* is often determined by the area from which it is collected. The range in retail price of *shihu* in Singapore is 14-fold, from cheap to

pricey. Table B1 shows the distribution of *shihu*
sources in China according to species (Figs. 4.12,
4.13, 4.14, 4.15, 4.16 and 4.17).

In the 1960s, Traditional Chinese Medicine
(TCM) divided *shihu* into four main groups. A
similar classification is employed today, the only
difference being additional species:

Fig. 4.6 *Dendrobium jenkinsii* (©Teoh Eng Soon 2019.
All Rights Reserved.)

1. *Herba Dendrobii nobilis* constituted by *Chin
   chai shihu*, golden hairpin *Dendrobium*
   (*Dendrobium nobile*), and *Chin shihu*—golden
   Dendrobium (*Dendrobium linawianum*). The
   best quality should display a golden sheen; the
   body is long and compact. Sichuan Province
   was the main source of *Chin chai shihu*. It was
   an export item.
2. *Herba Dendrobii rotundicaulis* which includes
   Tieh pi shihu, iron skin *Dendrobium*
   (*Dendrobium officinale*); *Ko chuang shihu*—
   hooklike *Dendrobium* (*D. aduncum*); *Wang
   mei chun shihu*, nettled nerve lip *Dendrobium*
   (*D. hercoglossum*); *Chin chai hua*, golden hairpin flower (*D. hancockii*); *Tung pi lan*, copper
   skin orchid; *Kwangtung shihu*, Guangdong
   Dendrobium (*D. kwangtungensis*); *Tung pi
   shihu*, copper skin *Dendrobium* (*D. crispulum*);
   Hsiao huang tsao, small yellow herb
   (*D. lohohense*); *Fen hua shihu*, pink flower

Fig. 4.7 *Dendrobium gratiotissimum* (©Teoh Eng Soon 2019. All Rights Reserved.)

Fig. 4.8 *Dendrobium aphyllum* (©Teoh Eng Soon 2019. All Rights Reserved.)

Dendrobium (*D. loddigesii*); *Hsiao chin chai*, small little hairpin *Dendrobium*; or *Wang shihu*, Wang's *Dendrobium* (*D. wangii*). Best quality *shihu* in this category possesses stems of uniform size and thickness and is of a golden colour and solid. Species with very slim stems are referred to as *Huo hu* or *Huo shihu*. In Kwangsi province, such stems are rubbed and twisted into small balls called *Huan tsao* (ring herb). Best quality *Huo hu* are curved, striated, with short internodes, golden coloured, powdery, flexible; on tasting, sweet, slimy and cooling and should come from Huo Shan in Anhui Province (the source of *Dendrobium lohohense*). Kwangsi also produces *Huang tsao shihu* (ring herb *Dendrobium*).

3. *Herba Dendrobii minima* consisting of *Dendrobium officinale* (syn. *Dendrobium catenatum*, *Dendrobium candidum*) (cover photo and Fig. 4.1) and *Ho chieh tsao*, black nodes herb, or *Hsiao mei shihu*—small

beautiful *Dendrobium* (*Dendrobium bellatulum*). In the best quality herbs, stems are thick, with few spirals, powdery but without any cracks, and they are complete with 'dragon head and phoenix tail'.

4. *Herba Dendrobium recens* containing fresh *Dendrobium* pseudobulbs. Stems and leaves should be stout, clear green with no yellow coloration (Figs. 4.18, 4.19, 4.20, 4.21 and 4.22).

The above species were employed throughout the country, with other species employed only locally. For instance, *Dendrobium moniliforme* (*Tung hu*, Eastern *Dendrobium*) imported as dry stems from Japan during the 1950s and 1960s was only available in the coastal cities. *Chin huang che*, golden yellow marsh (*Dendrobium jenkinsii*, syn. *D. aggregatum*); *Ma pien shihu*, horse whip *Dendrobium* (*Dendrobium chrysanthum*); and

Fig. 4.13 *Dendrobium hancockii* (©Teoh Eng Soon 2019. All Rights Reserved.)

Fig. 4.15 *Dendrobium chrysanthum* (©Teoh Eng Soon 2019. All Rights Reserved.)

Yiu kua shihu, gourd-bearing *Dendrobium* (*Flickingeria fimbriata* syn. *Dendrobium plicatile*) were only employed locally (Hu 1968). Currently, the ten most commonly employed species that constitute commercial shihu are *Dendrobium nobile, D. thyrsiflorum, D aduncum, D. aphyllum, D. chrysanthum, D. chrysotoxum, D. densiflorum, D. devonianum, D. fimbriatum* and *D. loddigesii* (Luo 2016). They do not include

the classic species, *D. cretenatum* and *D. moniliforme*.

For medicinal use, stems are best collected in the 7th or 8th lunar month (August to September). Roots, leaves and flowers are removed from the stems which are then cut into desired lengths. These are washed and soaked overnight in alcohol. After sun-drying they are steamed or baked to dryness. Alternatively, drying over a slow fire is followed by drying in the sun. In Kwangsi,

Fig. 4.14 *Dendrobium bellatulum* (©Teoh Eng Soon 2019. All Rights Reserved.)

Fig. 4.16 *Dendrobium fimbriatum* (©Teoh Eng Soon 2019. All Rights Reserved.)

Fig. 4.17 *Dendrobium chrysotoxum* (©Teoh Eng Soon 2019. All Rights Reserved.)

Fig. 4.19. *Dendrobium thyrsiflorum* (©Teoh Eng Soon 2019. All Rights Reserved.)

Fig. 4.18 *Dendrobium densiflorum* (©Teoh Eng Soon 2019. All Rights Reserved.)

Fig. 4.20 Old *Dendrobium* pseudobulbs could either be sold as shihu or kept to provide a source of new plants. (©Teoh Eng Soon 2019. All Rights Reserved.)

Fig. 4.21 Stick *shihu*. Pseudobulbs have been cut into sections and left to retain their form after drying. (©Teoh Eng Soon 2019. All Rights Reserved.)

fresh material is dipped in boiling water and dried by rubbing. To produce earring *shihu*, fresh stems, about 10 cm in length, are washed clean,

Fig. 4.22 Earring *shihu*, this is the commonest presentation of *shihu*. Pseudobulbs are twisted into coils. (©Teoh Eng Soon 2019. All Rights Reserved.)

put into an iron container and heated until they are soft and malleable. Sheaths are then are moved by rubbing and the stems are allowed to dry in a well-ventilated setting. Two days later, they are collected into a lead sieve and heated over a low fire. The stems are twisted into spirals and cooled. The process may be repeated until the stems retain their spiral shape, and they are dry (Hu 1968). *Shihu* is sometimes left in their original form. Stored in bottles, *shihu* has a shelf life of several years. For use, dried *shihu* is soaked in water until they have an 80% water content. They are the cut into 2 cm sections.

The taste of *shihu* is slightly sweet and sour. It is listed among the non-toxic medicines. In the *Shennong Ben Cao Jing*, it is mentioned that *shihu* improves tired eyes, enhances *yin*, benefits *jing* and strengthens stomach and intestines. The *Mingyibielu* (Ming Dynasty Records of Famous Doctors) states that '*shihu* should be administered to people who are weak to balance stomach *qi* and develop muscles, chase away evil *qi* from skin. It may also be used to treat painful knees, numbness arising from cold. It has the effect of calming passion, makes one feel lighter and it promotes longevity'. *Rihuazhjiabencao* (Chinese Compendium of Herbs) adds that 'it strengthens tendons and bones, protects kidneys from cold and benefits the brain (promotes clear thinking)'. Li Shizhen mentioned that it enhances spleen and kidney meridians, benefits the legs and corrects oligospermia and difficult micturition. For instance, to treat an enlarged prostate, one could boil two *chern* of *shihu* with a slice of raw ginger and consume it as though it were tea. The drink is also good for the lungs and spleen.

According to contemporary Traditional Chinese Medicine (TCM) practice, the functions of *shihu* are to strengthen the stomach, improve one's appetite, enhance liver functions, brighten the eyes and strengthen bones and tendons. It enhances yin in the body. *Shihu* is used to treat patients affected by internal heat and sweating. In modern terminology, it may be regarded as a health supplement.

Wherever they may reside, *shihu* is a popular tonic among Chinese who resort to Traditional Chinese Medicine (TCM) for alternative cures

Fig. 4.23 Worker planting *Dendrobium officinale* on a cliffside. Photo: courtesy of Luo Yibo

or for supplements to enhance their quality of life and promote longevity. *Records of Imperial Chinese Customs* from 1884 and 1888 revealed that enormous amounts of *shihu* collected from various Chinese provinces passed through the numerous ports along the Yangtze and the Chinese coast destined for distribution in Chinese and overseas. Exporters employed various names for their *shihu*, calling it *chin cha, huang tsao, hsien tuo, hsien hu tou, hsien shihu, mu hu, mu hu pi, huay tsan, ya tuo, huan chai*. The English equivalent in the Customs Records was *Dendrobium ceraia*: but this is an old name for *Dendrobium crumenatum*, the pigeon orchid, a species that is familiar to many Englishmen at that time because it is widely distributed throughout Southeast Asia and Bengal. In fact, *Dendrobium crumenatum* does not occur in China, and it is hardly ever employed in the country as *shihu*, albeit it might be a substitute outside.

Demand for *shihu* is so strong that sometimes an entire forest well known for its *Dendrobium* was been stripped of the flourishing orchid communities leaving sparse populations that are unlikely to sustain the species in the area. Awareness of this threat has been brought home to botanists in China, and currently there is a massive effect to propagate premier medicinal *Dendrobium* species and return plants to the wild. Professor Luo Yibo who is president of the Orchid Society of China and president of the Orchid Specialist Group Asia Branch, Species Survival Commission of IUCN (International Union for the Conservation of Nature) reported that over 7000 ha of agricultural land was devoted to cultivating medicinal *Dendrobium* derived from seed in 2016. When plants were large enough to survive in the wild, they were transferred to forest trees, and some were even attached to rocks in remote Danxia formations. Collection was permitted every 3 or 4 years from any single area. Such 'wild plants' fetched a better price than plants that had spent their entire life in nurseries (Luo 2016) (Fig. 4.23).

A significant proportion of *shihu* is now grown in nurseries because of increasing demand, diminishing supply from the wild, an awareness of the need for conservation and good financial returns. Total annual cultivated *shihu* production now exceeds 19 million kg, with *Dendrobium officinale* accounting for slightly over 9 million kg (including *Dendrobium huoshanense*) and *Dendrobium nobile* 4.5 million kg. The latter species is cultivated mainly in Guizhou Province, whereas *Dendrobium officinale* is cultivated in nine provinces (Zhejiang, Yunnan, Guangdong, Hunnan, Guangxi, Fujian, Jiangxi, Guizhou and Anhui). Luo Yibo reported that a farmer cultivating 1000 m^2 of *Dendrobium officinale* can expect to earn 10000 RMB annually after 2 years (Luo 2016).

Dendrobium Usage in Other Medical Traditions

Dendrobium species are also employed medicinally in Ayurveda and other traditions. Anderson who worked with the Hmongs in northern Thailand during the early 1990s reported that they chewed and ate the leaves of a *Dendrobium* to relieve bone pain. Alternatively Hmongs use a paste prepared by pounding *Dendrobium* with another orchid, *Coelogyne trivernis*, to apply over broken bones (Anderson 1993). Native healers in Thailand interviewed by Chuakul in 2002 identified five Thai *Dendrobium* species which they employed to treat various conditions. Stems of *Dendrobium cumulatum* were employed to treat asthma and *Dendrobium draconis* and *D. trigonopus* to treat fever and correct anaemia. Whole plants of *Dendrobium indivisum* and the related *Dendrobium leonis* both of which belong to the section *Aporum* (which is characterized by equitant leaves and small flowers) were used for headache (Chuakul 2002). In the Malay Peninsula, a Malay medicine man told Burkhill and Haniff (1930) that a poultice made with leaves of *Dendrobium subulatum* was applied to relieve headache (Figs. 4.24, 4.25, 4.26, 4.27 and 4.28).

Anyone who has lived in Malaysia or Singapore would be familiar with the beautiful, fragrant white pigeon orchid, *Dendrobium*

Fig. 4.24 *Dendrobium indivisum* (©Teoh Eng Soon 2019. All Rights Reserved.)

crumenatum which flowers gregariously several days after a thunderstorm. The orchid is ubiquitous on roadside trees in Singapore and blooming plants lend enchantment along major roads. In the early years of the twentieth century, Henry Ridley reported that juice from fresh, crushed, boiled or roasted pseudobulbs was dropped into the ear to relieve earache in Malaysia and Indonesia (Ridley 1906). Malays in Kelantan used a more elaborate method: they stuffed the orchid stem with onion and *jintan manis* seeds before roasting the preparation in hot ashes. The heated juice was then squeezed into the ear (Gimlette and Thomson 1939). It must have been a painful remedy, but if the original pain was caused by a boil, this was one way to rupture and drain the abscess, a rather novel way to administer the ancient Greek dictum: *Ubi pus, ibi evacuat* (if there is pus, it must be drained). A poultice made with the residue was then applied around the base of external ear.

In prewar Malaya, Burkill reported that the pigeon orchid was highly regarded as a magical plant, employed in ceremonies to induce the return of beneficent spirits (Burkill 1935).

Fig. 4.27 *Dendrobium purpureum* (©Teoh Eng Soon 2019. All Rights Reserved.)

Fig. 4.25 *Dendrobium leonis* (©Teoh Eng Soon 2019. All Rights Reserved.)

Fig. 4.26 *Dendrobium monticola* (©Teoh Eng Soon 2019. All Rights Reserved.)

Fig. 4.28 *Dendrobium draconis* (©Teoh Eng Soon 2019. All Rights Reserved.)

Gimlette and Thomson (1939) mentioned that it was used to ward off evil spirits after a recent demise.

Dendrobium crumenatum has the distinction of being the only orchid included in the list of 194 local medicinal plants compiled by Kwan Koriba and Watanabe for the Japanese Occupation Army in Singapore during the Second World War. This *Compilation of Medicinal Plants of the Malay District* published in July 1944 is probably the result of an independent effort because publications on economic and medicinal plants already in print before this period carried far more extensive listings of medicinal orchids of the Malay Peninsula. In his *Dictionary of Economic Products of the Malay Peninsula, Volume II*, Burkill (1935) noted that Malays referred to orchids as *anggreks*, as it were a generic term without bothering about species distinction, and they employed one of several common orchid species to treat a similar condition. Thus, *Dendrobium purpureum* which was also common in East Malaysia might be employed instead of *Dendrobium crumenatum*, whereas in West Malaysia, they might employ *Dendrobium planibulbe* instead. In the state of Perak, *Dendrobium subulatum* served as another possible substitute. This substitution went beyond the genus: to treat earache, Malays also employed *Acriopsis liliifolia* (syn. *A. javanica*), or *Cymbidium finlaysonianum*, or *Grammatophyllum species*, whichever species was on hand.

This practice of using the orchid to treat earache and other medicinal usages probably had its origins in India. Adventurers from that country in search of a fabled 'Land of Gold' arrived in Peninsular Malaysia 2000 years ago, and archaeological remains indicate that there was a large ancient Indian settlement in northwestern part of the peninsula around 500 C.E (Miksic 2013). Valmikis, a tribe living in Visakhapatnam in Andra Pradesh still treat earache with juice from tender growing tips of *Dendrobium herbaceum* or *D. macrostachyum* (Reddy et al. 2005). The Nukadoras use only *Dendrobium herbaceum* for this purpose. In Uttar Pradesh, *Dendrobium crumenatum* is valued as an antiseptic. It is applied to boils and pimples, whereas in Nepal *Dendrobium densiflorum* is employed. Before the

widespread availability of modern medicine and antibiotics in Malaysia, *Dendrobium planibulbe*, which is related to *Dendrobium crumenatum*, is pounded into a poultice and applied to sores and infected wounds in Peninsular Malaysia (Burkhill and Haniff 1930). Stems of *Dendrobium purpureum* was used to treat whitlow (Rumphius, second half, seventeenth century) A poultice prepared with young canes of *Dendrobium discolor* is used by Australian aborigines living in Queensland to draw a boil. Old canes are made into a liniment for treating ringworm (Lawler and Slaytor 1970).

Dendrobium crumenatum enjoys the honorific of *Jivanti*, a remedy for affections of the brain and nerve in India. A conserve of its flowers and leaves were fed to people suffering from cholera. Nicobar Islanders in the Bay of Bengal employ the orchid to treat fever, headache, giddiness, chest pain and body aches (Fig. 4.29).

It may be noted that as early as 1935, Burkill already suspected that Jacobus Bontius was

Fig. 4.29 *Dendrobium crumenatum* (©Teoh Eng Soon 2019. All Rights Reserved.)

referring to *Dendrobium crumenatum* when the Dutch physician observed that 'Malays thought nothing to be its equal for affections of the brain and nerves and that a conserve of its flowers and leaves was used for cholera'. Bontius visited Java during the seventeenth century. He was the first European to describe the symptoms and signs of cholera. He was a pioneer in tropical medicine.

In India, *Dendrobium crumenatum* and *D. monticola* are rendered as an emollient to treat pimples, boils and other skin lesions. *Dendrobium monticola* is sold in Kanpur's herb markets in the northeastern state of Uttar Pradesh as a nerve tonic and antiphlogistic. It is applied externally to relieve rheumatic pain (Trivedi et al. 1980).

Dropsy is treated with *Dendrobium quadrangulare* and *D. pumilum* in Malaya (Ridley 1906; Burkhill and Haniff 1930), but only *Dendrobium pumilum* is used in India (Caius 1936). Indians employ *Dendrobium normale* as an aphrodisiac (Sood et al. 2005). *Dendrobium ovatum* had multiple uses in India. In his monumental *Hortus Malabaricus*, Hendrik Adriaan van Rheede tot Drakenstein who was the Colonial Administrator of the Dutch East Indies, a military man but also a naturalist, recorded that juice prepared freshly from the plant is used to treat all sorts of pain and colic. It is commonly used to relieve tummy ache (van Rheede 1703). Later, Father Caius who made a careful study of medicinal orchids in Kerala during the 1930s explained that the native orchid was used in Bombay (Mumbai) as an emollient. It acted as a laxative and was administered to patients suffering from stomachache (Caius 1936). In the adjacent state of Karnataka, tribals at the Kudremukh National Park still employ the orchid to treat the similar complaints (Rao 2007). In Uttar Pradesh it is a tonic, stomachic, pectoral and antiphlogistic. It is used to treat rheumatism. Traces of alkaloids are present in the leaves (Trivedi et al. 1980).

Leaves of the startlingly beautiful, golden *Dendrobium densiflorum* mixed with salt are grounded into a paste and applied to fractures to help set bone in India and Bangladesh (Musharof Hossain 2009).

When Nepalese children have fever, healers advise bathing them with lukewarm water to which juice of *Dendrobium longicornu* has been added. In this country boiled root of the orchid is fed to livestock suffering from cough (Manandhar and Manandhar 2002).

It is evident that wide gaps exist on the usage of *Dendrobium* species in the various traditional medicinal practices. Additionally, the species in use are different. Although many *Dendrobium* species employed as *shihu* occur in abundance in the Indian Himalaya (e.g. *Dendrobium nobile*, *D. catenatum*, *D. transparens*, *D. crepidatum*, *D. chrysanthum*, *D. fimbriatum*, *D. primulinum*, *D. pulchellum*), none of them are employed in India or Nepal in the manner that people in the Far East use *shihu*, and many such species are entirely ignored for medicinal purposes.

Clinical data in support of any of the medicinal usage of *Dendrobium* is scarce (Teoh 2016). The alkaloid dendrobine has weak pain-relieving activity, but it has no effect on fever. Denbinobine, another alkaloid isolated from *Dendrobium nobile*, exhibits anti-inflammatory effects in vitro. Similar studies have not been conducted on *Dendrobium* species employed to treat pimples boils and whitlow in India and Southeast Asia.

Polysaccharide haemopoietic growth factors, GM-CSF and G-CSF, were isolated from *Dendrobium huoshanenese* (Hsieh et al. 2008); however, no study has been reported on *Dendrobium draconis* which is used by Thai native healers to boost blood production (Figs. 4.30, 4.31, 4.32 and 4.33).

Rats rendered diabetic by injections of streptozotocin eventually develop diabetic cataract. When such streptozotocin-injected rats were treated simultaneously with polysaccharides of *Dendrobium huoshanense* at doses of 50–200 mg/kg daily, they gained weight, but their blood sugar levels were suppressed. Lens opacity was reduced in polysaccharide treated rats (Luo et al. 2008). iNOS gene expression is increased three- to fivefold in Zucker diabetic fatty (ZDF) rats prone to developing cataracts (Kim et al. 2007). *Dendrobium huoshanense* polysaccharide treatment produces an inhibitory

response on iNOS gene expression in a dose-dependent manner (Luo et al. 2008). These studies provide some support for the claim that *shihu* improves eyesight. (Kew's World Checklist of Plant Names consider *Dendrobium huoshanense* to be synonymous with *Dendrobium officinale*.)

Research on the polysaccharides of *shihu* is recent, but *shihu* has been studied for its phytochemicals for well nearly a century. The first crystalline alkaloid, dendrobine was isolated by Suzuki in 1932. Several decades passed before Bjorn Luning undertook a screening for alkaloids

in *Orchidaceae*. In his study, out of 2044 orchid species in 281 genera tested, 214 species (10.5%) in 64 genera (29.4%) tested positive for alkaloids, i.e. they had an alkaloid content of 0.1% or higher. Only 8.3% of 384 *Dendrobum* species gave a positive test. Large amounts of alkaloid were present in those species included in *shihu* (*Dendrobium nobile, D. liniawanium, D. finlayanum, D. wardianum, D. crepidatum, D. aphyllum, D. chrysanthum, D. lohohense, D. primulum, D. parishii*) and other species such as *D. hildebrantii* and *D. friedericksianum* that belong to the northern clade. On the other hand, *Dendrobium* species in the section *Spathulata* which are distributed in Indonesia and Papua New Guinea do not contain any alkaloid (Luning 1974).

By 1980, 15 alkaloids had been identified in *Dendrobium* (dendrobine, 2-hydroxydendrobine, nobilonine, dendrine, dendrowardine, dendroxine, dendroprimine, dendrochrysine, dendrocreptine, crepidine, crepidamine, hygrine, dendroparine, pierardine and shihunine). Thirty-two alkaloids had been isolated from 42 species of *Dendrobium* by 2003 (Zhang et al. 2003), and two more were added in 2003 and 2007. To date over a hundred compounds have been isolated from medicinal species of *Dendrobium*. In

Fig. 4.33 *Dendrobium pulchellum* (©Teoh Eng Soon 2019. All Rights Reserved.)

addition to alkaloids, these consist of bibenzyls, phenanthrenes, lignin, flavonoids and polysaccharides. It is a major task to work out if any of them relate to and empower the medicinal usage of the specific orchid species.

Contemporary research workers are focusing on the search for new drugs to treat cancer and diseases associated with aging. In this respect, several compounds isolated from *Dendrobium* show promise. Denbinobin, a phenanthrene isolated from *Dendrobium nobile* and *Flickingeria xantholeuca (= Dendrobium xantholeucum)*, kills A549 human lung cancer, Sk-OV-3 human ovarian adenocarcinoma and HL human promyelocytic leukaemia cells in vitro (Lee et al. 1995; Kuo et al. 2008; Magwere 2009). Denbinobin also exhibits potent anti-inflammatory effects in vitro (Lin et al. 2001). Two dimeric phenanthrenes and denthyrsinin isolated from *Dendrobium thyrsiflorum* are cytotoxic against HeLa, K-562 and MCE cell lines (Zhang et al. 2005). Moscatilin present in stems of *Dendrobium loddigesii* prevented cell division in cancers of the placenta, stomach and colon but not the liver (Ho and Chen 2003). Intraperitoneal injection of a neutral polysaccharide from *Dendrobium denneanum* resulted in higher tumour inhibition and improved immunomodulatory response in rats (Fan et al. 2010).

There is currently no data from clinical (human) studies, so it is not possible to know whether denbinobin can be developed into a useful drug for treating tumours that are resistant to other cytotoxic drugs.

Chrysotoxine protected human nerve cells derived from a neuroblastoma (SH-SY5Y) from 6-hydroxydopamine-induced programmed cell death (Song et al. 2010). This is a possible avenue for research into the development of a new drug for neuroprotection in old age, to delay onset of Parkinsonism and dementia.

Two novel sesquiterpenoids named Dendrowardols A and B were isolated from *Dendrobium wardianum* Warner, a species employed as *shihu*. When tested on human lens epithelial cells (which are responsible for ensuring the health of the lenses in the eyes), Dendrowardol B slightly enhanced the promoting effect induced by D-galactose on such cells (Fan et al. 2013). Further experiments may determine whether *shihu* enhances eyesight (Figs. 4.34 and 4.35).

Fig. 4.34 *Dendrobium findlayanum* (©Teoh Eng Soon 2019. All Rights Reserved.)

Fig. 4.35 *Dendrobium parishii* (©Teoh Eng Soon 2019. All Rights Reserved.)

References

Administration Department of the Japanese Army, Syonan (1944) Compilation of medicinal plants in the Malay district. Japanese Imperial Army, Singapore

Anderson EF (1993) Plants and people of the Golden Triangle. Ethnobotany of the hill tribes of Northern Thailand. Dioscorides Press, Portland

Bao X, Shun Q (1999) Investigation and identification of 'shihu' medicinal materials from Shanghai. Zhong Yao Cai 22(2):61–63

Burkhill IH, Haniff M (1930) Malay village medicine. Gard Bull Straits Settl 6:165–321

Burkill IH (1935) A dictionary of economic products of the Malay Peninsula, Vol. II. Crown agents for the colonies, London (1966 reprint, 2nd ed, with contributions by Birtwistle W, Foxworthy FW, Scrivenor JB, Watson IG). Ministry of Agriculture & Co-operatives, Kuala Lumpur

Caius JF (1936) The medicinal and poisonous plants of India. J Bombay Nat Hist Soc 38(4):791–799

Chuakul W (2002) Ethnomedical uses of Thai Orchidaceous plants. Mohidol Univ J Pharm Sci 29(3–4):41–45

Fan YJ, Chun Z, Luo A et al (2010) In vivo immunomodulatory activities of neutral polysaccharide (DPP1-1) from *Dendrobium denneanum*. Chin J Appl Environ Biol 16(3):376–379

Fan WW, Xu FQ, Dong FW et al (2013) Dendrowardos A and B, two new sesquiterpenoids from *Dendrobium wardianum* Warner. Tetrahedron Lett 54 (15):1928–1930

Gimlette JD, Thomson HW (1939) A dictionary of Malayan medicine. Oxford University Press, London

Ho CK, Chen CC (2003) Moscatilin from the orchid *Dendrobium loddigesii* is a potential anticancer agent. Cancer Invest 21(5):729–736

Hsieh YS, Chien C, Liao SK, Liao SF et al (2008) Structure and bioactivity of the polysaccharides in medicinal plant *Dendrobium huoshanense*. Bioorg Med Chem 16 (11):6054–6068

Hu SY (1968) *Dendrobium* in Chinese medicine. Econ Bot 24:165–174

Kim JH, Kim KH, Kim JH et al (2007) Homoisoflavanone inhibits retinal neovascularization through cell arrest with decrease of cdc2 expression. Biochem Biophys Res Commun 362(4):848–852

Kuo CT, Hsu MJ, Chen BC, Chen CC, Teng CM, Pan SL, Lin CH (2008) Denbinobin induces apoptosis in human lung adenocarcinoma cells via Akt inactivation, Bad activation, and mitochondrial dysfunction. Toxicol Lett 177(1):48–58

Lawler IJ, Slaytor M (1970) Uses of Australian orchids by aborigines and early settlers. Med J Aust 2:1259–1261

Lee YH, Park JD, Baek NI, Kim SI, Ahn BZ (1995) In vitro and in vivo antitumoral phenanthrenes from the aerial parts of *Dendrobium nobile*. Planta Med 61 (2):178–180

Lin TH, Chang SJ, Chen CC, Wang JP, Tsao LT (2001) Two phenanthraquinones from *Dendrobium moniliforme*. J Nat Prod 64(8):1084–1086

Luning B (1974) Alkaloids of the Orchidaceae. In: Withner CL (ed) The orchids. Scientific studies. Wiley, New York

Luo YB (2016) The development of traditional Chinese medicinal *Dendrobiums* business. Abstracts/3rd Shanghai Chenshan International Orchid Symposium. Shanghai, p 8

Luo JP, Deng YY, Zha XQ (2008) Mechanism of polysaccharides from *Dendrobium huoshanense* on streptozotocin-induced diabetic cataract. Pharm Biol 46(4):243–249

Magwere T (2009) Escaping immune surveillance in cancer: is denbinobin the panacea? Br J Pharmacol 157 (7):1172–1174

Manandhar NP, Manandhar S (2002) Plants and people of Nepal. Timber Press, Portland

Miksic JN (2013) Singapore and the silk road of the sea. NUS Press, Singapore

Musharof Hossain M (2009) Traditional therapeutic uses of some indigenous orchids of Bangladesh. Med Aromat Plant Sci Biotechnol 3:100–106

Reddy KN, Reddy CS, Raju VS (2005) Ethno-orchidology of orchids of Eastern Ghats of Andhra Pradesh. EPRTI Newsl 11(3):5–9

Ridley HN (1906) Malay drugs. Agric Bull Straits Settl Fed Malay Staes 5: 245, 277

Rao TA (2007) Ethnobotanical data on wild orchids of medicinal value as practised by tribals in Kudremukh National Park in Karnataka. Orchid Newsl 2(2):1–7

Rumphius GE (second half 17th cent.: published posthumously 1741–1750) Amboinsch Kruidboek (The Amboinese Herbal, vol 1–6). Translated and annotated into English, with an introduction by E M Beekman (2011). Yale University Press, New Haven

Song JX, Shaw PC, Sze CQ et al (2010) Chrysotoxine, a novel bibenzyl compound, inhibits 6-hydroxydopamine apoptosis in SH-SY5Y cells via mitochondria protection and NF-kappa B modulation. Neurochem Int 57(6):676–689

Sood SK, Rana S, Lakhanpal TN (2005) Ethnic aphrodisiac plants. Scientific Publishers (India), Jodhpur

Suzuki H, Keimatsu I, Ito M (1932) Alkaloid of the Chinese drug 'Chin-Shih-Hu'. II. Dendrobine. J Pharm Soc Jpn 52:1049–1060

Teoh ES (2016) Medicinal orchids of Asia. Springer, Cham

Trivedi VP, Dixit RS, Lal VK (1980) Orchids in the drug markets of Bareilly, Kanpur and nearby districts. Nagarjun (Calcutta) 23(8):157–163

van Rheede HA (1703) Hortus Indicus Malabaricus, vol 12. Dutch East India Company, Kerala

Zhang GN, Bi ZM, Wang ZT, Xu LS, Xu GJ (2003) Advances in studies on chemical constituents from plants of *Dendrobium* Sw. Chin Tradit Herb Drug 34: S5–S8. (Appendix)

Zhang GN, Zhong LY, Bligh SW, Guo YL, Zhang CF, Zhang M, Wang ZT, Xu LS (2005) Bi-cyclic and bi-tricyclic compounds from *Dendrobium thyrsiflorum*. Phytochemistry 66(10):1113–1120

Metamorphosis

5

Modern Medicine Finds New Uses for an Ancient Herb

Over the last two decades, the ancient medicinal herb *baiji* has been attracting attention from Chinese scientists and physicians because new uses have been found for it, and more are being investigated.

Alternative Names

Baiji is prepared from the roots of the beautiful orchid, *Bletilla striata* (hyacinth orchid). It was a drug of inferior class when it first appeared in the *Shen Nong Bencao Jing* (first century CE) under the name *Lianjicao*. The tuberous roots of the orchid plant are clustered, hence *baiji* or 'white and attached'. Its alternative name was *Gangen* (sweet root) which is paradoxical because it is bitter and somewhat astringent. Wu Pu, a disciple of the famous physician-surgeon Hua Tuo (Three Kingdoms Period), called it *Baigen* (white root). It was also conferred a Buddhist name, *Wangdaluohexiduo* in the sutra *Jingguangming Jing*, first translated by Dharmasema during the Liang Dynasty (414–421): this name is a transliteration, and it is not in common usage. Two additional uncommon names are *Baijiertou* (white hen's head) and *Shantianji* (mountain frog). In Hong Kong it is called *Bak-kup*. The Vietnamese name is similar, *Bach cap*, but in this country, it is also known as *Hua lan tia*. The Japanese name *Shiran* (purple orchid) is shared with another species, *Bletilla formosana*.

This does not seem to matter in *Kanpo* Medicine because the two species are employed in a similar manner. To Japanese gardeners *Bletilla striata* is *No Ran* and Korean *Jaran* (Fig. 5.1).

The Herb

Bletilla striata is a small- to medium-sized herb varying in height from 15 to 50 cm. It has irregular, compressed, ovoid white pseudobulbs and short stems each carrying four or five linear-lanceolate leaves, 8–30 by 4–5 cm. The inflorescence is terminal, bearing 3–8 purple flowers with an attractive lip marked with purplish, longitudinal striations over a white and yellow background. There is also an alba form that bears white flowers. The species is terrestrial or lithophytic, growing on grassy slopes or scrub at 1100–3200 m. It is distributed from Sichuan, Gansu and Shaanxi in the west to Guangdong and Jiangsu in the east of China. It also occurs in Korea and Japan. Flowering season is April to June (Chen et al. 2009). In the temperate region, *Bletilla striata* is an easy plant to grow and maintain outdoors (Fig. 5.2).

Old herbal texts stated that the plant could be found in many places on rocky mountains. It was a common roadside plant. In spring, around March or April, the interconnected rhizomes send out new shoots that eventually reach a height of 1 m, bearing linear leaves about two fingerbreadths in

© Springer Nature Switzerland AG 2019
E. S. Teoh, *Orchids as Aphrodisiac, Medicine or Food*, https://doi.org/10.1007/978-3-030-18255-7_5

Fig. 5.1 *Baiji.* From: Li Shizhen, *Bencao Gangmu* (1593). Not the emphasis on the clustered, compressed pseudobulbs in the woodcut illustration

Fig. 5.2 *Bletilla striata* [as *Cymbidium hyacinthinum* Sw.] From: *Curtis Botanical Magazine* vol. 56: t. 492 (1812) [artist: ST Edwards]

width. Plants are mature by July. Li Shizhen (1518–1593) reported that various physicians had recommended that the rhizomes be collected in July, August or September, but, in one instance, either in February or July. The data provided by Perry and Metzger (1980) who authored *Medicinal Plants of East and Southeast Asia* differed in some respects from the foregoing. They reported that the rhizomes were collected from August to November with a non-metallic tool. In the case of planted crops, harvesting took place in September and October, 3–4 years after planting (Perry and Metzger 1980).

To meet the demand from herbalists, *Bletilla striata* is cultivated in Guizhou, Yunnan, Jiangxi and Guangxi. Although the species is not native in Guizhou, this province produces the largest quantity of top quality *baiji*. This is not surprising, given Guizhou's geography and

salubrious climate. A vigorous clump left undisturbed can carry a hundred pseudobulbs.

Alternatives or Substitutes

In addition to *Bletilla striata*, another three species which occur in China are employed as *baiji* in TCM, namely, *Bletilla formosana* (Chinese names: *Taiwan Baiji*; also *Xiaobaiji*), *B. ochracea* (Chinese name: *Huanghua Baiji* or yellow flower white mucilaginous root) and *B. sinensis* (Chinese name: *Xiaobaiji* or small *Baiji*; small white root). Their medicinal applications are similar.

The Medicinal Preparation

Tubers of *Bletilla striata* are described as bitter, sweet and acerbic in taste. For consumption the tubers are cut into thin slices or crushed into a fine powder after cleaning and drying. These slices are irregular in shape. The cut surface is translucent, white, with a suggestion of veins. They are brittle and somewhat sticky, and they emit a faint odour. Slices become gluey when they are chewed. *Baiji* is stored in dry containers in well-ventilated stores (Figs. 5.3 and 5.4)

Baiji powder is slightly yellowish, odourless and bitter. When mixed with water, it turns gluey. The powder is stored in tightly sealed containers to protect against moisture.

Fig. 5.4 *Bletilla ochracea* from Shanxi Province, China. Photo: Huang Weichang

The Evolution of Traditional Uses

The *Shen Nong Bencao Jing* stated that it can be used to treat swellings and carbuncles, festering sores and injured muscle. It dispels harmful factors from the stomach and cures paralysis.

Over the course of centuries, new uses were discovered by herbalist-physicians and published in various Chinese *Materia Medica*. *Mingyi Beilu* (Records of Famous Physicians, circa second to third century CE) advocated its use to treat tinea or ringworm, athlete's foot and other fungal infections of the skin and scabies. Zhen Quan

Fig. 5.3 *Bletilla striata* from Hubei Province, China. Photo: Huang Weichang

(541–643) who made notable contributions to acupuncture during the reign of Emperor Tang Taizong (r. 627–649) observed that it cleared freckles and acne, gave a glossy appearance to the countenance and corrected impotence. In the *Xinxiu Bencao* (New Revised Pharmacopoeia) compiled in 659 the principal editor, Su Gong (Su Jing, d. 674) stated that mountain dwellers chewed *baiji* and then applied the sticky paste on their hands when they had chapped skin. This state-sponsored *Tang Bencao* (an alternative name for *Xinxiu Bencao*) commented on and expanded Tao Hongjing's Liang Dynasty (502–557) southern *Herbal* to include all the drugs employed throughout the Tang Empire. Brought to Japan in 721 the *Tang Bencao* exerted an influence on *Kanpo* Medicine. Fragments of the *Xinxiu Bencao* have been discovered in Dunhuang, but original copies of the complete pharmacopoeia are no longer extant in China. An original copy of the *Tang Bencao* survives in Japan.

Four centuries of incessant internal strife in China prevented further development of herbal medicine, and it was only when peace was restored during the Song Dynasty that new pharmacopoeia were compiled. In Fujian Province, Su Song (1020–1101) and his colleagues produced the *Jin Yu Bencao* in 1057, and *Tu Jing Bencao*

(Illustrated Materia Medica) in 1062, the latter pharmacopoeia sourcing information from nearly 200 texts and containing over 900 illustrations. The preface of the *Tu Jing Bencao* stated that the illustrations were prepared to replace the illustrations from Tang Dynasty's *Xinxiu Bencao* which had been lost. Unfortunately, the *Tu Jing Bencao* is also no longer extant, but some illustrations are preserved in the *Daguan Bencao* (1108) (Sterckx 2008). Su Song reported that physicians often employ *baiji* to treat incised wounds, carbuncles and phlegmons.

The *Rihuazi Bencao* also compiled during the Song Dynasty stated that *baiji* '…regulates blood disorders and stops dysentery with bloody discharge….It is effective for treating hematochezia (passage of fresh blood per rectum), physical injury, sores due to incised wounds, burns and scalds' (Fig. 5.5).

Anecdotal accounts of miraculous cures are a common feature in folk medicine. Not confining himself to reading only medical tomes, Li Shizhen (1518–1593) came across a story of an extraordinary cure achieved with *baiji*. It was related in Hong Mai's *Yijian Zhi* which was a massive collection of fantastic tales recorded during Southern Song (1127–1279) by a court literatus and historian:

At one time, a condemned man in a Taizhou prison told his jailer that on seven occasions he had been

Fig. 5.5 *Bletilla formosana.* Photo: Huang Weichang

arrested and charged with capital crimes for which he was severely tortured. His lungs were so badly damaged he coughed out blood. Seeing his plight, someone advised him to consume *baiji* powder suspended in thin rice water. That stopped the hemoptysis. Later, when the curious executioner examined his lungs he saw dozens of cavities containing a white substance. (Comment: It was probably miliary tuberculosis, a common disease in old China.)

When Hong Mai became prefect of Yangzhou, one of his subordinates was suddenly afflicted with haemoptysis and the boss offered him treatment with *baiji*. The subordinate stopped coughing up blood almost immediately.

Li Shizhen also mentioned the work of Li Gao (Li Dong Yuan 1180–1251) who stated that *baiji* stopped haemoptysis caused by diseases of the lungs. This immediately boosted the status of *baiji* because in old China pulmonary tuberculosis was a common fatal disease for which there was no cure. Any herb that could stop the bleeding would be inordinately valuable.

Li Gao, however, came from a rich family and he had no interest in profiting from herbal medicine. He paid 1000 taels of gold to study medicine from the renowned doctor Zhang Yuansu whom he followed for many years. Afterwards, he only practised medicine as a hobby. He did not depend on the practice for his livelihood and was incorruptible. However, he was available for urgent cases and managed many 'miraculous cures'.

Zhu Zhenheng (1281–1358) who lived during the Mongol Dynasty took an eclectic, syncretic approach to the practice of medicine after observing that generally one master was an expert in treating only one type of illness. Therefore, he picked ideas from numerous sources. Zhu proposed that *baiji* should be added to prescriptions used for treating persistent hematemesis (vomiting of blood). This proposal survives in contemporary indications for using *baiji*.

Li Shizhen's *Bencao Gangmu* published posthumously in 1593 (Ming Dynasty) is the most authoritative historical pharmacopoeia that is still widely referred to today. It contains nine prescriptions that employ *baiji*. The first deals with its use to treat nose bleed: the other prescriptions are for fungal skin infections, furuncles, sores and abscesses, open wounds and fractures, burns and scalds and chapped skin. Two prescriptions are curious—the one that utilizes *baiji* to treat chest pain and the second that uses it to treat a prolapsed womb, presumably one so badly prolapsed as to result in abrasions.

Contemporary Usage

In TCM, *baiji's* property is stated as slightly cold and its meridian tropisms are the lung and stomach meridians. The alleged effects of the medicine are: it is haemostatic, it reduces swelling and it promotes regeneration of muscle and other tissues (Anonymous 2004). *Baiji* is used to treat swelling and inflammation, injuries related to *yin*, inflamed muscles or harmful *qi* in the stomach. It smoothens the skin; relieves anxiety; cleanses the blood; clears injected eyes; removes heat; heals burns, cuts and penetrating wounds; promotes tissue regeneration; relieves pain; and stops pulmonary haemorrhage (Li Shizhen 1578).

In practice, *baiji* is used to treat visible bleeding from any source—when a person coughs up or vomits blood, has a nosebleed, blood in his stools, bleeding from the anus, or is suffering from external wounds, carbuncles, burns, frostbite or cracked skin at the extremities (Anonymous 1974; Bensky et al. 2004).

When it is confirmed that a patient was coughing up blood, the recommended TCM treatment is to drink a soup prepared by boiling goat's lung, liver and heart with *baiji* powder added at the final stage. This soup should be consumed daily until the patient is well.

To relieve chest pain, pills are prepared by blending 2 qian (10 g) of *baiji* with equal amounts (by weight) of pomegranate skin, and three of such pills were consumed a day. When it is necessary to treat fissured skin at the extremities, powdered *baiji* is mixed with a small amount of water to produce a cream which is then applied on the affected parts. Afterwards, the hands and feet should be kept dry. Swollen and inflamed parts need to be treated with a dressing. This is prepared in the following manner: add powdered *baiji* to water; stir; decant excess water; transfer the gluey *baiji* onto a sheet of thick paper; and stick this on the affected part (Li Shizhen 1578).

A contemporary adaptation of this usage is the *Chuangyiling* dressing employed to assist the healing of wounds. Designed by the Orthopaedics Department of Union Hospital in Wuhan, it is made with an extract of traditional medicines mixed into a scaffold of gelatin and *Bletilla striata* gum to form a spongy porous material. The dressing has met the requirements for safety (absence of acute toxicity, skin irritation, sensitization and cytotoxicity) set by the Ministry of Health of China (Peng et al. 2005).

Several laboratory studies support the use of *baiji* as a wound dressing (Takagi et al. 1983; Luo et al. 2010). A natural polysaccharide (BSP) hydrogel prepared by oxidation and cross-linking methods was shown to promote excellent wound healing in a full-thickness trauma mouse model. Eleven days after surgery, wound area in the treated group was reduced by 67–80% compared to the untreated controls (Luo et al. 2010). Alcoholic extract *Bletilla striata* tubers kills *Staphylococcus aureus*, the bacterium that commonly infects wounds (Tagaki et al. 1983). *Bletilla ochracea* which is also employed as *baiji* contains phenanthrenes that act against gram-positive bacteria (Yang et al. 2012).

Oil is used in place of water to produce an ointment with powdered *baiji* to treat burns and scalds (Li Shizhen 1578). In Vietnam, an emollient containing *Bletilla striata* is employed to treat burns (Nguyen Van Doung 1993). This is similar to Woodville's eighteenth century emollient that makes use of *salep* which is a polysaccharide powder prepare with terrestrial orchid tubers.

The recommended amount of *baiji* for oral decoction is 3–10 g of *baiji*: as an oral powder, 1.5–3 g (Yang et al. 1999). *Baiji* is said to be incompatible with *Radix aconite* and allied drugs, and they may not be used together (Guo 1996).

Baiji has also been used with sepium and also on its own to treat mucosal damage of bowel and bleeding peptic ulcers (Anonymous 1959). *Baiji* is one of the several components of *Xiao Wei Yan Powder* that is used for treating intestinal metaplasia and atypical hyperplasia of the gastric mucosa (Liu et al. 1992). There is some evidence from laboratory experiments that *baiji* promotes gastric healing. *Baiji* has antibiotic activity against *Streptococcus mutans* which is responsible for dental caries.

In 1982 a team of Korean scientists reported that an organic extract from *baiji* killed liver flukes, but there has been no follow-up on the research (Rhee et al. 1982). Liver fluke infestation occurs when people eat uncooked freshwater fish that are infested with the parasite. Millions of people in East Asian countries are infected. Infestation results in blockage of the bile duct and causes cancer of the bile duct and liver. Korea has the highest liver cancer rate in the world (Shin et al. 1996; Wang et al. 2004).

An Embolizing Agent

At Tongji Medical University in Wuhan, *baiji* was employed to treat inoperable liver cancer by a process known as embolization which involves blocking the blood supply to the tumour. In the United States, doctors employ Spherex® for the same purpose. Spherex® are biodegradable starch microspheres prepared from potato. It was patented in 1982. During the 1980s, doctors in Wuhan performed animal experiments to determine how intraarterial *baiji* could affect blood supply to the liver (Feng et al. 1995, 1996, 2003). These experiments have now been repeated in Germany. *Baiji* coagulates in the bloodstream and blocks the blood supply to a specific site when it is introduced into the appropriate artery. Body enzymes digest starch so *baiji* is biodegradable, but embolization with *Bletilla striata* particles was found to produce extensive and permanent vascular obstruction.

In a study conducted at Tongji Medical Centre in Wuhan, China, 56 patients with inoperable primary liver cancer (hepatic carcinoma) were treated with *Bletilla striata* embolization, and 50 by conventional gelfoam embolization. Embolization with *Bletilla striata* resulted in permanent vascular obstruction, marked shrinkage of tumour size, decrease in serum alpha-fetoprotein (AFP, a marker of primary liver cancer). Survival at 1, 2

and 3 years were 81.9%, 44.9% and 33.6% with a median survival of 19.8 months. By comparison, the 50 patients treated with conventional gelfoam had poorer survivals: 48.9%. 31.1% and 16.0% at 1, 2 and 3 years (Zheng et al. 1996, 1998). The result of embolization with *Bletilla striata* is superior to reported series of embolization performed in the United States.

Experiments are now being conducted to determine whether *baiji* might be a useful vehicle for drug delivery to the eye or as a carrier to deliver genes into the human body (Dong et al. 2009).

References

Anonymous (1959) Preliminary observations on the treatment of haemorrhage of digestive ulcers with sepium and *Bletilla hyacinthina*. Zhonghua Nei Ke Za Zhi 7 (1):9

Anonymous (trans) (1974) Barefoot doctors manual. NIH, Bathesda

Anonymous (2004) The new century Chinese-English dictionary of traditional Chinese medicine. People's Military Medical Press, Beijing

Bensky D, Clavey S, Stoger E, Gambie A (2004) Herbal medicine materia medica, 3rd edn. Eastland Press, Seattle, WA

Chen XQ, Gale SW, Cribb PJ (2009) *Bletilla* Rchb. f. In: Chen XQ, Zj L, Zhu GH et al (eds) Flora of China—Orchidaceae. Science Press, Beijing

Dong L, Xia S, Luo Y et al (2009) Targeting delivery oligonucleotide into macrophages by a cationic polysaccharide from *Bletilla striata* successfully inhibited the expression of TNF-alpha. J Control Release 134 (3):214–220

Feng XS, Qiu FZ, Xu Z (1995) Experimental studies of embolization of different hepatotropic blood vessels using *Bletilla striata* in dogs. J Tongji Med Univ 15 (1):45–49

Feng G, Kramann B, Zheng C et al (1996) Comparative study of long term effect of permanent embolization of hepatic artery with *Bletilla striata* in patients with primary liver cancer. J Tongji Med Univ 16 (2):111–116

Feng GS, Li X, Zheng CS et al (2003) Mechanism of inhibition of tumor angiogenesis by *Bletilla* colloid—an experimental study. Yo Xue Za Zhi 83 (5):412–416

Guo JX (1996) *Bletilla striata* (Thunb.) Reichb.f. In: Kimura et al (ed) International collation of traditional and folk medicine, vol 1. World Scientific, Singapore, p 205

Li Shizhen (1578 date of completion) Bencao Gangmu (published posthumously, 1596. Hu Chenglong, Nanjing)

Liu XR, Han WQ, Sun DR (1992) Treatment of intestinal metaplasia and atypical hyperplasia of gastric mucosa with Xiao Wei Yan powder. Zhongguo Zhong Xi YiJie He Za Zhi 12(10):602–603

Luo Y, Diao HJ, Xia SH et al (2010) A phyiologically active polysaccharide hydrogel promotes wound healing. J Biomed Mater Res 94A(1):193–204

Nguyen Van Doung (1993) Medicinal plants of Vietnam, Cambodia and Laos. Nguyen Van Duong, Hanoi

Peng R, Zheng Q, Hao J et al (2005) Biological evaluation of ChuangYuLing dressing—a multifunctional medicine carrying biomaterial. J Huazhong Univ Sci Technolog Med Sci 25(1):72–74

Perry LM, Metzger J (1980) Medicinal plants of East and Southeast Asia: attributed properties and uses. MIT Press, Cambridge, MA

Rhee JK, Kim PG, Baek BK et al (1982) Isolation of anthelmintic substance on *Clonorchis sinensis* from tuber of *Bletilla striata*. Kisaengchunghak Chapchi 20(2):142–146

Shin HR, Lee CU, Park HJ et al (1996) Hepatitis B and C virus, *Clonorchis sinensis* for the risk of liver cancer: a case-control study in Pusan, Korea. Int J Epidemiol 25 (5):933–940

Sterckx R (2008) The limits of illustration: Animalia and Pharmacopeia from Guo Pu to Bencao Gangmu. Asian Med 4(2):357–394

Tagaki S, Yamaki M, Inoue K (1983) Antimicrobial agents from *Bletilla striata*. Phytochemistry 22:1011–1015

Wang KX, Zhang RB, Cui YB et al (2004) Clinical and epidemiological data of patients with clonorchiasis. World J Gastroenterol 10(3):446–448

Yang YL, Wu KJ, Lu G (eds) (1999) Traditional Chinese materia medica. Wuhan University Press, Wuhan

Yang X, Tang C, Zhao P, Shu G et al (2012) Antimicrobial constituents of *Bletilla ochracea*. Planta Med 78 (6):606–610

Zheng C, Feng G, Zhou R (1996) New use of *Bletilla striata* as embolizing agent in the intervention treatment of hepatic carcinoma. Zhonghua Zhong Gu Za Zhi 18(4):305–307

Zheng C, Feng G, Liang H (1998) *Bletilla striata* as a vascular embolizing agent in interventional treatment of primary hepatic carcinoma. Chin Med J 111:10060–11063

Spiritual Tibetan Medicine: Popular *Wangla*

6

Taxila (now in Pakistan) was a famous medical centre during the sixth-century BCE. The city belonged to the Achaemenid Empire and was populated by many ethnicities that included Greeks, Persians and Northern Indians. Here, Jivaka Komarabhacca studied medicine and surgery with a seer, Atreya before returning to Rajagrha to look after the royal family, Buddha and his monks. Physicians, artisans and soldiers left behind by Alexander in Northern India and Central Asia in the fourth century B.C.E. contributed to the further spread of Greek medicine in Taxila and areas north of the Punjab. Today, Unani or Greek Medicine is one of five schools of medicine practised in India.

Interest in the fabled aphrodisiac properties of orchid tubers has survived in this region spreading even further eastwards to Mongolia following the Mongol conquest during the fourteenth century. Most prized are the palmate tubers of *Dactylorhiza* species (*D. hatagirea* and *D. viridis*) and *Gymnadenia conopsea* from the northern highlands, all shaped like a hand, and further south, pseudobulbs of *Eulophia dabia* and *Flickingeria fugax* that do not possess bulbous roots. *Dactylorhiza hatagirea* is exceptionally popular, to an extent that in several Himalayan nations and provinces, the orchid has become critically endangered (Figs. 6.1, 6.2, 6.3 and 6.4).

Meanwhile, Unani medicine was amalgamated with Ayurveda, Traditional Chinese Medicine and Bon traditions (ancient Sowa-Rigpa medicine) to form Tibetan medicine which developed its own unique characteristics. This knowledge was transmitted by oral tradition. It was esoteric. In 1696, with the support of the powerful Fifth Dalai Lama, Desi Sangye Gyatso founded the pioneering Chagpori College of Medicine in Lhasa. The College taught the classic *Four Tantras* and Desi Sangye Gyatso's extensively illustrated *Blue Beryl* (*Baidurya sngon po*) which contains drawings of medicinal herbs. Graduates from the College remained in monasteries. Care of the public was left to other lamas and healers. Tibetan medicine continues to be practiced at two levels—an esoteric spiritual form intended for adepts and yogins and a simpler form which is more generally available to the common people. Public medicine improved after 1916 when a secular college of Tibetan medicine and astrology was started by Khenrab Norbu to teach people who would return to their rural areas to work as doctors and educators. Khenrab Norbu is physician to the 13th Dalai Lama.

The Spiritual Tibetan *Materia Medica*

An aphorism from Hippocrates may well be applied to Tibetan Buddhism:

> Life is short, the art is long.

To cultivate *bodhicitta* and reach enlightenment, one needs to have a long life, else

© Springer Nature Switzerland AG 2019
E. S. Teoh, *Orchids as Aphrodisiac, Medicine or Food*, https://doi.org/10.1007/978-3-030-18255-7_6

Fig. 6.1 *Dactylorhiza hatagirea*. Photo: Bhaktar B Raskoti

mentioned the following benefits of *bcud len*: long life, enhanced body complexion, improved functioning of the sense organs, a sharp mind, melodious voice and virility (Gerke 2012a).

Bcud len has its origin in legend. Original transmission was oral, from a divinity to a Master. Pha dam pa Sangs rgyas (d. 1117), who received the *Bcud len gyi gdams pa rim pa lnga pa* (*Essence extraction in five stages*) from the goddess Vajrayogini, is said to have lived for 572 years by practising this flower *bcud len*. Other hermits practising it were also reported to have enjoyed long life. Several illustrious Masters were named in various texts, such as Nagarjuna (c. 150–c. 250 CE), Padmasambhava (c. eighth century), Milarepa (1052–1135) and Tsongkhapa (1357–1419) (Oliphant 2018) (Figs. 6.5 and 6.6).

Constituents in *bcud len* generally consist of several herbs, with calcite, honey, sugar, butter or milk as binders. Collection of a herb may involve a spiritual component, i.e. the sex and age of the collector, the lunar date and time are specified. Unique preparations may contain spices, saffron, stones, mercury, semen, menstrual blood, urine, and flesh. The famous Tibetan poet and yogin, Milarepa, is commonly portrayed in *thangkas* (Tibetan Buddhist religious paintings) with a greyish green body because during the early years of his meditation practice; his diet was based on nettles growing outside his cave. Milarepa ate so much nettle that when his terracotta pot broke, he discovered an inner nettle shell had formed from the nettle extract. In sculptures he is often featured holding this nettle bowl in his left hand (Rhie et al. 1991).

In his exhaustive Ph.D., dissertation on *bcud len*, Charles Jamlang Oliphant reported that the orchid, *Dactylorhiza hatagirea* (Tibetan *dbang lag*), was present in 10 out of the 72 *bcud len* recipes which he managed to collect. *Flower bcud len* always included *dbang lag*. However, the other plant materials were not orchidaceous: they were derived from *Polygonatum cirrhifolium* (Tibetan *ram mnye, ra mosha*), *Terminalia chebula* (Tibetan *a ru ra*), *Rhododendron* species (Tibetan *ba lu*), *Juniperus communis* (Tibetan *shug pa*), *Asparagus adscendens* (Tibetan *nye shing*), *Drosera peltata* (Tibetan *rtag tu ngu pa*),

innumerable lives. Consequently, Tibetan medicine as practised by the clergy delves into the preparation of *bcud len* (pronounced *chew len*) or 'essence extraction', pills that promote vitality and longevity. *Bcud len* has also been translated as 'elixir' or 'elixir of rejuvenation'. *Bcud len* practice is conducted with the objectives of rejuvenation, prevention of aging and spiritual realization. The *Four Tantras* of Tibetan medicine

Fig. 6.2 *Gymnadenia conopsea*. From: Baxter W. *British phaenogamous botany*, t. 401–509, vol. 6: t. 409 (1834–1843)

Ephedra (Tibetan *mtshe*) and *Artemisia* species (Tibetan *mkhan pa*). *Terminalia chebula* is a popular Ayurvedic medicine for preventing or treating disorders of the digestive system, whereas *Juniperus communis* contains fragrant essential oils that may also exert an effect on the digestive system. *Ephedra* (in Chinese Traditional Medicine, *ma huang*) contains ephedrine which is a stimulant that could counter the tendency to fall asleep (*nacrolepsy*). *Artemisia*

Fig. 6.3 Tubers of *Dactylorhiza hatagirea* Photo: SK Ghimire

annua (in TCM *qinghao*) is the source of *artemisinin*, an effective remedy for deadly *falciparum* malaria. *Asparagus adscendens* is employed as a tonic in Northern India. Polypharmacy is almost a universal feature of traditional medicine.

However, it is not necessary for *bcud len* to contain many ingredients. In fact, the earliest

Fig. 6.5 Milarepa (©Teoh Eng Soon 2019. All Rights Reserved)

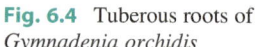

Fig. 6.4 Tuberous roots of *Gymnadenia orchidis*

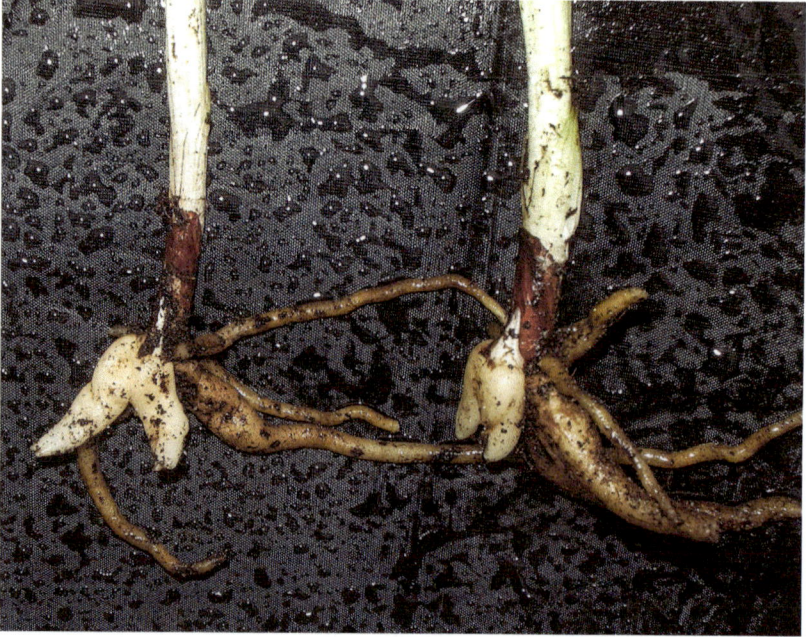

recorded mention of *bcud len* dating from the eleventh century stated that food could be transformed into *bcud len* by reciting an Avalokitesvara mantra (Oliphant 2018).

One of the earliest *bcud len* compilations was written by Rgyal mtshan dpal bzang (1310–1391). Entitled *Bcud len sna tshogs,* it is a work in three folios. It contains several recipes for making *bcud len* which were consumed to enhance the attainment of transcendental wisdom through *prajna* (wisdom) meditation.

The work begins with a recipe for a *Dbang lag bcud len* (Extracted Essence of *Dactylorhiza hatagirea*) that was alleged to be transmitted by a *nagarajah* (serpent king) to Nagarjuna (150–250 CE), the philosopher who wrote commentaries on the *Prajnaparamita Sutras* (*Discourse on Transcendental Wisdom*) and founded the Madhyayamaka School of Mahayana Buddhism. *Dbang lag* (*Dactylorhiza hatagirea* tuber) is grounded into powder, then cooked in milk and sugar and made into pills. If grounded female organs of a deer or sheep are added to these pills, they will 'help to remove obstacles'. The pills have an expiry date of 1 month (Oliphant 2016).

Another recipe uses the bark of *Asparagus adscendens*, beaten into coarse chunks, then boiled with one's own urine. To this is added refined butter, beaten molasses and three chunks of *Dactylorhiza hatagirea* tubers; the mixture is made into pills with the addition of calcite powder. The pills are consumed before performing *prajna* (wisdom) meditation (Oliphant 2016).

The third prescription states the following: flowers of *rtag tu ngu* (*Drosera peltata*) are boiled with wild yak meat, *dbang lag* (*Dactylorhiza hatagirea* tuber) and calcite in butter or milk to liquefaction. Jagger, garlic and *chang* (barley) are then added. After drying, the paste is rendered into powder. Heated with *bri* butter and after the addition of honey, it is made into pills, each of the size of deer's droppings. These are to be taken on an empty stomach. (Oliphant 2016). (*Drosera peltata* is a medicinal flower growing at high altitudes in Tsari, the most easterly district of Central Tibet. Dew is invariably present on this flower; hence it is said to be always in tears, like the perfect bodhisattva, Sadaprarudita, the 'Perpetually Weeping' bodhisattva of a past aeon who exemplifies unwavering devotion and perseverance).

Spiritual practice is not explained in the *Bcud len sna tshogs*, but it is specified in Pha dam pa Sangs rgyas's *Bcud len gyi gdams pa rim pa lnga pa*. Here, one is required to visualize oneself as Chenrezig (Avalokitesvara, bodhisattva of compassion) and to recite the *Om Mani Padme Hum* mantra (Teoh 2003) (Fig. 6.7) when one enters a field to gather flowers of *ba lu* (*Rhododendron*) and *rtag tu ngu* (*Drosera peltata*). After drying, the flowers are grounded and mixed with *a ru ra* (*Terminalia chebula*), *dbang lag* (*Dactylorhiza hatagirea* tubers), the 'six precious substances' and other ingredients. The practitioner then makes offerings at the shrine, seeks refuge (in the Buddha, his teachings and the community of bodhisattvas), refreshes the *bodhicitta* commitments (repeats the bodhisattva vows), and visualizes himself as Vajrayogini (Vajravarahi), a female tantric deity or *dakini* whose several iconographic features symbolize non-duality, an important concept in Buddhism. Masters of the lineage give their blessings. A *mantra* is recited 1000 times. When

Fig. 6.7 *Mani mantra* increased on stone. Presence of such imply sacred ground such as the vicinity of a monastery (©Teoh Eng Soon 2019. All Rights Reserved)

empowered, the pills are 'a nectar of wisdom'. While engaged in a meditative state of bliss, daily one recites the *mantra* 21 times and consumes two to three pills over a period of 21 days. Breathing exercise is not included. This practice is for the elimination of disease and protection against illnesses. It is claimed that it also improves intelligence, enhances clarity, assists the practitioner to comprehend the way and becomes attractive to dakini. This *bcud len* is practiced by hermits in Eastern Tibet (Oliphant 2016).

Whereas several *bcud len* treatises instruct the practitioner to visualize himself as Avalokitesvara or Vajrayogini, in one or two, the practitioner is required to visualize himself as Amitayus (Buddha of Infinite Life) (Fig. 6.8) or Heruka (a wrathful deity). It is claimed that by consuming a particular long-life pill which contains *dbang lag*, practising breath retention and meditating on Buddha Amitayus, the practitioner 'may attain the life power of Vidyadhara (in Indian religions, a semi-divine being with magical powers) and (he) can live for 16 months without food' (Oliphant 2016).

This spiritual *Materia Medica* is not intended for ordinary people who are not deeply involved in the tantric practice of Tibetan Buddhism. It is

Fig. 6.8 Amitayus, Buddha of Infinite Life. Photo: Kwa Bee Hua

included here to provide a glimpse of the role of the orchid, *Dactylorhiza hatagirea* in the advanced tantric practice of Tibetan yogins.

'Instructional *bcud len* texts describe how the practitioner's meditative experiences and subtle energies can be reinvigorated through sexual practice, both actual and visualized in the mind of the yogi' (Oliphant 2016). Thus, iI is indeed curious that no sexual practice is mentioned in any of the ten *bcud len* texts that specified *dbang lag* (*Dactylorhiza hatagirea*) (Oliphant 2016) whose alleged aphrodisiac quality is so highly valued in Southern Himalaya. However, in some Yogacara lineages, sexual abstinence (*chags spang*) is absolutely essential during yogic practice. Lama Yeshe informed Dr. Barbara Gerke that 'experiencing strong sexual pleasure during this retreat is extremely dangerous; rather, the revitalized strength from the aphrodisiac component of the *bcud len* pill should be completely directed to gain 'penetrative concentration' in meditation. Sangye Gyatso advised that in addition to being clean, quiet and pleasant, the place for *bcud len* practice should be 'inaccessible for snakes, women and stupid people'' (Gerke 2012b).

In both this spiritual and also the plebian form for ordinary people, Tibetan medicine requires maintenance of a healthy lifestyle with attention to diet, meditation and the recitation of mantras. Visualization of deities is a requisite in the spiritual practice.

Contemporary Tibetan medicine as practiced in Dharamsala, India conveniently interprets *bcud len* as tonic, health supplements and/or aphrodisiacs for people above 50. Like most people, ordinary Tibetans prefer popping a pill daily to long struggle with fasting and meditative practice bcud len practice. The pills on sale are packaged accordingly. Sometimes, constituents of the pills may be blessed or consecrated by a high standing lama. *Gaay pa Siwae chulen* (Elixir of Rejuvenation) or *rgas pa gso ba bcud len* (lit/exlixir to heal the aged) are marketed as aphrodisiacs. *Gcong chen bcud len* (chronic wasting elixir) while not described as an aphrodisiac claims to boost the 'activities of the sexual organs' while simultaneously treating chronic

illnesses, respiratory problems, lack of appetite and general weakness. They are tonics to 'promote kidney heat', an euphemism for sexual stamina (Gerke 2009, 2012a, b). The Indian (here, also Tibetan) concept of rejuvenation couples with virility; the final, eight branch of the Indian and Tibetan medical traditions both focus on restoring virility (Sanskrit *vajikarana*; Tibetan *ro tsa bar by aba*). At least two World Heritage temples, the Konark Sun Temple in Orissa and Khajuraho in Madhya Pradesh, illustrate that notion.

Dactylorhiza hatagirea (2018)

Dbang lag (*Dactylorhiza hatagirea*) is a Eurasian temperate, alpine orchid which shares characteristics with *Dactylorhiza viridis*, other *Dactylorhiza* species and the European genera *Anacamptis* and *Orchis*. They were, and still are, alleged to possess tonic and aphrodisiac properties. *Dactylorhiza hatagirea* enjoys a wide distribution from Heilongjiang to the British Isles and southwards to Northeastern Tibet, Bhutan, Nepal, Pakistan and Afghanistan. It occurs in marshes, damp meadows and grassy slopes. Plants are 30–90 cm tall. Leaves are shaped like a lance, and they are green, blotched with purple. The flower spike is erect, 15 cm tall, shaped like a hyacinth and bears numerous purple (sometimes flesh pink, yellow or white) flowers. Flowering season is May to July. As its tubers are palmate, they easily substitute for *Dactylorhiza viridis* as *wangla* (Fig. 6.9).

However, *Dactylorhiza hatagirea* enjoys a far greater reputation as an invigorating agent and aphrodisiac than almost any other Asian orchid. It is especially in high demand in India where besides being an aphrodisiac and tonic (Vij 1995; Sood et al. 2005), it is used to treat a long list of ailments, including diabetes, chronic diarrhoea, dysentery, hoarseness of voice, coughs, paralysis, cuts and wounds, fractures and paralysis (Teoh 2016). In the Dolpa, Humla and Mustang districts within the alpine zone of Northwestern Nepal which borders Tibet, *Dactylorhiza hatagirea* is one of the four

Fig. 6.9 Himalayan meadow with luxuriant growth of *Dactylorhiza hatagirea*

commonest herbs employed by the people. Here, juice extracted from the tubers are used for cuts, wounds and gastritis (Kunwar et al. 2006). *Dactylorhiza hatagirea* is consumed in Bhutan as aphrodisiac or tonic and to promote longevity (Wangchuk 2009) (Figs. 6.10 and 6.11).

Chinese herbal medicinal texts state that the herb 'benefits the kidney, stomach and spleen', relieving thirst and improving appetite. It is employed as a cardiac stimulant or to treat dizziness, anaemia and irregular menstruation (Wu 1994).

Despite this extensive medicinal usage of the orchid tubers, scientific studies on *Dactylorhiza hatagirea* are scarce, and most are concerned with conservation of the orchid species. Five new compounds (dactylorhins A, B, C, D and E) and two new natural compounds (dactyloses A and B) have been isolated together with 12 known compounds from tubers of *Dactylorhiza hatagirea*. Additionally two lipid mixtures were also obtained (Kizu et al. 1999). A small study involving six treated albino rats and six controls observed that the treated rats gained weight and were more sexually active. Testosterone levels were more than three times higher in the treated animals. The authors concluded that

administration of *Dactylorhiza hatagirea* causes a rise in testosterone levels resulting in weight gain, increased libido and enhanced sexual performance (Thakur and Dixit 2007).

Wangla

Wangla applies to two orchid species, *Gymnadenia conopsea* and *Dactylorhiza viridis* (syn. *Coeloglossum viride*), that are widely distributed across the highlands of Central China. These fragrant herbs are attributed with health-promoting properties which do not seem to be very much inferior to those of *bcud len*, except that they do not help to further spiritual advancement. They are consumed with the intent to overcome illness and malaise, gain strength and vigour, as aphrodisiac, and to promote longevity. Some of these properties are also attributed to fragrant herbs with tuberous roots, e.g. *ginseng* (*Panax ginseng*) and *danggui* (also spelt *dong quai*; *Angelica sinensis*).

Apart from taking herbs, when a person is ill, Tibetan medicine advocates breathing meditation and dietary restrictions.

Fig. 6.10 Amchi Tangyal Sangbo Baiji of Pungmo collecting *Dactylorhiza hatagirea.* Photo: SK Ghimire

Fig. 6.11 Amchi Tangyal Sangbo Baiji of Pungmo harvesting *Dactylorhiza hatagirea* Photo: SK Ghimire

Gymnadenia conopsea

Gymnadenia conopsea is a widely distributed Eurasian terrestrial orchid species growing in open forests, grasslands, water-logged meadows and rocky slopes. In China it occurs at 200–4700 m, from Tibet and Yunnan through Sichuan, Gansu, Shaanxi, Shanxi, Hebei, Inner Mongolia, Liaoning, Heilongjiang and Jilin onwards to Korea, Russia and Japan (Chen et al. 2009). Plants are slender or robust, reaching up to 60 cm, with 3–5 long, narrow, pointed leaves, 5–11 cm long and 1–2.5 cm wide, sheathing at their base. Roots are tuberous. In summer, numerous small, fragrant and purple flowers appear clustered around a tall inflorescence (see Fig. 6.2).

This fragrant herb enjoys wide usage in China, Russia and Mongolia. Its Chinese herbal name is *Shou Zhang Shen* (ginseng-like palm) because its tubers resemble a human hand and like the tubers of ginseng (*Panax ginseng*) which has been fancifully likened to a human figure. The herb is employed as a tonic and to some extant as a panacea. Also going by the Tibetan name, *wangla*, it is an important herb in Tibetan medicine. It was recorded in *Sibuyidia*, the eighth century classic on Tibetan medicine (Shang et al. 2017), and it is likely that the usage originated in the Tibetan region. Tibetans conquered and occupied parts of Sichuan, Shaanxi, Gansu and Xinjiang in the mid-seventh century, and Tibet retained control over parts of this region for almost 200 years. During the Mongol Period, Tibetans influenced the Mongol court. *Wangla* is thus widely employed as a tonic in Traditional Chinese Medicine and in the folk medicine of the Mongols, Koreans, Bai and Naxi. It is not employed in European folk medicine.

Wangla is employed in Tibetan medicine to treat weakness of the lungs (perhaps tuberculosis, bronchiectasis, bronchitis, black lung, silicosis, etc.) by 'moisturizing the lungs and invigorating the kidney'. Basically, it strengthens and nourishes a weak body.

Sometimes it is employed as a solitary herbal preparation. But more commonly, it is employed in combination with other herbs. For example, powdered *Wangla* (*Gymnadenia conopsea* 30 g), mixed with equal amounts by weight of *Tianma*, *Radix Phlomii*, *Drosera peltata* (Tibetan *rtag tu ngu*) and *Rhododendron parvifolium* (Tibetan *ba lu*), is mixed in 40 g of honey to prepare pills of which 3 g is consumed twice a day to treat impotence, spermatorrhoea, anaemia and insomnia (Shang et al. 2017).

In a survey of 4500 Tibetan prescriptions provided in Xizang a decade ago, *wangla* was prescribed 104 times (2.3%), principally as a tonic to invigorate the body, strengthen *Yang* and promote longevity (33 cases). It was also employed to treat kidney or lung disease, gout, disorders of the eye and parasites (Ji and Li 2009) and hepatitis. At Mount Taibai in Shaanxi Province, *Gymnadenia conopsea* is employed to treat weakness, dizziness, wounds and liver complaints (Chang et al. 2017). It is also relished as a vegetable. The dish, *Shiguo Ji*, is 'chicken cooked with *wangla* in a stone hotpot' (Shang et al. 2017).

Chinese scientists who are busy studying the constituents of *Gymnadenia conopsea* have managed to isolate over 120 compounds that include predominantly glycosides which account for more than a third. Other compounds are diethystilbenes, phenanthrenes, alpha-tocopherol and polysaccharides, but alkaloid was not detected. Some compounds exert an antiallergic effect and others an antioxidant action (Teoh 2016). The antioxidant effect on cultured brain cells from the old rats provided the basis for a claim that *Gymnadenia conopsea* possesses an antiaging property (Shang et al. 2017).

When aqueous extracts of *Gymnadenia conopsea* was fed to mice at the rate of 10, 20 or 40 g/kg for 6 days, their swimming time increased two and half to threefold, albeit this fatigue tolerance was exceeded in mice fed with 20 g/kg *ginseng* (*Panax ginseng*) (Zhao and Liu 2011). Assuming that humans could achieve similar fatigue tolerance by taking a proportionately similar amount of *wangla*, a 60 kg man would need to consume at least 600 g of *Gymnadenia conopsea* daily for 6 days. In the medicinal recipes, the usual amount is 60 g daily.

How could one reconcile the difference in required dosage? Traditional Mongolian medicine requires *Gymnadenia conopsea* (*Erihaoteng*) to be decocted in goat's or cow's milk. When the goat's milk decoction of *Erihaoteng* was fed to mice at the rate of 2 g/kg, it trebled their swimming time compared to animals fed in a saline decoction of *Erihaoteng*. Stamina enhancement was minimally improved by feeding with cow's milk decoction (Jin and Wang 2009). So a 60 kg man might only need to consume 120 g of *Gymnadenia conopsea* daily. However, the authors discussion overlooked the nutritional content of goat's milk.

There are at least two patent applications in China for the use of *Gymnadenia conopsea* to treat diabetes. In one, the claim is based on a traditional prescription that advocates the addition of another orchid, *Pecteilis susannae*, and 16 other herbal products, including *Astragalus* (not an orchid).

Dactylorhiza viridis

Coeloglossum viride or *Coeloglossum viride* var. *bracteatum* were names assigned to this orchid, once regarded as the sole species in the genus *Coeloglossum*. Ribosomal DNA studies now place *Coeloglossum* within the genus *Dactylorhiza*, so the correct name for this herb is *Dactylorhiza viridis* (Fig. 6.12).

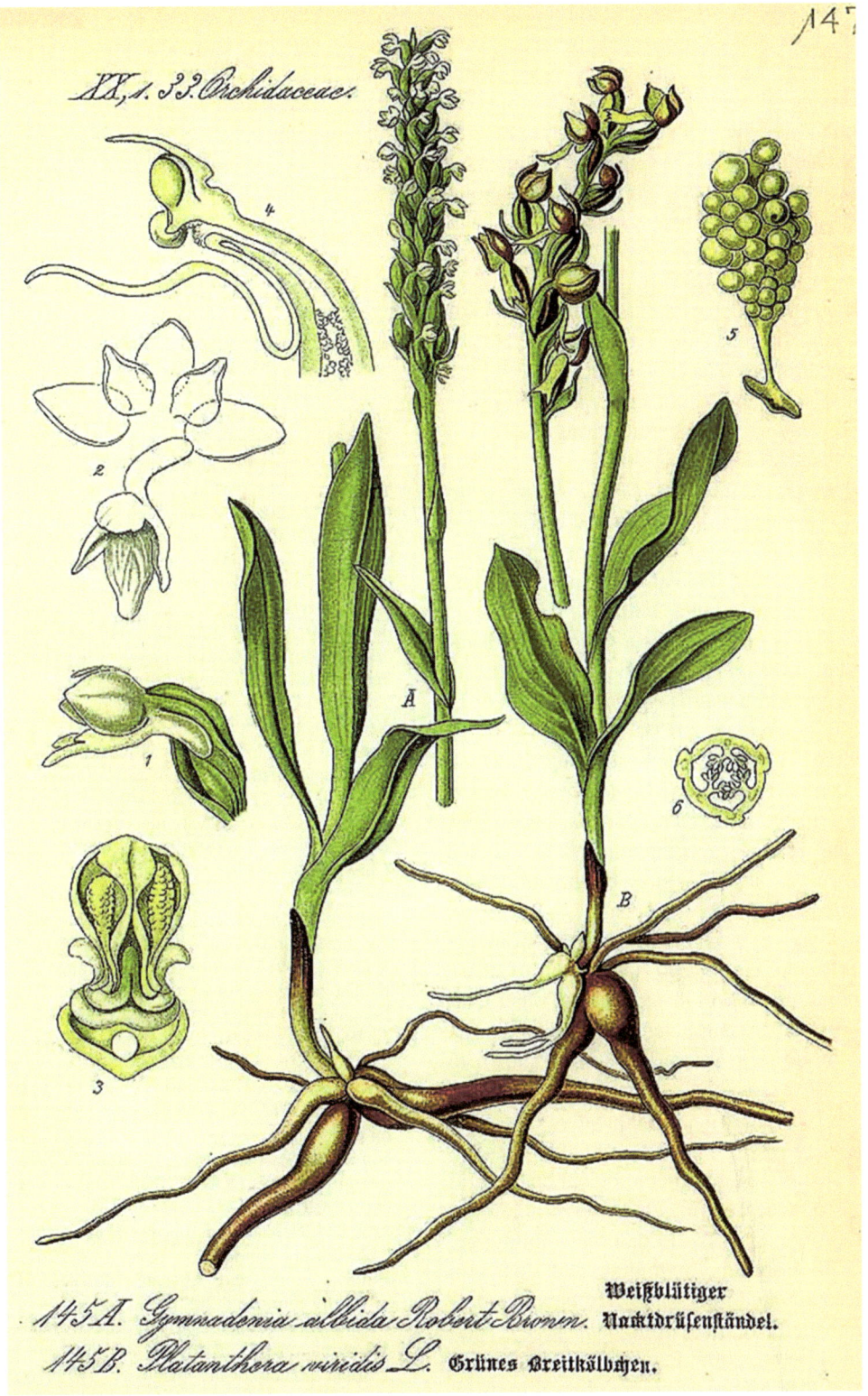

Fig. 6.12 *Dactylorhiza viridis* (syn. *Coeloglossum viride*) [as *Platanthera viridis*] From: Thome OW. *Flora von Deutschland Osterreich und der Schweiz, Tfein*, vol. 1: t. 145, Fig. B. (1855)

This robust terrestrial herb is found in forests, bogs, thickets and wet grasslands or steep slopes at 1200–4300 m across China (Gansu, Hebei, Heilongjiang, Henan, Hubei, Jilin, Liaoning, Nei Mongolia, Ningxia, Qinghai, Shaanxi, Shanxi, Sichuan, Xinjiang, NE Xizang, NW Yunnan). It also occurs in Korea, Japan, Taiwan, Mongolia, Russia, Kazakhstan, Kyrgyzstan, Turkmenistan, Kashmir, Nepal, Bhutan, SW Asia and North America (Chen et al. 2009).

Plants are slender or stout, up to 45 cm tall, with widely spaced, elliptic leaves and palmate, three-lobed tubers. The inflorescence is 3–15 cm with few to many small, greenish yellow or greenish brown flowers. Flowering season is June to August (Chen et al. 2009).

For centuries, tuber of *Dactylorhiza viridis* (*wangla*) has been widely employed in traditional medicine in Tibet and in other Chinese provinces that have a strong Tibetan influence (Qinghai, Gansu, Inner Mongolia and Shanxi) and in Mongolia. It is employed as a tonic, aphrodisiac and tranquillizer, to improve memory and also to treat asthma and hepatitis (Zhang et al. 2006; Li et al. 2009; Guo et al. 2013).

Chinese scientists investigating *wangla* (*Dactylorhiza viridis* syn. *Coeloglossum viride*) initially isolated 7 new compounds, coelovirins A–G, and 14 known compounds that included Dactylorin A and B, loroglossin, militarine, coelovirin, gastrodin, thymidine, quercetin, gastrodin and hydroxybenzyl alcohol from the tubers (Huang et al. 2004). Several of these compounds are present in *Tianma* and other Chinese herbs that are employed to treat disorders of the nervous system [these herbs were shown to exert a neuroprotective effect in vitro and in animals (Teoh 2016)]. Subsequently, extract of *Dactylorhiza viridis* was shown to correct memory deficits and prevent pathological changes in the brain of senescent mice (Zhang et al. 2006). Dactylorin A and B, loroglossin, militarine, coelovirin, gastrodin, thymidine and quercetin were demonstrated to be antioxidants which protect brain cells of rats from damage by beta-amyloid (Li et al. 2009; Pan et al. 2017). Nevertheless, in the absence of proper clinical trials on human subjects, it is not possible to determine whether *wangla* sharpens the mind, improves memory and prevents dementia. Well-designed clinical trials should be conducted to determine whether such benefits can be derived from consumption of *wangla*.

References

Chang N, Luo ZW, Li DW, Song HY (2017) Indigenous uses and pharmacological activity of traditional medicinal plants in Mount Taibai, China. Evid Based Complement Alternat Med 2017:8329817. https://doi.org/10.1155/2017/8329817

Chen XQ, Gale SW, Cribb PJ (2009) *Dactylorhiza* Necker ex Nevski. In: Chen XQ, Liu ZJ, Zhu GH et al (eds) Flora of China—Orchidaceae, vol 25. Science Press, Beijing

Gerke B (2009) Chulen (bcud len)—understanding Tibetan essence extractions, elixirs and tonics. Paper read at the Seventh International Conference on Traditional Asian Medicine (ICTAM) in Thimphu, Bhutan, September 2009

Gerke B (2012a) 'Treating the aged' and 'Maintaining health'—locating bcud len practices in the four Tibetan medical tantras. J Intern Assoc Buddhist Studies 35 (1–2):329–362

Gerke B (2012b) Treating essence with essence: re-inventing bcud len as vitalizing dietary supplements in contemporary Tibetan medicine. Asian Med 7(1):196–224

Guo Z, Pan RY, Qin XY (2013) Potential protection of *Coeloglossum viride* var. bracteatum extract against oxidative stress in rat cortical neurons. J Anal Methods Chem 2013:326570

Huang SY, Li GQ, Shi JG, Mo SY (2004) Chemical constituents of the rhizome of *Coeloglossum viride* var. bracteatum. J Asian Nat Prod Res 6(1):49–61

Ji N, Li Y (2009) Effect of different strains of Amillaria mellea on the yield of Gastrodia elata f. glauca. J Fungal Res 6(4):231–233

Jin L, Wang XL (2009) Study on the strengthening with tonics activity of the different processed products of *Gymnadenia conopsea*. J Med Pharm Chin Minor 1:28

Kizu H, Kaneko EI, Tsuyoshi Tomimori T (1999) Studies on Nepalese crude drugs. XXVI. Chemical constituents of Panch aunle, the roots of *Dactylorhiza hatagirea* D. DON. Chem Pharm Bull 47 (1):1618–1625

Kunwar RM, Nepal BK, Kshhetri HB et al (2006) Ethnomedicine in Himalaya: a case study from Dolpa, Humla, Jumla and Mustang districts of Nepal. Ethnobiol Ethnomed 2:27

Li M, Guo SX, Wang CL, Xiao PG (2009) Quantitative determination of five glycosyloxybenzyl 2-isobutylmalates in thetubers of *Gymnadenia conopsea* and *Coeloglossum viride* var. bracteatum by HPLC. J Chromatogr 47(8):709–713

Oliphant CJ (2016) 'Extracting the essence': bcud len in the Tibetan literary tradition. Ph.D. Thesis. Oxford University

Oliphant CJ (2018) Subject: Rasayana (chulen) history & literature. Himalayan Art Resources 2018

Pan RY, Ma J, Wu HT et al (2017) Neuroprotective effects of *Coeloglossum viride* var. bracteatum extract in vitro and in vivo. Sci Rep. https://www.nature.com/articles/s41598-017-08957-0

Rhie MM, Thurman RAF, Taylor JB (1991) The sacred art of Tibet. Thames & Hudson, London

Shang XF, Guo X, Liu Y et al (2017) *Gymnadenia conopsea* (l.) R.Br.: a systemic review of the ethnobotany, phytochemistry and pharmacology of an important Asian Folk Medicine. Front Pharmacol: 1–28

Sood SK, Rana S, Lakhanpal TN (2005) Ethnic aphrodisiac plants. Scientific Publishers (India), Jodhpur

Teoh ES (2003) Lotus in the Buddhist art of India. Teoh Eng Soon, Singapore

Teoh ES (2016) Medicinal orchids of Asia. Springer Nature, Switzerland

Thakur M, Dixit VK (2007) Aphrodisiac activity of *Dactylorhiza hatagirea* (D.on) Soo in male albino rats. Evid Based Complement Alternat Med 4(Suppl 1):29–31

Vij SP (1995) Orchid genetic diversity in India: conservation and commercialization. Proc. 5th Asia Pacific Orchid Conference and Show, Fukuoka, pp 20–39

Wangchuk P (2009) High altitude medicinal plants of Bhutan. An illustrated guide for practical use. Pharmaceutical and Research Unit, Institute of Traditional Medicine Services, Ministry of Health, Thimphu

Wu XR (1994) A concise edition of medicinal plants in China. Guangdong Higher Education Publication House (in Chinese), Guangdong

Zhang D, Liu GT, Shi JG, Zhang JJ (2006) Effects of *Coeloglossum viride* var. bracteatum extract on memory deficits and pathological changes in senescent mice. Basic Clin Pharmacol Toxicol 98(1):55–60

Zhao L, Liu G (2011) Experimental study of Shaoshang Shen liquids on antifatigue effects in mice. Chin J Chin Med 22:17

Indian tradition upholds the infinity vine as the symbol of abundance, fertility and wealth. It features the lotus (*Nelumbium nucifera*) whose rhizome traverses lakes and streams for miles shielding the waters with its shimmering, emerald leaves and pink flowers throughout summer (Teoh 2003, 2019). In the orchid world, *Vanilla* is the infinity vine. A single plant is capable of climbing to a height of 20 m or higher and spreading over an area of 2000 m^2 (Soto-Arenas and Cribb 2010; Cameron 2011). Such a large *Vanilla* plant would be a suitable challenger to a 2-ton tiger orchid (*Grammatophyllum speciosum*) for the title, 'Largest Orchid Plant in the World'.

Vanilla is the orchid with which the world is most familiar because of its usage in food, beverages and cosmetics, albeit many people are unaware of its orchidaceous identity. In 1520, Aztecs whose nobles enjoyed drinking vanilla-flavoured chocolate offered the drink to Cortez and his Spanish conquistadors. Moctezuma allegedly consumed inordinate amounts of vanilla-flavoured chocolate daily. An Aztec Herbal of 1552, the *Codex de la Cruz-Badianus*, described its medicinal usage. This was elaborated by Francisco Hernandez after 1575 (de Toledo 1615). Aztecs accepted *Vanilla* pods as tribute from the Totonacs and other Indian tribes as early as during the reign of Itzcoatl (1427–1440). The Franciscan friar Bernardino de Sahagun reported that vanilla pods were sold in Mexican markets for use to flavour drinking chocolate (de Sahagun 1529).

The earliest mention of *Vanilla* in a *Herball* is in the *Codex de la Cruz-Badianus* (commonly shortened as the *Badianus Manuscript* or alternatively referred to as the *Barberini Codex*) written in Nahuatl by an Aztec physician whose given Catholic name was Martinus de la Cruz. In 1552 (30 years after the Spanish conquest of Mexico), the *Herball* was translated verbatim into Latin, without comment, introduction or personal bias, by another native whose Catholic name was Juannes Badianus. Both author and translator were residing and teaching at the Colegio de Santa Cruz de Tlatelolco. The *Codex de la Cruz-Badianus* described the use of 183 native Aztec herbs many of which were for pain relief, trauma, swellings and skin lesions (Fig. 7.1).

Taken to Spain as a gift for King Charles V, it stayed unread in the royal library and then mysteriously found its way into the hands of an obscure Spanish apothecary, Didacus Cortavila. Then in 1625 it was presented to the visiting representative of Pope Urban VIII, Cardinal Francesco Barberini, who was a prominent patron of the arts and a collector of rare manuscripts but, simultaneously, Grand Inquisitor of the Roman Inquisition (a post he held from 1633 to 1679). Francesco Barberini was an unprejudiced and compassionate man who argued for leniency towards Galileo during the latter's trial and in

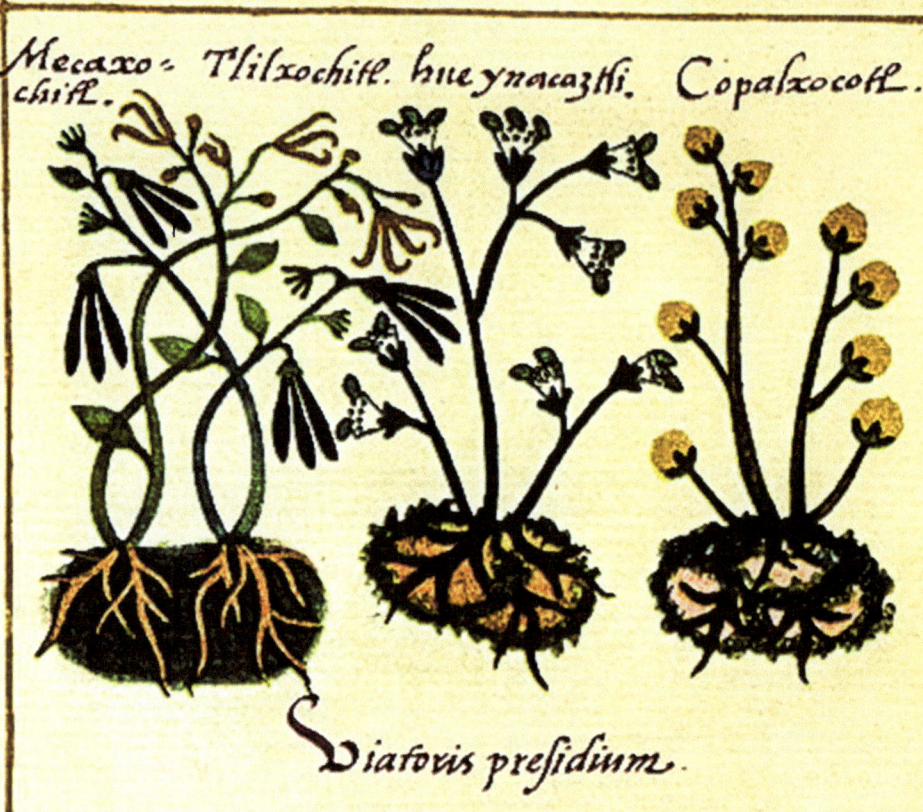

Fig. 7.1 *Vanilla planifolia* (left), *Nahuatl tlilxochitl*, featured in *Codex de la Cruz-Badianus* (1552). Drawing and original Nahuatl text by de la Cruz

the early seventeenth century. He refused to sign Galileo's sentence, and afterwards he arranged for the scientist to be transferred to the custody of the Archbishop of Siena instead of doing penance at a monastery. Barberini loaned the *Codex de la Cruz-Badianus* to the Linneans who made a copy that is now housed at the Royal Library in Windsor Castle. However, the cardinal did not seem to appreciate the priority of the *Codex*. He was impressed by the enormously more voluminous and superbly illustrated Mexican Treasury produced by Leonardo Antonio Recchi's adaptation of Francisco Hernandez's *Quatro Libros de la Naturaleza* (1615) and *Rerum Medicarum Novae Hispaniae Thesaurus* (1651). Furthermore, at that point in time, church leaders in Europe were not prepared to accept that native civilizations which they had set out to destroy or educate and dominate had anything worthwhile to offer. Barberini merely kept the *Codex* in his library and did not bring it to public attention. In 1902 the Barberini Library was incorporated into the Vatican Library, and there the *Codex* lay hidden until 1929 when it was discovered by Charles U Clark. *Codex de la Cruz-Badianus* contains beautiful colour drawings of the herbs (Fig. 7.1). It was reprinted by Johns Hopkins Press in 1940. The Maya Society published an English translation by William Gates in 1939. Emily Walcott Emmart rendered another English translation in 1940 that was published by Johns Hopkins Press. Spanish translations are also available.

Mexican Legends on the Origin of *Vanilla*

Two Mexican legends pertaining to the origin of the vine possibly promoted the consumption of vanilla for its alleged aphrodisiacal properties. Totonacs who were the first people in the world to enjoy vanilla told this legend about the origin of the plant:

> During an ancient era when the country was ruled by Tinitzli III, the queen gave birth to a beautiful daughter, named Tzacopontziza (Morning Star) whom they consecrated as priestess to the goddess of fertility. Her job was to collect flowers and other offerings from the forest for the goddess. During one of these trips her beauty aroused the carnal passions of Prince Zkatan-Oxga (Running Deer) who had chanced upon her. Every morning thereafter, the prince would hide in the thickets to await her arrival just to get a glimpse of her. He knew that they could not be together because priestesses were required to be celibate, and if caught, the amorous couple would be beheaded.
>
> One day when the sky was overcast, Running Deer cast his fear aside and ran to Morning Star whom he caught by the hand. Surprised at this approach by the rash but handsome young man, and unable to react appropriately, she followed in his flight. Their escapade was short-lived for the alerted priests caught up with them and the inevitable followed. On the ground where their blood was spilled, a bush sprouted, and clinging to it was a delicate vine. Scattered around were dry black pods which emitted a mesmerizing fragrance. When the priests of Tonoacayohua visited the site, they concluded that this exceptional vine and its pods came about through the intervention of their goddess, so vanilla was collected and made into an offering for Tonoacayohua. They called it zanat, 'nectar of the gods', although it was derived from the blood of Morning Star.

A different story also originating from the Mexican city of Papantla, the 'city which perfumed the world', is told by V P Cano that appeared on the web on 3rd February 2009. In her story,

> Xanath was a young and beautiful daughter from a noble family who was fond of making offerings to the gods. One day when she entered the temple to make her offering to the Messenger God, she heard a lovely tune coming from the inner courtyard. Tzarahuin, an artist who was decorating the temple was whistling to himself. It was love at first sight. In the days that followed, the couple used the temple for their meetings. One day, the fat God of Happiness caught sight of Xanath and also fell in love with her. Failing to win her love, he approached her father and conferred so much wealth and status on the latter that the man could not avoid wanting to please his god. When ordered to marry the god of happiness, Xanath stood her ground. She declared that she would not abandon Tzarahuin. In a fit of rage the god transformed Xanath into a plant, and poor Tzarahuin committed suicide. He became a bee. Every year when Vanilla came into bloom, this humble melipona bee would buzz around the flowers for hours, as it were, making love with sweet Xanath. (Fig. 7.2)

Fig. 7.2 Mexican god of happiness

Fig. 7.3 Mexican postage stamp featuring *Vanilla planifolia*

From Mexico to Europe

In 1570, King Phillip II of Spain commissioned an expedition to study the medicinal resources in the New World. The expedition was led by the king's personal physician, Francisco Hernandez de Toledo (1514–1587), who evinced an interest in medicinal botany having earlier made a Castilian translation of the work by Pliny. Hernandez spent 7 years in Mexico with his son and three artists. In his publications Hernandez described 3000 plants. He dispatched 16 leather-bound volumes to King Phillip II, and around 22 volumes of his works were rendered in Latin. On *Vanilla* (Nahuatl *tlilxochitl*, 'black flower') Hernandez reported:

> …vanilla beans are small like musk or balsam of the Indies and they are black—hence the name. It grows in hot, moist places. They are hot in the third degree and they are usually mixed with cocoa as well as mecaxochitl. … Two vanilla beans dissolved in water will provoke urine and menstruation.[1] It strengthens the stomach, and expels flatulence. It heats and thins the humours. It invigorates brain and heals fits of the mother.[2] It is said that

these beans are a similar remedy against cold poisons and against cold poisonous animal stings. It is also said to be one of the most aromatic plants in this region (Varey 2000).

Hernandez, however, never saw the *Vanilla* flower. He described it as *flore nigro aromatic*, the perfumed black flower, i.e. merely translating the Nahuatl name, *tlilxochitl* (Ecott 2004), whereas the ephemeral flower is green ad scarcely aromatic (Fig. 7.3). The Nahuatl name described the fermented black pods discovered on the ground around the plant.

In Spain the dark pods were named after their shape, *vaynilla*, a diminutive of *vayna* derived from Latin, *vagina*, meaning sheath. Thus 'orchid' and 'vanilla' both share an erotic terminology, 'orchid' being derived from Latin *orchis*, which means testicle. In the sixteenth century,

---

[1] Ecott (2004) translates this as abortion. The term probably refers to vaginal bleeding which could imply menstruation or a miscarriage.

[2] Possibly referring to eclampsia, the deadly fits are preceded by severe high blood pressure of pregnancy and fluid retention.

Hieronymus Tragus was promoting the belief that Mediterranean terrestrial orchids did not arise from seed but sprung from the semen of copulating goats, hence 'goat stones' or *Satyrion*. From this arose the belief that orchids and vanilla were aphrodisiacs.

The word *vanilla* spelt in this manner was originally a Dutch, and later an English, corruption of *vaynilla*. *Vanilla* made its first appearance in *De Indiae utriusque re naturali et medica* by Willem Piso published in Amsterdam in 1658 (Ecott 2004). Piso was an outstanding physician, a pioneer of tropical medicine. Travelling to Brazil where he remained from 1637 to 1644, he concluded that many sailors had fallen ill because of a deficiency of fresh food. Instead of treating them with medication, he prescribed fresh fruit, vegetables and fish. Piso and Georg Markgraf co-authored *Historia Naturalis Brasiliae* published in 1648. Markgraf contributed the eight books on botany and zoology, and one wonders whether he might have suggested the term *vaynilla* to Piso (Figs. 7.4 and 7.5).

Phillip II was not happy with the academic tone of Hernandez's work, and only fragments of the writings began to appear in print late in 1615 in a Mexican tome, *Plantas y Animales de la Nueva Espana, y sus virtudes por Francisco Hernandez, y de Latin en Romance por Fr. Francisco Ximenez*. It was published in Latin in 1628 and again in 1648. Hernandez's works inspired the British through the writings of Sir Hans Sloane who quoted him extensively in the *Natural History of Jamaica* (c. 1725).

From 1560 onwards, Spanish ships plying between Spain and Central-South America returned with pearls as their principal cargo, together with cocoa, vanilla, indigo, deerskin and annatto (Leal and de Clavijo 2012). However, Spanish nobility failed to appreciate vanilla because it was presented as a drink with chocolate, without sugar and milk. 'Better thrown to hogs than presented to man,' one conquistador remarked (Ecott 2004).

Queen Elizabeth I liked the drink. Hugh Morgan, apothecary to the queen, suggested to his monarch that vanilla could be added to other food and drink besides chocolate, and she became so fond of its fragrance that in the final year of her life (1602–1603), unfortunately only then, she insisted that all her food and drink should be enhanced by the addition of vanilla. The gutsy queen might have relished the flavour because it was loot from the Anglo-Spanish maritime wars and evidence of British naval supremacy over Spanish merchantmen!

When Morgan proposed that vanilla could have other culinary uses besides just the flavouring of chocolate, he gave the item a new, independent life of its own. Morgan also sent vanilla pods to the French physician-botanist, Charles de l'Ecluse (Carolus Clusius 1526–1609), court physician to Emperor Maximilian II and overseer of the imperial gardens in Vienna, who was to remark that 'anyone who sniffed the pods would soon develop a headache' (Harkin-Franke and Belanger 2011). Nevertheless, the French soon found many uses for vanilla. The addition of milk and sugar to chocolate and vanilla resulted in a delightful drink which became fashionable with the aristocracy within a decade, perhaps in no small measure due to the fact that Anne of Austria, who was married to Louis XIII, also expressed her fondness for the drink. Madame de Pompadour (Jeanne Antoinette Poisson), mistress of Louis XV whose name was given to a historically outstanding *Dendrobium*, served chocolate that was flavoured with vanilla and ambergris at her dinners.

Casanova (1725–1798) flavoured his wine with vanilla. In his memoirs, he related having a lover's lock of hair grounded to fine powder by a Jewish confectioner and stored in the form of a paste in a crystal sweetmeat box. Ingredients of the paste included amber, angelica, sugar and vanilla. Marquis de Sade (1740–1814) served dessert chocolate containing vanilla and Spanish fly which resulted in his guests being 'seized with a burning sensation of lustful ardour' (Ecott 2004). In 1847, an American volume on *Medicinal Botany* by R.E. Griffith, MD, mentioned that vanilla acts 'powerfully on the generative system as an aphrodisiac', and he advocated a dose as 8–10 grains (0.52–0.65 g). In the twentieth century, the sensual propensity of chocolate was

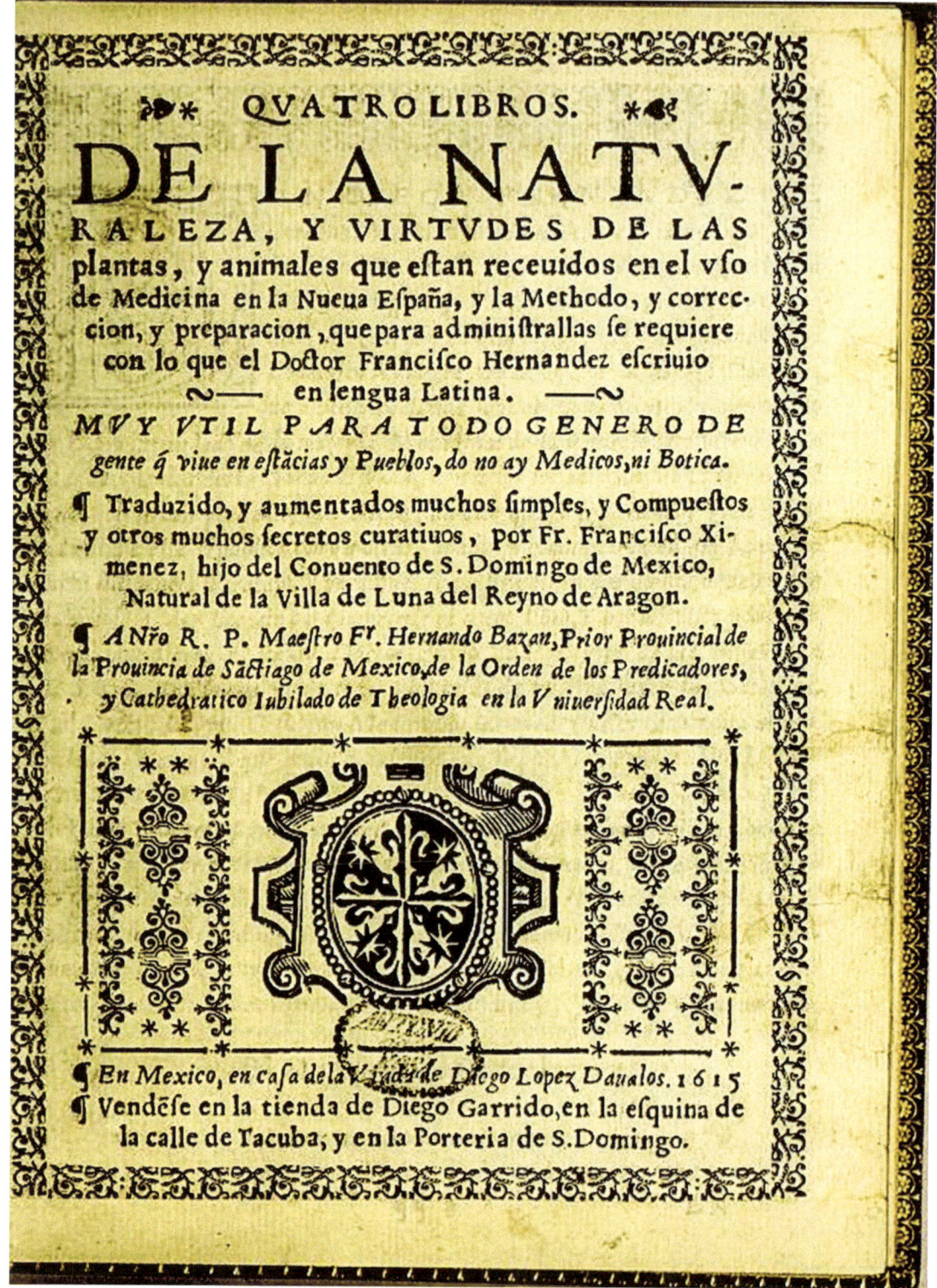

Fig. 7.4 Hernandez F, *Quatro Libros, De la naturaleza y virtudes deslas plantas y animals que estan receuidos en la vso de medicina en la Neuna Espana, etc. (1615)*. Mexico, title page

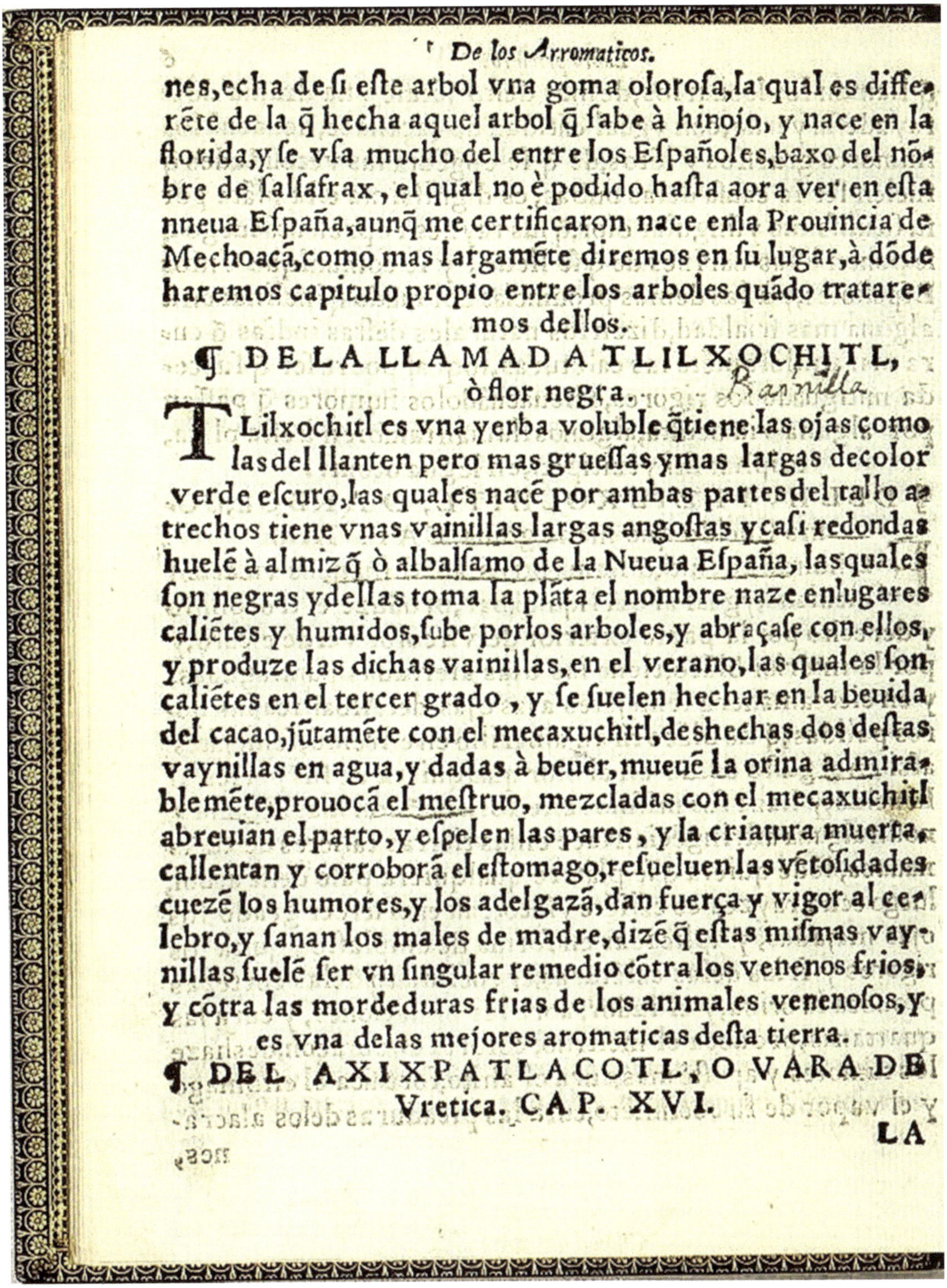

De los Arromaticos.

nes,echa de ſi eſte arbol vna goma oloroſa,la qual es diffe‑
rēte de la q̃ hecha aquel arbol q̃ ſabe à hinojo, y nace en la
florida,y ſe vſa mucho del entre los Eſpañoles,baxo del nō‑
bre de ſalſafrax, el qual no è podido haſta aora ver en eſta
nneua Eſpaña,aunq̃ me certificaron nace enla Prouincia de
Mechoacā,como mas largamēte diremos en ſu lugar,à dōde
haremos capitulo propio entre los arboles quádo tratare‑
mos dellos.

¶ DE LA LLAMADA TLILXOCHITL,
ò flor negra.

Tlilxochitl es vna yerba voluble q̃tiene las ojas como
las del llanten pero mas grueſſas ymas largas decolor
verde eſcuro,las quales nacē por ambas partes del tallo a
trechos tiene vnas vainillas largas angoſtas ycaſi redondas
huelē à almiz q̃ ò albalſamo de la Nueua Eſpaña, lasquales
ſon negras ydellas toma la plāta el nombre nace en lugares
caliētes y humidos,ſube porlos arboles,y abraçaſe con ellos,
y produze las dichas vainillas,en el verano,las quales ſon
caliētes en el tercer grado , y ſe ſuelen hechar en la beuida
del cacao,jūtamēte con el mecaxuchitl,deshechas dos deſtas
vaynillas en agua,y dadas à beuer,mueuē la orina admira‑
blemēte,prouocā el meſtruo, mezcladas con el mecaxuchitl
abreuian el parto,y eſpelen las pares , y la criatura muerta,
callentan y corroborā el eſtomago,reſueluen las vētoſidades
cuezē los humores,y los adelgazā,dan fuerça y vigor al ce‑
lebro,y ſanan los males de madre,dizē q̃ eſtas miſmas vay‑
nillas ſuelē ſer vn ſingular remedio cōtra los venenos frios,
y cōtra las mordeduras frias de los animales venenoſos,y
es vna delas mejores aromaticas deſta tierra.

¶ DEL AXIXPATLACOTL, O VARA DE
Vretica. CAP. XVI.

LA

Fig. 7.5 Hernandez F, *Quatro Libros, De la naturaleza y virtudes deslas plantas y animals que estan receuidos en la vso de medicina en la Neuna Espana, etc. (1615).* Mexico, page 6. On line 4 of his description of *tlilxochitl.* Hernandez used the term '*vanillas*' to describe the orchid pod. He is the originator of the name for the orchid genus

implied when it was sent as a gift together with red roses to lovers, a practice much featured in American movies.

Cultivation of Vanilla

Spurred by their queen's fondness for exotic *tlilxochitl*, the English moved quickly to gather knowledge on vanilla. In 1630, Thomas Gage reported that it was grown in Guatemala along with *achiote* (*Bixa orellana*, a small shrub whose seeds are a source of orange food-colouring), *mecasuchitl* (*Flor de cuerda*) and other herbs that were commonly added to chocolate. English merchantmen collected wild vanilla from Northeast Nicaragua and Honduras during the 1640s. In the 1660s, the English buccaneer, William Dampier, discovered wild populations of *Vanilla* being tended to in Mexico and the Caribbean. He made extensive comments on its curing and trade but did not manage to process the beans personally (Varey 2000; Ecott 2004).

One Father Labat imported three *Vanilla planifolia* vines to Martinique in 1697 and to Guadeloupe in 1701, both in the West Indies. Although the vine is grown in the two countries, on the world market, they are minor producers today.

Following rising popularity of vanilla in Europe, Totonac Indians established vanilla plantations in the Veracruz region in 1767. They were the world's first producers of commercial vanilla. Plants were brought to Reunion in 1793. Upon the discovery of hand pollination in 1841, vanilla plantations sprouted rapidly in Reunion, Madagascar and other islands in the Indian Ocean. *Vanilla planifolia* was reintroduced to Reunion in 1819, 1820, 1822 and 1875. Most of the cultivated *Vanilla* in Reunion belongs to a clone derived from the Mexican plant brought back by a merchant who obtained it from Jardin du Roi in Paris in 1822. *Vanilla planifolia* was introduced to Mauritius in 1827 and 1836 and to Seychelles in 1866. Cultivation commenced in Mauritius in 1865. The French also introduced it to their Asian and Pacific colonies, Indochina in

1865 and New Caledonia in 1861, but in both places, it did not become a significant crop.

Marchal introduced *Vanilla planifolia* to Buitenzorg (Bogor Botanic Gardens) in 1819, and commercial cultivation commenced in Java in 1846 spreading to Bali, Sulawesi and Sumatra only in the twentieth century. Production in Java in 1918 was a mere 67 kg, but the following year, it rose to 3678 kg, then more than doubling to 8516 kg the following year and 20,293 kg in 1923. Heyne (1927) commented that as plantations were productive 3 years after planting with yield increasing rapidly thereafter, vanilla could be considered a highly profitable crop in Batavia, despite competition from artificial vanillin. At that time, despite vanillin content of Java vanilla being comparable to vanilla produced elsewhere, it was priced lower; one reason being that too rapid drying in the hot tropical sun prevented the full aroma from developing (Heyne 1927). Currently, Indonesia is the second largest source of vanilla. The British sent *Vanilla* to India in 1835 and to Sri Lanka in 1853 (Plucknett and Smith 2014). The crop is grown in both countries but not on as extensive a scale as tea: perhaps it was also overshadowed by opium which was then 'legal' and far more profitable!

Madagascar did not enter the picture until much later, the first introduction being made only in 1870, but over the next 20 years, there were four more importations of the orchid. Almost 90% of the vanilla is grown across the northeastern tip of the island near the coastline, between Vohemar and Antalaha, going inland to Andapa. By 1924, Madagascar had become the dominant producer of high-grade vanilla, a position it has retained to the present day. Madagascar produced about 3000 tons of vanilla in 2005 which accounted for half of the world's total vanilla crop. At peak production a top Madagascan establishment may hold as much as 500 tons of vanilla pods. In the same year, Indonesia produced 2400 tons, and China, a late-comer, produced 1000 tons (Cameron 2011). Although approximately 100 ha of vanilla reported were into cultivation in the states of Kerala, Karnataka, Tamil Nadu and Lakshadweep in Southern India in 2000 (Bhattacharjee 2000), India is not noticeable on the vanilla map.

Cultivation of vanilla did not receive much attention in the British Straits Settlements. *Vanilla planifolia* and *Vanilla pompano* flowered and bore fruit at the Singapore Botanic Gardens. Plants on oil palm were remarkably vigorous, ascending to the crown then hung down in a circular cascade. Flowering season was November to March, one or two flowers per vine. Fruiting occurred, but because they grew so high above the ground, it was difficult to pollinate them by hand. Once or twice small lots were planted in the Straits Settlements, but Henry Ridley did not receive any report of their results. He commented that *Vanilla* might be successfully grown in the drier part of the Malay Peninsula, but the heavy rainfall in many parts militates against it. Although plants grow well and fruit setting occurs after pollination, the heavy rainfall are extremely liable to cause fruits to fall. Since the fruits take such a long time to ripen, the chances of their being destroyed by rain are great. Curiously wild *Vanilla griffithii* fruits in December and January during the period of heaviest rain. Unfortunately, the fruits, though sweet and edible, tasting like small bananas, contain no vanillin (Ridley 1897). Ridley's pessimistic view may not be correct because Hainan which also experiences heavy rainfall has emerged as a major producer of vanilla within a few decades (Cameron 2011).

The *Vanilla* Plant (Figs. 7.6 and 7.7)

Vanilla is an ancient orchid genus. Its 110 member species are distributed throughout the tropics (except Australia) between the 27th north and south parallels. Tropical America is home to 52 species, Southeast Asia with Papua New Guinea 31 species. Experts are divided as to whether it originated in the Indonesian region or in the Americas (Figs. 7.8 and 7.9).

There are altogether between 18 and 35 aromatic species of *Vanilla*, mostly distributed in America, but 95% of the world's supply of vanilla is derived from a single species, *Vanilla*

planifolia, now mostly cultivated in Madagascar. This species is naturally distributed from Oaxaca in Eastern Mexico to Guatemala and Belize. It is now rare in its natural habitat, with only one specimen being discovered per 4 km^2 in Oaxaca (Soto-Arenas 1999). However, *Vanilla planifolia* already shows wide genetic and phenotypic diversity in cultivation, and gene banks have been established to ensure survival of its varied forms (Roux-Cuvelier and Grisoni 2010).

Vanilla tahitensis is another popular aromatic species. It is widely cultivated in Tahiti and other Pacific Islands but is not native to the islands. It was introduced into Tahiti from the Philippines in 1848 by Admiral Hamelin (Correll 1953). The genetic status of this species was uncertain until several experts demonstrated that it was a natural hybrid between *Vanilla planifolia* and *Vanilla odorata*. Chloroplast DNA indicated that *Vanilla planifolia* was the pod-bearing parent; 93% of the markers were common to *Vanilla planifolia*, whereas *Vanilla odorata* was the pollen parent. Both parents are closely related, distributed in the same area, and the natural hybrid appeared 500–600 years ago, before the arrival of the Spanish (Bory et al. 2008; Lubinsky et al. 2008).

Vanilla pompano is a variable species with thicker, broader leaves, a larger plant, yellow flowers with a funneled lip and larger seed pods. In Peru the plants flourish in palm swamps and flower twice a year, in April and from September to October. Pods which are known locally as *pompon*, *gruesa* or *platanillo* vanilla beans are oily and emit a strong vanilla fragrance, but the aromatic profile is different from that of *Vanilla planifolia*. *Platanillo* vanilla is sold in Mexico, but it is not widely traded on the global market. *Vanilla planifolia* has formed hybrids with *Vanilla pompona*, and these are presently cultivated (Soto-Arenas and Dressler 2010) (Fig. 7.10).

The fourth fragrant species grown for its pods is *Vanilla odorata*. The species is distributed from Mexico southwards to Central America, Columbia, Ecuador, Bolivia and Peru, but it is only cultivated on a small scale in Chiapas, Mexico, and only for the local market. Mexican Indians add it to rum. They might have employed the

Vaillant pinxit.   Vanille a feuilles planes ┆ *Vanilla planifolia*

N. Rémond imp.

pods as a vermifuge in the past. The species is more difficult to handle because its pods have a tendency to split on the vine and are more susceptible to fungi. However, it is probably the pollen parent of *Vanilla tahitensis*, and therefore it may be useful as a parent for breeding new varieties of *Vanilla*.

Vanilla abundiflora which is native to Indonesia was also reported to produce fragrant fruits, but they are less fragrant than those of *Vanilla planifolia* (Heyne 1927). It is not a commercial crop. In the Caribbean other aromatic species with local usage include *Vanilla mexicana* (syn. *V. aromatica*), *V. claviculata*,

Fig. 7.7 *Vanilla planifolia* (©Teoh Eng Soon 2019. All Rights Reserved)

Fig. 7.8 *Vanilla aphylla* tastes like banana, but it is not aromatic enough to be commercialized (©Teoh Eng Soon 2019. All Rights Reserved)

V. poitoei (syn. *V. eggersii*), *V. palmarum*, *V. Pphaeantha* and *V. decaryi* (Teoh 2019).

In its native Mexico, sexual reproduction is rarely observed in *Vanilla planifolia*. Flowers are pollinated by *Euglossine* bees (*Euglossa viridissima*) but do not reward the pollinator (Soto-Arenas 2006). Between 4 and 6% of cultivated crops in Mexico self-pollinate, but some varieties self-abort after 3 months, and those that survive eventually led to inbreeding depression, loss of vigour and lack of resistance to disease.

Fragrance collection by orchid bees have been observed for *Vanilla grandiflora*. *Vanilla planifolia* is adapted for bat dispersal. Alternatively, it could also be dispersed by birds or by wind and water. Seeds do not readily germinate, and vegetative propagation is the mainstay for the orchid's survival in nature. One plant of *Vanilla planifolia* can cover an area of 0.2 ha, albeit not densely (Soto-Arenas 1999). Commercial *Vanilla* are derived from stem cuttings.

Vanilla planifolia has been bred to *Vanilla tahitensis* and *Vanilla pompano* in Madagascar.

In India it has been successfully hybridized with the Asian species, *Vanilla aphylla*.

Hand Pollination of *Vanilla*

Natural pollinators of *Vanilla* are specific species of bees that are native to Central and South America, and since they were not introduced into Europe and other parts of the world together with the vine, imported *Vanilla* flowered but did not bear seed. Commercial production of vanilla started in Bourbon (renamed La Reunion when granted its independence from France in 1858). This only came about because a young negro slave named Edmond Albius (1829–1880) discovered the process of hand pollination which he demonstrated to his master, the planter Ferreol Bellier-Beaumont, one morning in 1841. Prior to that discovery, *Vanilla* had been growing in the plantation for 20 years without bearing any fruit (Fig. 7.11).

Fig. 7.9 *Vanilla pompona.* Mulsant E, Verreaux E, *Histoire naaturelle des oiseaux-mouches*, t. 8 (1877)

EUCEPHALA SUBCAERULEA

(Vanillon de Antilles)

Bellier-Beaumont was a generous man who not only invited other landowners to learn and make use of Edmond's technique, he also granted Edmond a surname, Albius (white), and freedom from slavery in mid-1848. In that year Bourbon exported 50 kg of vanilla pods to France. Ten years later the amount increased to 2 tons, then 20 tons in 1867 and finally 200 tons in 1898 (Ecott 2004). In London, vanilla pods (*vanilloes*) were always highly priced, and varied in value according to quality and size, in 1896 as follows:

5.5–6 inches—19s. 6d. to 20 shillings per pound
7–7.5 inches—29 shillings per pound (Ridley 1897)

Several authors questioned the originality of Edmond's discovery. In 1836 *Vanilla* was successfully hand-pollinated by Charles Morren at the Liege Botanical Gardens in Belgium. The fruits took a year to ripen, and during a lecture to the Institut de France, Morren showed a vine carrying three pods. His discovery was well

Fig. 7.10 *Vanilla odorata* [as *Vanilla ensifolia*]. From: *Transactions of the Linnean Society of London, 2nd Series: Botany*, vol. 4(3): t. 33 (1895) [R. Morgan]

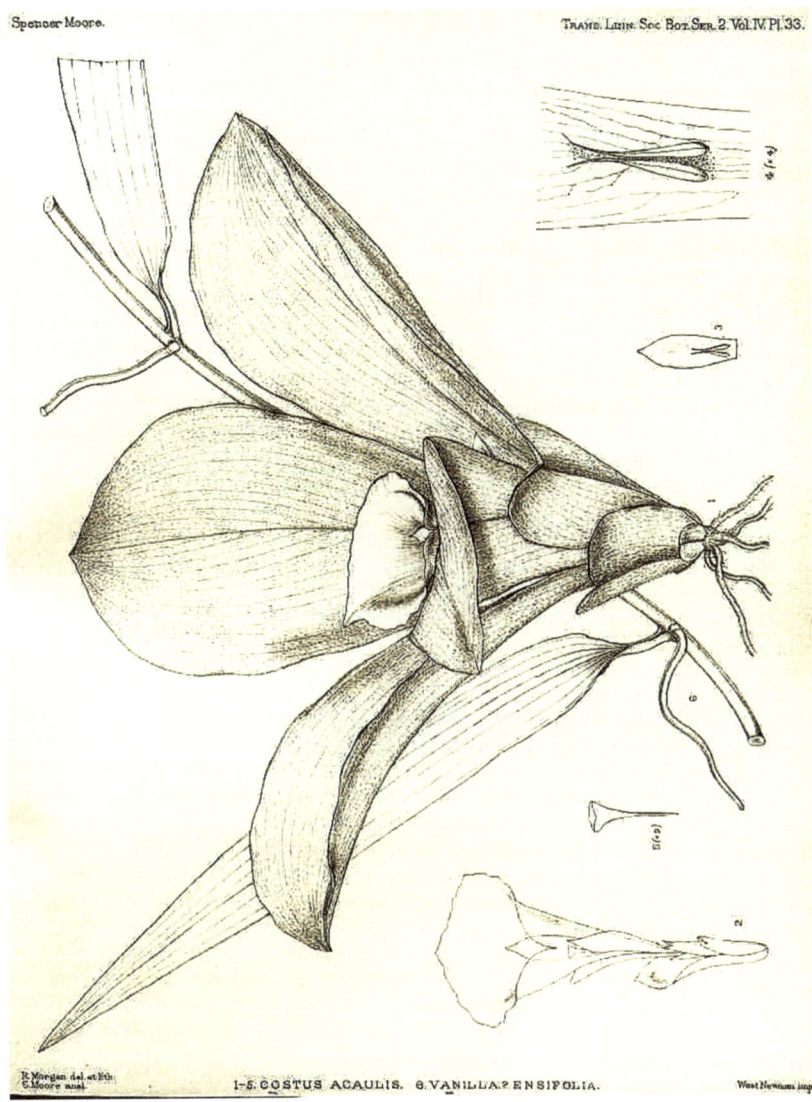

Spencer Moore.

TRANS. LINN. SOC BOT.SER. 2. Vol.IV. Pl. 33.

R. Morgan del. et lith. C. Moore anal.

1–5. COSTUS ACAULIS. 6. VANILLA? ENSIFOLIA.

West Newman imp.

documented in various publications of the period. Shortly thereafter, Joseph Neumann at the Museum National d'Histoire Naturelle Jardin des Plants claimed that he had discovered the process in 1830. However, a publication in 1899 raised two questions: why would Neumann wait it 8 years to announce such an important discovery, and why did it take 10 years for the French to introduce it to Reunion (Arditti et al. 2009)? In all likelihood, 12-year old Edmond Albius made an independent discovery, one which led to the actual commercialization of vanilla. Whereas the Belgian and French discoveries could claim priority, they remained academic: neither brought about commercial exploitation.

At the Singapore Botanic Gardens, fertilization was effected by cutting through the small plate in the column that separated the male and female parts of the flower. The pollen was then pressed down into the stigma (Burkill 1935).

The young pod grows rapidly and within 2 months it reaches maximum size. However,

brand their pods when they are large enough to protect against theft. At curing plants, female staff who handle the beans are body-searched by male guards when they go to the toilet and at the end of the day when they leave the establishment. Curfews prohibit trucking of vanilla after 6 in the evening. Due to the high price of vanilla, dealing in vanilla at source can be a dangerous business. Buyers have been maimed or murdered (Ecott 2004).

Postharvest Handling: Sweating and Curing

There is no fragrance in ripe vanilla pods. At that stage, pods are loaded with glucovanillin at the placental site and at the tiny white hair (trichomes) to which the seeds are attached. When cells in the pod perish, a catalytic enzyme is released from cells comprising the outer wall of the pod. This enzyme is beta-glucosidase: it cleaves glucovanillin thereby producing the aromatic vanillin and glucose. This is the predominant chemical process. However, natural vanilla contains over 200 substances, their composition varying with the cultivated variety, source of the pods, their stage of maturation, seasonal influences, handling, etc. To obtain high-quality vanilla, the rate of drying during the curing process has to be carefully controlled such that the desired chemical changes proceed without interference from fungal contamination and other decompositions.

Fig. 7.11 Edmond Albus (1829–1880), the slave who launched the commercial cultivation of *Vanilla* by showing a simple and quick method of hand pollination to his owner Fereol Bellier Beaumont. Albius was only 12 when he discovered the method. Beaumont in turn taught other plantation owners

the pod does not become mature until 4–6 months later. Best Bourbon vanilla comes from pods that have remained on the vine for the longest time. They require to be individually hand-picked when the tips turn yellow. Buyers pay premium prices for the best quality. Meanwhile, pods get stolen, and to pre-empt theft, some farmers resort to harvesting their pods before they are ripe (Ecott 2004). This has led to marked variation in the quality of exported vanilla even from the same source (Gassenmeier et al. 2008). Under duress farmers may even harvest their crops when the pods are very unripe. They are sold by weight, and pods do not become heavier when they ripen; they become lighter.

Madagascar has a law which prohibits buying and selling of stolen vanilla pods. Some people

The old Mexican method was to spread the plucked fruits on racks in a room for 24 h during which time they lose moisture and wilt. The next day they are spread on mats or blankets placed across a cement floor exposed to the sun, and in the afternoon, the mats are folded over the pods to prevent excessive drying. At night they are put into airtight cases to sweat. Castor oil may be applied if the pods are too ripe or have been injured to prevent accelerated drying through the fissures. The whole process is repeated over several days until the pods become dark and

Fig. 7.12 *Vanilla* pods on vine Photo: Chang Yoon Ching

shriveled. This may take up to 2 weeks. Should rain interfere with sunning an oven is employed to provide the heat that is essential for the enzymatic processes to proceed at the desired rate. Meanwhile the pods are gathered according to size and shape and steps are taken to prevent their dehiscence (Burkill 1935). A great deal has been written about the processing of vanilla pods, and details of the method vary from country to country, curing plants, and even among growers, with each preferring its own recipe (Fig. 7.12).

Curing today involves dipping bundles of pods in hot water to kill the cells forming the wall of the pods. Timing varies according to water temperature—15 s at 85 °C, 3 min at 65 °C. Critics argue that being tied into bundles, the pods are unevenly exposed to the hot water, pods in the centre being isolated from the heat, whereas the outermost pods are subjected to too much heat which may denature some of the enzymes required for the curing process. If this is true, the longer exposure at lower temperature should be the better option. In Peru the beans are dipped in boiling water before they are tied into bundles. Hot ash or an oven is employed in some places. Freezing at subzero temperature or snap freezing in liquid nitrogen have also been employed. Theoretically, the last method is less likely to inactivate enzymes than slow freezing at subzero temperatures.

Handling the pods during the curing process has not changed materially over time, even though one study found that the vanilla content reaches a peak of 6% within 3 days, and thereafter there is no further increase (Cameron 2011). However, a different study reported that a 90% conversion of glucovanillin to vanillin occurred in beans which were sweated continuously at 35 °C for 12 days, a rate that was much higher than the 70% conversion for beans blanched at 67° and sweated at 45° for 4 days or at 35° for 5 days. In both instances the beans had turned brown. The appearance of the blanched beans was more attractive, but such beans lost out on aroma (van Dyk et al. 2010).

As in the past, great care is taken over the handling of the pods. They need heat and drying, not too quickly nor too much, nor too slowly, lest they acquired mould or bacteria. Blankets play an important role in controlling the slow drying process. At night they act as wicks to drain off surface moisture: in sunlight such moisture from the blankets prevent pods from drying too quickly.

Once the pods are cured, they are placed on racks to be sun-dried. This prevents them from getting mouldy. Drying takes about 2 months, smaller beans requiring a shorter time than the larger beans. Dried vanilla pods are a dark brown, oily, fragrant and still pliable. They are now sorted out into classes according to size, appearance and fragrance. Large beans, 18–20 cm which are free of blemish, evenly coloured and most fragrant form the top class. Vanilla is packed in

Table 7.1 Twelve top producing countries of vanilla in 2008 (from Hays 2009)

Country	Metric tons	Price ($1000)
Indonesia	3700	18,191
Madagascar	2800	13,766
China	1400	6883
Mexico	637	3131
Turkey	170	835
Tonga	150	737
Uganda	70	344
Comoros	50	245
French Polynesia	30	147
Malawi	20	98
Reunion	12	52
Zimbabwe	10	49

Fig. 7.13 Processed *Vanilla* pods (©Teoh Eng Soon 2019. All Rights Reserved)

plastic or glass and vacuum sealed. At their peak, they have fetched US $500 per kilogram or more (Fig. 7.13).

The Vanilla Market

Prices of vanilla are highly volatile with precipitous ups and downs. Many farmers have suffered when they set their expectations too high. After a disastrous hurricane which wrought havoc in vanilla plantations hit Madagascar in 2000, prices of vanilla shot up to US$275 per pound but, according to the Statistics Division of the Food and Agricultural Organization (FAO), in the following year, it dropped to only 1% of that value. It sold at the lower price in 2008 (Hays 2009). By comparison it fetched US$10 per pound in 1997. The ups and downs of vanilla have inspired forex traders to coin the term 'vanilla option' for their deals.

In this setting, smallholders have a tough time maintaining their livelihood by growing vanilla, and Madagascar may not be able to maintain its position as the top producer of vanilla. The Chinese challenge and heavy taxation are additional factors. During the 1970s Madagascar commanded 70% of the world's trade in vanilla. By the 1990s, this dropped to 40%. The 12 top exporting countries in 2008 in terms of metric

tons and dollar values are summarized in Table 7.1 (Hays 2009).

Nevertheless, a saving grace is the fact that Bourbon vanilla sourced from Madagascar, Reunion (Bourbon being the French name before the island gained its independence) and the Comoros is traditionally the most sought-after type of vanilla. It has a rich flavour and aroma; it is thick, oily, creamy, hay-like and sweet and provides familiar vanilla notes.

Tahitian vanilla is fruity, smelling of cherry, liquorice, prunes and wine and retails at more than twice the price of Bourbon vanilla. The unique fragrance of Tahitian vanilla is due to its additional contents of piperonal (heliotropin, 3,4-dioxymethylenbenzaldehyd) and diacetyl (butanedione). Vanillin content is about 1.7% (National Tropical Garden Website, Vanilla 2012). Mexican vanilla is mellow, smooth and spicy, with a woody fragrance, and is slightly more costly than Bourbon (The Vanilla Company 2012). Indonesian vanilla is said to be woody, nutty and smoky, but it has greater staying power that gives it an edge with high heat preparations (Cameron 2011).

Today, vanilla is ubiquitous in western food, drinks and perfumes. The dairy industry makes the most use of the spice to flavour ice cream, yoghourt and related items. Confectionaries, chocolate, biscuits, cakes and soft drinks also make extensive use of vanilla. However, only premium products make use of natural vanilla. Most items, even ice cream or biscuits, make use of synthetic vanillin which is produced from eugenol or guaiacol and from lignin derived from wood, coal tar or tonka beans. Vanillin is mostly synthesized from wood or coal tar. Nevertheless, purified vanillin is as safe for human consumption as natural vanilla. Although it does not have the same range of fragrances as natural vanilla, it is a strong competitor in the food and drink market because of price.

Vanillin is a foundation base for perfumes, staying on after the initial fragrances have evaporated. The flavouring of cigars with vanilla is based on an ancient Aztec practice. Vanillin is also employed in industry to lessen the irritating smell of rubber, paint and cleansing agents. The United States is the principal consumer of vanilla taking 60% of the world's total consumption. However, vanilla pods make up only 10%, with 90% made up by synthetic vanillin. Even in France, the gastronomical nation, synthetic vanillin accounts for 50% of total 'vanilla' consumption (Hays 2009).

With synthetic vanillin, a single chemical, the concentration can be fixed, but this is not the case with natural vanilla. Generally, the fermented fruit of *Vanilla* contains about 2% vanillin, depending on provenance: Mexico 1.75%, Sri Lanka 1.5% and Indonesia 2.75%. Vanilla pods of exceptionally good quality may be covered with tiny white needles which are crystallized vanillin, the so-called *givre*, from the French word for 'frost'. Apart from vanillin which accounts for 85% of total volatiles, there are other important aroma components such as *p*-hydroxybenzaldehyde (up to 9%) and *p*-hydroxybenzyl methyl ether (1%). About 130 more compounds have been identified in vanilla extract: phenols, phenol ether, alcohols, carbonyl compounds, acids, esters, lactones, aliphatic and aromatic carbon hydrates and heterocyclic compounds. Two stereoisomeric vitispiranes (2,10,10-trimethyl-1,6- and methylidene-1-oxaspiro(4,5)dec-7-ene), although only occurring in traces, further influence the aroma (National Tropical Garden Website 2012). Other aromatic compounds present in traces can also significantly improve the flavour.

Piperonal was found in Tahitian vanilla but not in other sources by a few investigators, but an overwhelming majority could not discover any piperonal in the beans that they sampled. Several explanations are plausible: (1) piperonal is a typical compound of only some botanical variations of *Vanilla tahitensis*, (2) formation of piperonal in *Tahitian vanilla* is dependent on the harvest conditions, or (3) the vanilla beans analysed had been adulterated with piperonal. The German investigators who did not discover any piperonal in the beans they sampled concluded that the third explanation is the most likely (Ehlers et al. 1994). Piperonal is being employed as a substrate to prepare illegal street drugs used by addicts in Japan and the United States (Wood et al. 2015).

Variability of botanical products is caused by a whole range of genetic, geographic, climatic, cultural, processing and other factors, many of which are often beyond human control. This has been the greatest obstacle to maintaining consistency for herbal preparations. It is impractical to perform biological standardization by weight of the raw plant material because of batch-to-batch variation. An analysis of over 90 samples of cured vanilla bean from the 2006 to 2007 crop derived from worldwide sources undertaken by Gassenmeier et al. (2008) discovered huge differences in the analytical parameters both between origins and between batches. Nearly half the beans from a major supplier did not meet the desired ratios for concentrations of vanillin, vanillic acid, *p*-hydroxybenzaldehyde and *p*-hydroxybenzoic acid. Many samples contained less vanilla than the prescribed content (Gassenmeier et al. 2008). In practice, different batches of beans are pooled, and grading is based on the vanilla content of pooled batches. This is practical, but it would not meet pharmaceutical standards.

Medicinal Usage

When vanilla was first introduced into Europe, it carried many medical usages. It was highly regarded as an aphrodisiac. Vanilla was an item in the *London Pharmacopoeia* in 1721. It was employed to treat impotence, hysteria and gastric ulcers and to assist digestion, but by the end of the nineteenth century, such usage had become obsolete in England. Some vanilla workers developed 'vanillism' which is characterized by headaches and skin rash (Bentley and Trimen 1880). Allergic reactions have also been reported in young children exposed to high doses of vanilla in their diet (Kanny et al. 1994).

The *Vanilla planifolia* plant is considered to have aphrodisiac value in India (Sood et al. 2002). Roots are used as a stimulant and also to treat gonorrhoea and dysuria, the latter possibly resulting from the sexually transmitted disease. They are mixed with onions, cumin, sugar and butter to prepare a confection for easy consumption. An extract of the root together with cumin and sugar is added to cold milk to provide a remedy for spermatorrhoea (Nadkarni 1954). Alkaloids are not present in significant amounts in *Vanilla* (Luning 1974) (Figs. 7.14, 7.15 and 7.16).

An additional dozen species of *Vanilla* have been employed medicinally in various parts of the world (Teoh 2019). *Vanilla claviculata* which is endemic in the Caribbean was once employed as a folk remedy for wounds. The fruits were once believed to be a cure for syphilis (Griffith 1847; Duggal 1972). In Peninsular India, *Vanilla wrightii* was employed to treat syphilis (Lawler 1984). *Vanilla mexicana* (syn. *Vanilla aromatica*)

Fig. 7.14 *Vanillla madagascariensis* (2–7) and *V. decaryana* (8–11) with the smaller floral parts. From: *Flore de Madagascar et des Comoros, Orchidees*, vol. 49(1): Orchidees p. 197 (1939) [MJ Vesque]

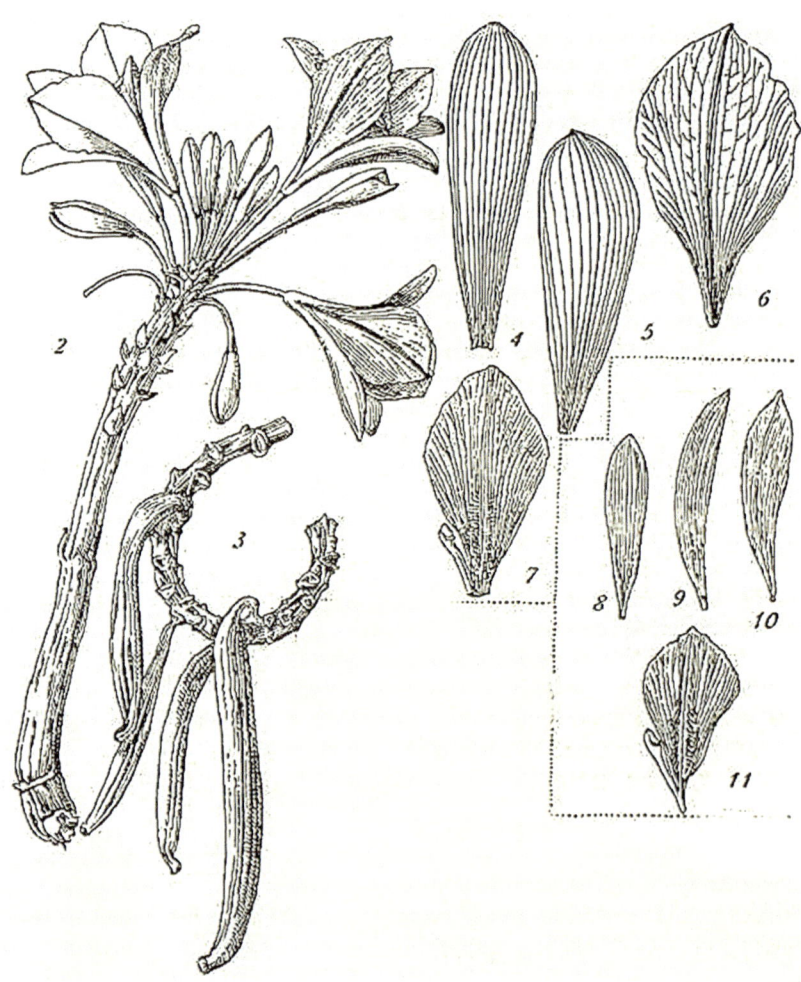

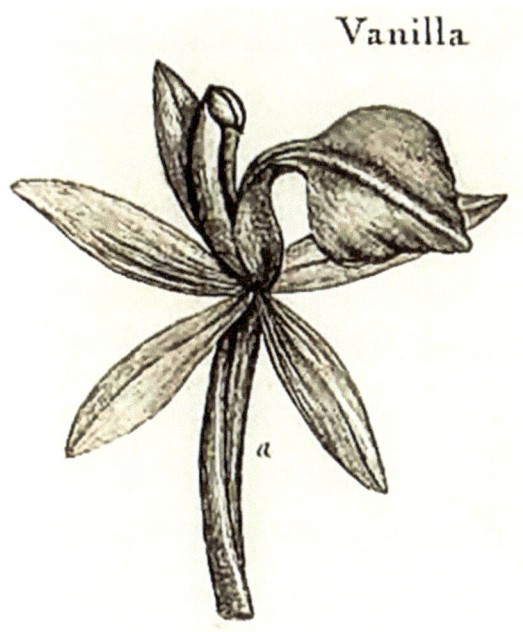

Fig. 7.15 *Vanilla claviculata*. From: *Nova Acta Regiae Societatis Scientiarum Upsaliensis*, vol. 6: t. 5. Fig. 1 (a) [O Swartz]

was used to treat hypochondria and hysteria, assist digestion and induce menstruation. Fresh juice extracted from the plant was applied on ulcers. Dried stems were kept in syrup for use as vermifuge. Root infusion was employed to treat syphilis in Cuba. Cubans in the west of the island employed juice from heated stems of *Vanilla poitoei* (syn. *Vanilla eggersii*) to treat worms. In Brazil *Vanilla palmarum* capsules were used to treat disorders of the psychiatric disorders and uterine dysfunction (Lawler 1984). *Vanilla phaeantha* was employed as a spice and also employed medicinally (Henelt 2001). *Vanilla decaryana* Perr. (local name: *Vahy amalona*) is an aboriginal aphrodisiac in Southern Madagascar (Uphof 1968). Another is *Vanilla madagascariensis* Rolfe (local name: *amalo*) (Cribb and Hermans 2009). In India, Irulas living at the Nilgiri Biosphere Reserve call *Vanilla walkerie* Wight by the name *Kundu pirandi*, and they use its stem as a veterinary medicine (Balasubramaniam and Prasad 1996). In Malacca, Alvins (c. 1884–1888) recorded that flowers of the native Malaysian *Vanilla griffithii* was pulped in water for application to the body to treat violent fever (Burkill 1935). Thai herbalists used *Vanilla aphylla* to treat liver dysfunction (Chuakul 2002). Leaves of *Vanilla crenulata* are employed medicinally in Liberia and the Ivory Coast (Teoh 2019). Recently it was reported that vanillin protects human keratinocyte (skin) stem cells against ultraviolet B irradiation (Lee et al. 2014). One should not be surprized that vanillin will find its way into skin products.

Other Forms of 'Vanilla'

Vainilla chica (little vanilla) and *vainilla en arbol* (vanilla on a tree) refer to the tall South American slipper orchid, *Selenipedium chica*, whose seed capsules are sometimes employed as a substitute by the Indians living in the mountains of Panama when vanilla is not available. In Brazil, fruits capsules of the native species, *Leptotes bicolor*, were used to flavour ice cream (Hawkes 1943).

Fig. 7.17 *Selenipedium chica*. From: Reichenmach HG, Arnott GAW, Xenia Orchidaceae vol 1: t. 2 (1900)

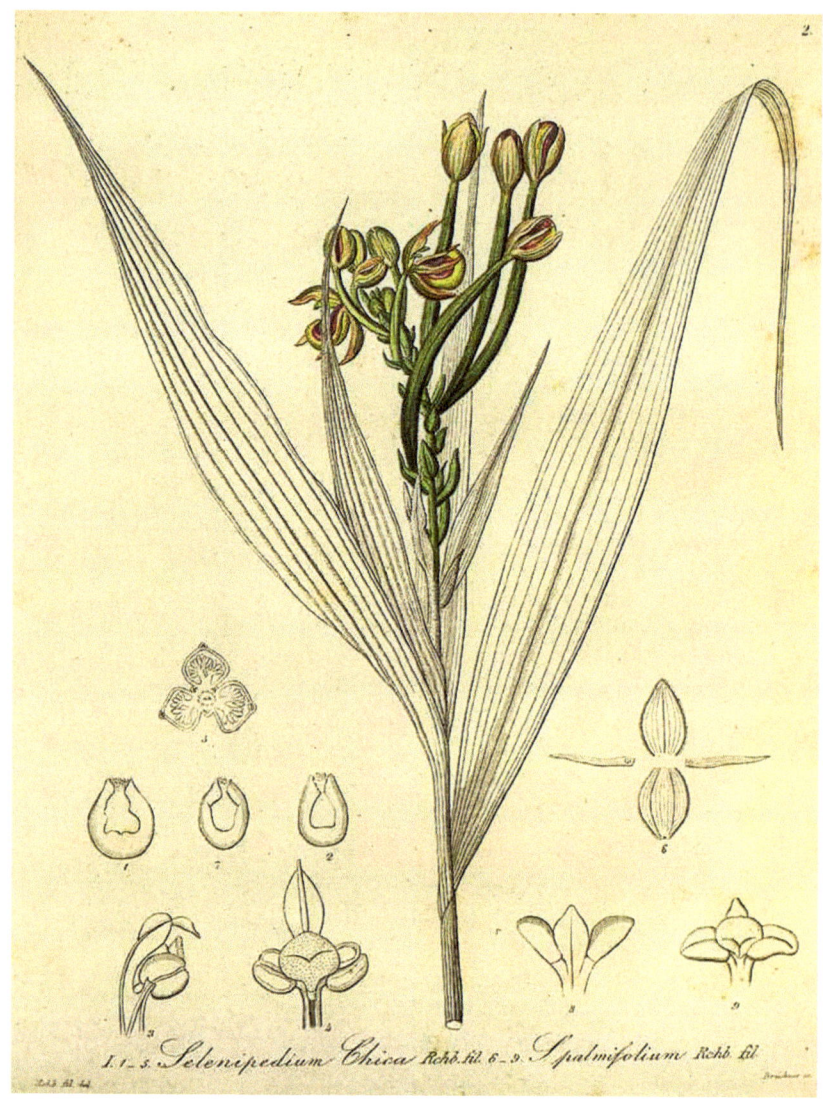

Fig. 7.18 *Leptotes bicolor*. From: *Lindenia Iconographie des Orchidees* [E. von Lindmann]. Plates 145–192, vol. 4: t. 157 (1888) [P. Pannemaeker]

Fig. 7.19 *Vanilla grandiflora* [as *Vanilla lutescens*]. From Houtte L. van, *Flora des serres et des jardin de l'Europe*, vol. 21: t. 0 (1845)

However, worldwide, vanilla is a dominant flavour in the dessert (Figs. 7.17, 7.18 and 7.19).

References

Arditti J, Rao AN, Nair H (2009) Hand pollination of vanilla: how many discoverers? In: Kull T, Arditti J, Wong SM (eds) Orchid biology and perspectives, X. Springer, Dordrecht, pp 233–249

Balasubramaniam P, Prasad SN (1996) Ethnobotany and conservation of medicinal plants by Irulas of Nilgiri Biosphere Reserve. In: Jain SK (ed) Ethnobiology in human welfare. Deep Publications, New Delhi

Bentley R, Trimen H (1880) Medicinal plants; being descriptions with original figures of the principal plants employed in medicine and an account of the characters, properties, and uses of their parts and products of medicinal value. Churchill, London

Bhattacharjee SK (2000) Handbook of aromatic plants. Pointer Publishers, Jaipur

Bory S, Grisoni M, Duval MF, Besse P (2008) Biodiversity and preservation of vanilla. Present state of knowledge. Genet Resour Crop Evol 55(4):551–571

Burkill IH (1935) A dictionary of the economic products of the Malay Peninsula, vol II. Crown Agents, London

Cameron K (2011) Vanilla orchids. Natural history and cultivation. Timber Press, Portland, OR

Cano VP (2009). http://storiesfromtheamericas.blogspot.sg/2009/02/legend-of-vanilla.html

Chuakul W (2002) Ethnomedical uses of Thai orchidaceous plants. Mahidol Univ J Pharm Sci 29(3–4):41–43

Correll DS (1953) Vanilla, its botany, history, cultivation, and economic import. Econ Bot 7:291–358

Cribb P, Hermans J (2009) Field guide to the orchids of Madagascar. Royal Botanic Gardens, Kew

de Sahagun B (1529) Florentine codex (manuscript). Repository: Biblioteca Medicea Laurenziana, Florence

de Toledo FH (1615) Quatro libros de la naturaleza y virtudes de las plantas y animals. Nardo Antonio Recchi, Mexico

Duggal SC (1972) Orchids in human affairs (a review). Pharm Biol 11(2):1727–1734

Ecott T (2004) Vanilla. Travels in search of a luscious substance. Michael Joseph, London

Ehlers D, Pfister M, Bartholoma S (1994) Analysis of Tahiti vanilla by high-performance liquid chromatography. Z Lebensm Unters Forch 199(1):38–42

Gassenmeier K, Reisens B, Magyar B (2008) Commercial quality and analytical parameters of cured vanilla bean (*Vanilla planifolia*) from different regions from the 2006–2007 crop. Flavour Fragr J 23(3):194–201

Griffith RE (1847) Medicinal botany of descriptions of the more important plants used in medicine with their history, properties and mode of administration. Lea and Blanchard, Philadelphia

Harkin-Franke D, Belanger F (2011) Handbook of vanilla science and technology. Blackwell, London

Hawkes AD (1943) Economic importance of the Orchidaceae. Amer Orchid Soc Bull 11:412–415

Hays J (2009) Vanilla. Its history, cultivation and processing. Factsanddetails.com

Henelt P (ed) (2001) Mansfeld's encyclopedia of agricultural and horticultural crops. Springer and IPK, Cham

Heyne K (1927) De Nuttige Planten van Nederlandsch Indie. Uitgave van Het Departement van Landbouw, Nijverheid & Handelin in Nederlandsche Indische

Kanny G, Hataret R, Monret-Vautrin DA et al (1994) Allergy and intolerance to flavouring agents in atopic dermatitis in young children. Allerg Immunol 26 (6):204–206

Lawler LJ (1984) Ethnobotany of the Orchidaceae. In: Arditti J (ed) Orchid biology. Reviews and perspectives, vol III. Cornell University Press, Ithaca, NY

Leal F, de Clavijo CM (2012) Annatto: botany and horticulture. Hortic Rev 39(1):389–419

Lee J, Cho JY, Lee S et al (2014) Vanillin protects human keratinocyte stem cells against ultraviolet B irradiation. Food Chem Toxicol 63:30–37

Lubinsky P, Bory S, Hernandez JR et al (2008) Origins and dispersal of cultivated Vanilla (Vanilla planifolia Jacks [Orchidaceae]). Econ Bot 62:127–138

Luning B (1974) Alkaloids of the Orchidaceae. In: Withner CL (ed) The orchids. Scientific studies. Wiley, New York

Nadkarni AK (1954) Dr. K M Nadkarni's materia medica. Bombay: Popular Book Depot

National Tropical Garden Website (2012) On Vanilla. https://www.edenproject.com/learn/for-everyone/plant-profile/vanilla

National Tropical Garden Website. Vanilla (2012) (online)

Plucknett DL, Smith NJH (2014) Gene banks and the world's food. Princeton University Press, Princeton, NJ

Ridley HN (1897) Agric Bull 6:124–126

Roux-Cuvelier M, Grisoni M (2010) Conservation and movement of Vanilla germplasm. In: Odoux E, Grisoni M (eds) Vanilla. CRC Press, Boca Raton, FL

Sood SK, Rana S, Lakhanpal TN (2002) Ethnic aphrodisiac plants. Scientific Publishers (India), Jodhpur

Soto-Arenas MA (1999) Systematics of vanilla. Royal Botanic Gardens, Kew

Soto-Arenas MA (2006) La Vainilla – los retos de un cultivo basado en uno especie amenazada con una historia de vida compleja. Congreso international de productores de vainilla. Papantla Ver, May 2000

Soto-Arenas MSA, Cribb P (2010) A new infraageneric classification and synopsis pf the genus Vanilla Plum ex Mill (Orchidaceae. Vanillinae). Lankasteriana 9 (3):355–398

Soto-Arenas MSA, Dressler RL (2010) A revision of the Mexican and Central American species of Vanilla Plumer ex Miller with a characterization of their ITS region of the nuclear ribosomal DNA. Lankasteriana 9 (3):285–354

Teoh ES (2003) Lotus in the Buddhist art of India. Teoh Eng Soon, Singapore

Teoh ES (2019) Vanilla. In: Hong H (ed) Healing orchids. World Scientific, Singapore

The Vanilla Company (2012) (online)

Uphof JCT (1968) Dictionary of economic plants. Verlag von J. Cramer, Lehre

van Dyk S, MGlasson WB, Williams M, Gair C (2010) Influence of curing procedures on sensory quality of vanilla beans. Fruits 65(6):387–399

Varey S (2000) The Mexican treasury: the writings of Dr. Francisco Hernandez. Chabran R, Chamberlain CL and Varey S (trans). Stanford University Press, Stanford, CA

Wood MR, Lalancette RA, Bernal I (2015) Crystallographic investigationsnof select cathinones: emerging illicit drugs known as 'bath salts'. Acta Crystallogr C: Struct Chem 71(Pt 1):32–38

Modern Treasure Hunters

At the height of orchidomania in the nineteenth century, orchid hunters were sent all over the tropics to look for new species of attractive orchids that would fetch handsome prices in the auction houses if they survived when they reached Europe. At times, tens of thousands of plants of a single species were stripped from the jungle to satisfy this lust. In one recorded incident, when Wilhelm Micholitz (1854–1932) lost a collection of Indonesian orchids through shipwreck, Frederick Sander wired him to go back and collect some more. Collectors were secretive about their sources to the extent that they gave misinformation about the original location of their plants. Some even went to the extent of destroying an entire population lest they met with competition when their plants were offered for sale.

Thankfully, those days are over. Well, almost. When new *Paphiopedilum* species were discovered in China during the mid-1970s, tens of thousands of plants were again removed from their habitats and trans-shipped through Hong Kong and Singapore to Europe and the United States. *Paphiopedilums* are now listed under CITES Category I which specifies that only plants proven to be cultivated from seed are allowed to cross borders; however, this does not entirely rule out collection from the wild. During the old colonial days, when Brazil banned the export of rubber seeds, Sir Henry Alexander Wickham went on a 'plant collecting expedition' in 1876 and took 70,000 rubber seeds out of the country, hidden among plant specimens as he passed through custom. Singapore, Malaya, Indonesia, Sri Lanka, India and England benefited from that theft. Much of the early prosperity of the Malay Archipelago was based on rubber (Figs. 8.1 and 8.2).

Treasure hunters still scour the jungles, now looking for plants which might contain a valuable chemical compound with possible usage to treat common illnesses that ravage mankind in old age. However, most treasure hunters work in the laboratory, and many have begun to examine orchids as a hitherto largely ignored source. Orchids are unique in that they require an association with fungus (mycorrhiza) for germination, and most, if not all, orchids continue to associate with various mycorrhiza species for life. Thus, there is a dual source when one is examining orchids. They may produce useful compounds: if they did not, their mycorrhiza might do so. Taxol (paclitaxel), commonly employed in cancer chemotherapy, was first isolated from the bark of the Pacific yew tree (*Taxus brevifolia*) by National Cancer Institute (NCI) researchers Monroe E Wall and Mansukh C Wani in 1966. Subsequently it was found that the compound was actually produced by endophytic fungi in the bark.

The story of Taxol is worth summarizing at this stage of our discussion because it illustrates the difficulties encountered in drug development when working with plant material. Monroe Wall

© Springer Nature Switzerland AG 2019
E. S. Teoh, *Orchids as Aphrodisiac, Medicine or Food*, https://doi.org/10.1007/978-3-030-18255-7_8

Fig. 8.1 *Paphiopedilum rothschildianum*, a rare species from Mount Kinabalu in Borneo (©Teoh Eng Soon 2019. All Rights Reserved)

Fig. 8.2 *Paphiopedilum micranthum* (©Teoh Eng Soon 2019. All Rights Reserved)

announced the discovery of Taxol at an American Cancer Society Meeting in Miami in 1967. The chemical structure was published in 1971. Although early testing with the crude extract during the mid-1960s had shown a cytotoxic effect, no further investigations were performed until 1978 when purified Taxol was available in sufficient quantity for in vivo studies in mice. This required a supply of more than 3 tons of bark. The studies showed that Taxol had a mildly beneficial effect in leukaemic mice. The following year a report by Susan Horowitz and Peter B. Schiff from the Albert Einstein University in New York, published in *Nature*, stated that Taxol stabilized microtubules. This announcement got many people excited and NCI proposed collecting 20 tons of bark, even more a little later, when tests began showing promising results. Clinical testing commenced in 1984. Four years later it was announced that Taxol produced remission in 30% of refractory, advanced ovarian cancer. Additionally, it was effective for treating melanoma. By extrapolation, calculations suggested that relying on Taxol to treat all cases

of ovarian cancer and melanoma in the United States alone would require harvesting bark from 360,000 trees of the Pacific yew every year. This immediately raised alarm among conservationists because the species would rapidly become extinct if so many trees were stripped of their bark. NCI then decided that it would be best to invite a large pharmaceutical firm to invest in the development of Taxol. Taxol was registered as a trade name in 1992, and *paclitaxel* was selected as the chemical name. I have a patient who has survived stage IV ovarian cancer for over 20 years today because of Taxol.

Paclitaxel has now become generic. Meanwhile, additional applications have been found. Drug-eluting stents employed to treat coronary heart disease contain paclitaxel which is an antiproliferative agent. It suppresses cell division. This property is being applied to prevent narrowing and blockage of coronary stents.

The story of paclitaxel illustrates the stages that a plant product must undergo in order for it to be part of a modern medical armamentarium. (1) It has to be isolated in pure form. (2) It needs to be tested against a panel of immortal cancer cell lines which are now readily available to research centres (Eagle and Foley 1958). For a compound to be considered cytotoxic, the concentration required to produce 50% cell death (IC50) should be less than 2 microgram/ml. If its LD50 is 90 microgram/ml or higher, the compound is classified as not cytotoxic. LD50 reading between 2 and 89 microgram/ml indicates that the compound is moderately cytotoxic. (3) Should the screening test prove positive, additional experiments are designed to determine the modes of action. (4) In vivo testing in animals is conducted to study its metabolism in the mammalian body, determines toxic dosage and ensures safety. (5) Once safety is ensured, the supply issue needs to be addressed. Sufficient material must be obtained, preferably through synthesis, before human trials are started. (6) Approval by National Ethics Committees and Institutional Review Boards are required before trials can be commenced on human subjects. (7) Such trials have to be carefully monitored for side effects and results. (8) Adequate funding and facilities are essential. (9) Even after the drug is approved for clinical usage, monitoring of possible side effects and complications continue, albeit this may be done on a voluntary basis, i.e. users and physicians report untoward effects. (10) Large-scale clinical trials and/or observational studies are necessary to convince clinicians of the efficacy and safety of a product.

Phytoalexins

Survival of the fittest is a law of nature. Plants being rooted at one spot cannot avoid danger by running away. They must compete successfully with other plants and defend themselves against competitors, invaders and predators if they are to survive and flourish. Thus, besides synthesizing their own food, plants produce secondary metabolites which are small molecules that exert a wide range of effects on the plant itself and on other living organisms. These compounds induce flowering, fruit set and abscission, maintain perennial growth or signal deciduous behaviour. They act as antimicrobials and perform the role of attractants or conversely as repellents. Many medicinal herbs and many modern medicines rely on such secondary plant metabolites for their actions. Over 50,000 secondary metabolites have been discovered in the plant kingdom. These are made up of flavonoids, alkaloids, polysaccharides, stilbenes and phenanthrenes.

Orchids enjoy a special relationship with fungus throughout their life. The tiny orchid seeds which are well adapted for wind dispersal do not carry their own supply of energy units sufficient for germination. In order to germinate and grow before they can photosynthesize, orchids need mycorrhiza to provide sugars for them, and mycorrhizal associations benefit orchids throughout their entire life. Heterotrophic species of orchids that spend a considerable part of their life in dark shade remain dependent on fungi all the time. Nevertheless, mycorrhiza and all species of fungi, especially when the latter are non-essential or parasitic, need to be controlled. This is where phytoalexins come in.

During the 1940s, Muller and Borgen recognized that plants produced antimicrobial compounds in response to infection by fungi. Such compounds suppressed the growth of fungi and bacteria without killing them, i.e. these compounds were fungistatic and bacteriostatic, but, unlike some synthetic compounds and antibiotics, they were not fungicidal nor bactericidal. They are known as phytoalexins (plant warding off compounds).

Orchinol was the first phytoalexin to be isolated from orchids. It was produced by *Orchis militaris* infected with *Rhizoctonia repens* (Boller et al. 1957). Orchinol was not labeled a phytoalexin at that time, but it was referred to as an *Abwehrstoff* (*Abwehr* 'defensive' or 'warding off'; *stoff* 'compound'), meaning much the same thing. Soon after, loroglossal was isolated from another Mediterranean orchid, *Loroglossum hircinum* (Hardegger et al. 1963). Hircinol was the third phytoalexin to be isolated, also from *Loroglossum hircinum*. Trace amounts of hircinol were present in *Orchis militaris*. Around the same time, orchinol was isolated from several species of *Serapias*, also from *Loroglossum longibracteatum*, but it was not found in *L. hircinum* (Stoessl and Arditti 1984). Dozens of phytoalexins have now been isolated from terrestrial and epiphytic orchids. Many have been synthesized and are being studied for a wide range of pharmacological effects. Numerous phytoalexins discovered in orchids are also produced by plants belonging to other families.

Healthy plant tissues, such as orchid protocorms in sterile culture, do not contain phytoalexin in any appreciable amount. However, the capacity to produce the compound is inherent in the plant tissues. Production is triggered by microbial invasion or when the plant is subjected to injurious stress and chemicals (heavy metals, pesticides, glycoproteins, protein-lipid-polysaccharide complexes, polypeptides, or glycans). Freezing injury has been shown to elicit phytoalexin production in beans. On the other hand, high, sublethal temperatures inhibit phytoalexin production, and the plant tissue temporarily loses its ability to resist fungal infection. However, this situation is reversible. When it is returned to normal temperatures, the plant regains its phytoalexin producing capability within a few days (Muller 1966).

Phytoalexin production triggered by cold and its suppression by high heat may be the reasons why it is claimed that Chinese medicinal orchids harvested from the wild in northern provinces (Sichuan or Gansu) are better than similar species cultivated in more southerly provinces. Now and then, dangerous levels of lead and other heavy metals have been discovered during surveillance screening of herbal products. Since heavy metals trigger phytoalexin synthesis one wonders whether in such instances, the heavy metals could have been present in the herbs selected for their better potency.

Alkaloids

Alkaloids are plant-derived compounds that contain nitrogen in addition to carbon and hydrogen, rarely, also oxygen, sulphur, chlorine, bromine or phosphorus. Like proteins, they are formed from amino acids. Alkaloids are bitter, already a warning to herbivores that the plant part should not be eaten. They have marked effects on animals and humans. They play a role in seed germination and

protect plants from predators. The term, *alkaloid* was introduced by Carl F. W. Meissner in 1819 in reference to its source and alkaline property, *alkaline* itself being derived from the Arabic *alqalwi* meaning 'ashes of plants'.

Alkaloids have been known to mankind from ancient times. Many alkaloids were used for healing, for pain relief and in rituals. The classic Chinese herbal remedy for asthma and hay fever, *Ma huang* (*Ephedra sinica*), contains the alkaloid, ephedrine. Many alkaloids act on the nervous system. Caffeine and nicotine are stimulants; cocaine is anaesthetic; morphine is narcotic; scopolamine induces 'twilight sleep'. Aminophylline is a bronchodilator. Papaverine is a vasodilator which had a role in treating erectile dysfunction before the discovery of Viagra ®. Reserpine which lowers blood pressure is an ancient Indian remedy derived from *Rauwolfia serpentina*, now totally replaced by a wide range of more potent and more reliable antihypertensives. Many alkaloid stimulants (e.g. morphine, cocaine, nicotine) are addictive. Improperly applied, some stimulants and sedatives are deadly. Strychnine is used as a rat poison (Fig. 8.3).

Many alkaloids continue to have important roles in modern medicine. Quinine and artemisinin are derived from ancient remedies for malaria, the latter being the first natural compound that was capable of destroying chloroquine-resistant *Plasmodium falciparum*. Malaxin, an alkaloid first isolated from the

Fig. 8.3 *Phalaenopsis lueddemanniana* (©Teoh Eng Soon 2019. All Rights Reserved)

common, tropical, terrestrial orchid, *Malaxis latifolia*, is dihydroartemisinin. It also has the ability to destroy *Plasmodium falciparum*, the organism responsible for the deadly cerebral malaria. Synthetic malaxin is employed to treat uncomplicated *falciparum* malaria. Curare, the alkaloid which played a vital role in the development of general anaesthesia and surgery, was found in an Amazonian arrow poison.

The search for alkaloids in orchids began when de Waldemann (1892) and de Droog (1896) screened European terrestrial species for the presence of such compounds, de Droog examining 104 species in 78 genera. The studies were not fruitful, but de Droog subsequently discovered a high alkaloid content in *Dendrobium nobile* and *Phalaenopsis lueddemanniana*. *Dendrobium*, *Malaxis* and *Phalaenopsis* and various genera of monopodial orchids are rich in alkaloids. Members of the first three genera are generally large and common plants and therefore easy to work with. Screening of 2044 species in 281 genera by Bjorn Luning showed that 214 species had an alkaloid content equal to or exceeding 0.1% of the plant's dry weight (Luning 1974). *Chin Shih-Hu* (*Dendrobium moniliforme*) from Sichuan Province had an alkaloid content of 0.5% (Chen and Chen 1935a).

Dendrobine was the first alkaloid discovered in an orchid. It was isolated in crystalline form from *Dendrobium moniliforme* in 1932 by Japanese scientists, H Suzuki, I Keimatsu and M Ito. In China, KK Chen and AL Chen found that it was present in *Chin Shih-Hu* (*Herba Dendrobii*) from Sichuan Province but absent in the 'Chin Shih-Hu' (also known as *Herba Dendrobii*) from

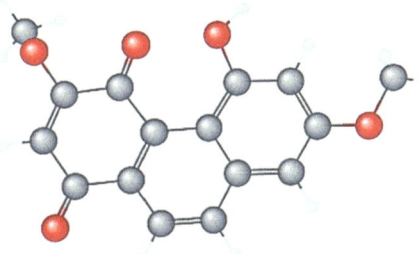

Fig. 8.4 Denbinobin

Guizhou province (Chen and Chen 1935a, b). Alkaloids are usually present in deciduous *Dendrobium* species belonging to the northern clade, but they are not present in detectable amounts in *Dendrobium* species occurring in Indonesia, Papua New Guinea and Australia (Fig. 8.4).

Phenanthrenes, Stilbenoids and Polysaccharides

Hydrocarbons with benzene ring structures and polysaccharides have taken over from alkaloids in attracting the interest of scientists working on the biochemistry of orchids, one reason being the fact that several such compounds possess unique anti-cancer effects on classic cancer cell lines.

Members of *Dendrobium* and related genera (*Flickingeria*, *Eria*) produce compounds which destroy human cancer cells in the test tube. *Dendrobium chrysotoxum* contains three fluorenones—dendroflorin, denchrysan A and 1,4,5-trihydroxy-7-methoxy-9H-fluoren-9-one—which exhibit cytotoxic against human liver cancer cell line BEL 7402. Denbinobin, a phenanthrene occurring in *Dendrobium nobile* and *Flickingeria fimbriata*, show in vitro cytotoxic activity on several types of cancers (A549 human lung cancer, SK-OV-3 human ovarian cancer, HL-60 human promyelocytic leukaemia and human pancreatic cancer). Testing was conducted on cultured immortal cell lines derived from human cancers which are available to research scientists (Figs. 8.5, 8.6 and 8.7).

Erianin also from *Dendrobium chrysotoxum* stops cell division and promotes cell death in human leukaemia HL-60 cells (Li et al. 2001), moderates growth delay in xenografted human hepatoma Bel7402 and melanoma A375 and induces significant vascular shutdown to the tumours (Gong et al. 2004). ZJU-6, a novel derivative of Erianin, shows potent anti-tubulin polymerization and anti-angiogenic activities. It is even more potent than erianin (Lam et al. 2012). An additional 23 isoerianin analogues were synthetized in France, and some show promising anti-angiogenesis activity, but proof of their

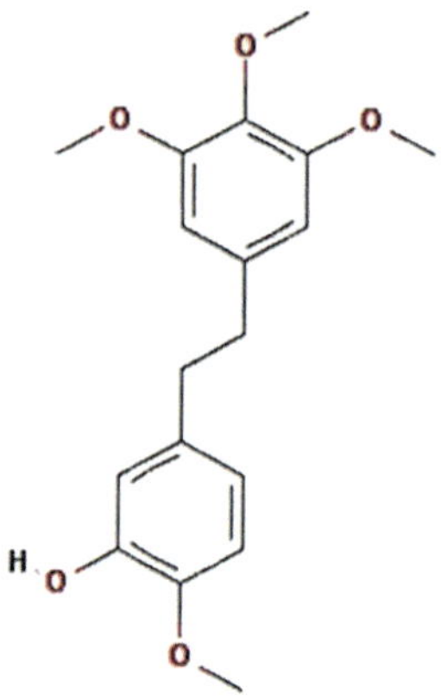

Fig. 8.5 Erianin

Fig. 8.7 Chrysotoxine

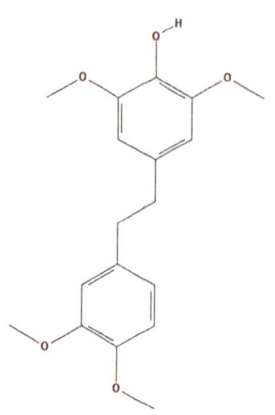

efficacy in a patient setting has to await clinical trials (Massaoudi et al. 2011).

Calanthe arisanensis produce calanquinone A which is cytotoxic for breast cancer cells that are resistant to paclitaxel (Taxol) and also vincristine-resistant nasopharyngeal carcinoma. Calanquinone A was synthesized in 2008 (Lee et al. 2008). Selective extracts of *Cremastra appendiculata* show potent anti-angiogenic activity (Shim et al. 2004; Hur and Kim 2009a, 2009b; Hur et al. 2010) and moderate cytotoxicity (Xia et al. 2005).

In theory denying blood supply to tumours appears to be a reasonable way to slow down their growth and division. However, many anti-angiogenesis compounds (anti-VEGF) were unable to halt progress of advanced cancers. They might prove useful for age-related macular degeneration. For instance, the A homoisoflavanone, 5,7-dihydroxy-3-(3-hydroxy-4-methioxybenzyl)-6-methoxychroman-4-one, reduces retinal neovascularization. There is a

possible application for the treatment of vaso-proliferative retinopathies, a major cause of blindness (Hur and Kim 2009a, 2009b; Hur et al. 2010).

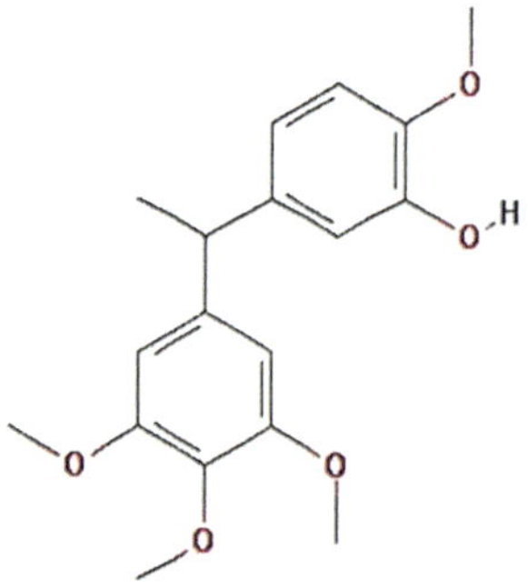

Orchids have been employed to treat painful joints, swellings and fractures. They possibly contain anti-prostaglandins. *Vanda coerulea* anti-prostaglandin action on skin cells protects the cells against UV damage. Some cosmetic firms have already added compounds derived from *Vanda, Cymbidium* and *Cattleya* to their products (Teoh 2016) (Fig. 8.8).

Orchids have been admired for their fragrance, *Cymbidium ensifolium* in China, *Vanda falcata* (syn. *Neofinetia falcata*) in Japan and *Phalaenopsis bellina* in Southeast Asia, but it is only recently that cosmetic firms in Japan and France began to include orchid fragrances in their products. Apart from the work by Roman Kaiser (1993), this promising area has not been extensively researched, and more can be done. Among the Asian orchids, numerous species of *Aerides, Arachnis, Calanthe, Coelogyne, Cymbidium, Dendrobium, Phalaenopsis* and *Vanda* hold promise (Fig. 8.9 and 8.10).

Micropropagation

The use of cultivated rather than wild-harvested medicinal plants not only alleviates pressure on natural populations but facilitates standardization and increased safety, as inconsistencies in the quality and composition (due to genetypic and

Fig. 8.6 Isoerianin

Fig. 8.8 *Vanda coerulea* (©Teoh Eng Soon 2019. All Rights Reserved)

phenotypic variation) are reduced, risk of adulteration lowered and yields are raised by management practice (Grace et al. 2002). The controlled

Fig. 8.10 *Jumellea fragrans* [as *Angraecum fragrans*]. From *Curtis Botanical Magazine*, t. 7153–7211, vol. 117 [ser. 3 vol. 47] t/7161 (1891) [M Smith]

environments of tissue culture systems and cell suspension cultures offer benefits of optimized plant uniformity and continuous and consistent natural products extraction. In vitro plant cell and callus cultures first described by Nickell (1962) are a firmly established means to produce secondary metabolites from rare or threatened plants (Bajaj 1999), and a range of products have been produced and patented. For high volume secondary metabolite production, cell suspension cultures are favoured for their rapid growth cycles and the totipotency of plant cells.

In vitro cultures tend to accumulate secondary metabolites when subjected to chemical stress or stimuli such as fungal elicitors and varied environmental conditions. Indeed these factors may

Fig. 8.9 *Cymbidium ensifolium* (©Teoh Eng Soon 2019. All Rights Reserved)

even elicit de novo synthesis of novel compounds not usually present in the plant taxon. The growth, production and excretion of secondary metabolites from cell suspension cultures in a continuous process have been studied by numerous emerging technologies. These include immobilization of plant cells, permeabilization, two-phase technology and the use of bioreactors and elicitors (Bourgaud et al. 2001).

Comment

Longer-life expectancy has shifted current pharmaceutical research to focus on the afflictions of old age, such as hypertension, heart disease, strokes, neurological disorders including dementia, cancer, connective tissue deterioration and diabetes. Numerous compounds isolated from orchids show promise that they might be useful in treating these conditions, but whether they can be translated into clinical usage is another matter. A compound may remain on the shelf for several decades before someone takes an interest in it, and then, with luck, amazing discoveries might follow.

References

Bajaj YPS (1999) Medicinal and aromatic plants XI, Biotechnology in agriculture and forestry, vol 43. Springer, Switzerland

Boller AH, Corrodi F, Gaumann E et al (1957) Uber induzierte Abwehrstoffe bei Orchideen Pt. 1. Helv Chim Acta 40:1062–1066

Bourgaud F, Gravot A, Milesi S, Gontier E (2001) Production of plant secondary metabolites: a historical perspective. Plant Sci 161:839–851

Chen KK, Chen AL (1935a) The alkaloid of Chin-shih-hu. J Biol Chem 111:653–658

Chen KK, Chen AL (1935b) The pharmacological action of dendrobine, the alkaloid of chin-shih-hu. J Pharamacol Exp Ther 55:319–325

Eagle H, Foley GE (1958) Cytotoxicity in human cell cultures as a primary screen for the detection of antitumor agents. Cancer Res 18:1017–1025

Gong YQ, Fan Y, Wu DZ et al (2004) In vivo and in vitro evaluation of erianin, a novel anti-angiogenic agent. Eur J Cancer 40(10):1554–1565

Grace OM, Prendergast HDV, Jager AK, van Staden J (2002) The status of bark in South African health care. S A J Bot 68:21–30

Hardegger E, Schellenbaum M, Corrodi H (1963) Uber onduzierte Abwehrstoffe bei Orchideen. Part 2. Helv Chim Acta 46:1171–1180

Hur S, Kim T (2009a) Homoisoflavanone, an extract from Cremastra appendiculata Makino, inhibits inflammatory and allergic response in mast cell. J Invest Dermatol 129(Suppl. 1):S125

Hur S, Kim T (2009b) Homoisoflavanone, an extract from Cremastra appendiculata Makino, inhibits UVB-induced skin inflammation. J Invest Dermatol 129(Suppl. 1):S23

Hur SG, Lee YS, Yoo H et al (2010) Homoisoflavanone inhibits UVB-induced skin inflammation through reduced cyclooxidagenase-2-expression and NF-kappa B nuclear localization. J Dermatol Sci 59 (3):163–169

Kaiser N (1993) The scent of orchids: olfactory and chemical investigations. Editiones Roche, Basel

Lam F, Bradshaw TD, Mao H et al (2012) ZJU-6, a novel derivative of Erianin, shows potent anti-tubilin polymerization and anti-angiogenic activities. Investig New Drugs 30(5):1899–1907

Lee C-L, Nakagawa-Goto K, Yu D et al (2008) Cytotoxic Calanquinone A from Calanthe arisanensis and its first total synthesis. Bioorg Med Chem Lett 18 (15):4275–4277

Li YM, Wang HY, Liu GQ (2001) Erianin induces apoptosis in human leukemia HL-60 cells. Acta Pharmacol Sin 22(11):1018–1012

Luning B (1974) Alkaloids of the Orchidaceae. In: Withner CL (ed) The orchids. Scientific studies. Wiley, New York

Massaoudi S, Hamze A, Provot O et al (2011) Discovery of isoerianin analogues as promising anticancer agents. ChemMedChem 6(3):488–497

Meissner W (1819) II. Ube rein neues Pflanzenalkali (Alkaloid). J Chem Phys 25:377–381

Muller CH (1966) The role of chemical inhibition (allelopathy) in vegetational composition. Bull Torrey Bot Club 93(5):332–351

Nickell LG (1962) Submerged growth of plant cells. Adv Appl Microbiol 4(1962):213–236

Shim JS, Kim JH, Lee J et al (2004) Anti-angiogenic activity of a homoisoflavanone from Cremastra appendiculata. Planta Med 70(2):171–173

Stoessl A, Arditti J, 1984): Orchid phytoalexins. In: J Arditti Orchid biology. Reviews and perspectives, III. Ithaca: Cornell University Press

Suzuki H, Keimatsu I, Ito M (1932) Alkaloid of the Chinese drug 'Chin Shih Hu'. II. Dendrobine. J Pharm Soc Japan 52:1049–1060

Teoh ES (2016) Medicinal orchids of Asia. Springer, Cham

Xia WB, Xue Z, Li S, Wang SJ, Yang YC, He DX, Ran GL, Kong LZ, Shi JG (2005) Chemical constituents from the tuber of Cremastra appendiculata. Zhongguo Zhong Yao Za Zhi 30(23):1827–1830

Barbarism characterized the Spanish conquest of Mexico, and sadly, during the early colonial period and for a long time afterwards, the indigenous people continued to face discrimination, brutality and despair. In 1542, responding to a plea by Bartholome de las Casas (1484–1566) to forbid racism, Emperor Charles V promulgated the *Nuevas Leyes* prohibiting slavery in New Spain, but the decree was largely ignored. Immigrants continued to look down and mistreat the natives. However, there were notable exceptions among the better educated. Among the educated Spanish elite, not the conquistadors, there were people who recognized that the Aztec and Inca civilizations were well organized and sophisticated in areas where the invaders were really quite deficient.

Wanting to demonstrate that the natives had a long and admirable culture and were thus worthy of respect by Spain and Spaniards, Francisco de Mendoza, son of Antonio de Mendoza, first viceroy to New Spain, proposed a compilation of the local traditional medicine. The job was assigned to a native healer from Xochimilco, Martinus de la Cruz, who was working at the Colegio de Santa Cruz de Tlatelolco that was founded by Franciscan friars in 1536. De la Cruz wrote the original Aztec text and was possibly responsible for the accompanying illustrations. Juannus Badianus, another resident at the college who was also from Xochimilco, was assigned the task of translating the text from Nahuatl to Latin. The manuscript was completed in 1552 and titled in Latin *Libellus de Medicinalibus Indorum Herbis* (*The Book of Indian Medicinal Herbs*).

Subsequently referred to as the *Badianus Manuscript* after its translator, the book carries the first written record and the earliest illustration of a Central American orchid, *Vanilla planifolia*. The proper name for the *Badianus Manuscript* should be *Codex de la Cruz-Badianus*. This *Codex* has a long and interesting history (Figs. 9.1 and 9.2).

The original Nahuatl text compiled by de la Cruz has not been found. It disappeared with the College which closed when the local government withdrew its financial support in 1606. Meanwhile, the translated text reached the court of Spain, but it was kept under wraps and quietly disappeared into the royal library without attracting any scholar's attention. Eventually the manuscript fell into the possession of Don Diego de Cortavila y Sanabria, apothecary to Phillip II, who promptly inscribed his name on its frontispiece, *ex Libris didaci Cortauils*. Later, when Cortavila received Francesco Barberini, Nunzio of the Vatican and nephew of the Pope, at his botanical garden, he gave the book to the cardinal. Barberini was a collector of books. However, like Cortavila, Barberini did not promote the *Codex* but merely deposited it in the Library of the Palazzo Barberini and thence into the Vatican Library. There it remained hidden until 1929 when it was discovered by the Smithsonian

© Springer Nature Switzerland AG 2019
E. S. Teoh, *Orchids as Aphrodisiac, Medicine or Food*, https://doi.org/10.1007/978-3-030-18255-7_9

Fig. 9.1 Francisco Hernandez de Toledo (1517–1580)

Fig. 9.2 Francisco Hernandez de Toledo's *Rerum Medicarum Novae Hispaniae Thesaurus* (1615) published in Mexico, title page

scholar, Charles Upson Clark. William E Gates and Emily Walcott Emmart produced English translations of the text that were published by the Maya Society in 1939 and Johns Hopkins Press in 1940, respectively.

Cardinal Barberini (1597–1679) was a member of the Accademia dei Lincei, and at one stage, he was even proposed for election as its president. However, when the Accademia contemplated publishing a *New World Materia Medica*, Barberini did not offer the *de la Cruz-Badianus Codex*. Since he loved Americana, Barberini should have remembered the *Codex*, and therefore it is difficult to understand his behaviour, given that he was a fair man because he refused to condemn Galileo Galilei when he served as grand inquisitor of the Roman Inquisition. Given the mood of the Inquisition, there is speculation that there was prejudice against the work of a native. Francisco Hernandez de Toledo's *Rerum Medicarum Novae Hispaniae Thesaurus* published in Mexico in 1615 thus became the only known major work for a long time, and it was hailed as the first scientific compilation from the New World although, in fact, a translated native Mexican treatise on Aztec medicine preceded it by 63 years (Figs. 9.3 and 9.4).

Altogether, the *de la Cruz-Badianus Codex* mentioned 249 medicinal plants but included only a single orchid that is positively identified, the famous *tlilxochitl* (*Vanilla planifolia*). This is the first account of *Vanilla* to appear in a European language. The description of the orchid was accompanied by a colour illustration which showed the plant-bearing flowers and the typical pods. Dried flowers of *tlilxochitl* (*mecaxochitl*) together with the fragrant bark of a few plants were rendered into powder and transferred to the hollow of dried flowers of *huacaxochitl* so that their scent could be intensified through fermentation. When this was accomplished, the powder was transferred to the hollow of dried flowers of *yolloxochitl* (*Talauma mexicana*, a member of the magnolia family) and given to travellers who wore them on the neck for protection during their journeys. The practice is reminiscent of another ancient practice mentioned in the Chinese *Book of Songs* (sixth century BC) which recorded

Fig. 9.3 *Stanhopea hernandezii* [as *Stanhopea devoniensis*] From: Houtte L van, *Flores des serres et des jardin de l'Europe*, vol. 10: p.17, t. 975 (1855)

Fig. 9.4 *Trichocentrum luridum* [as *Oncidium luridum*] From: *Edward's Botanical Register* vol. 9: pl. 727 (1869) [M. Hart]

that young people in springtime wore fragrant flowers (*lan* or *Cymbidium goeringii*) in their hair to ward off mountain spirits.

Direct evidence for the use of orchids as medicine in Mayan culture does not exist because in one of the great crimes against humanity, the Franciscan bishop, Diego de Landa (1524–1579), presided over the torture of 4500 Mayas and burnt the 27 rolls of Mayan codices 'since they only contained superstitions and perfidies of the devil'. It required only a single bigoted Inquisitor to destroy eight centuries of Mayan literature (Ossenbach, 2009; Encyclopaedia Britannica 2019). Mayan medicinal usage of orchids is generally surmised by what is known about Aztec usage.

Rerum Medicarum Novae Hispaniae Thesaurus (1615)

Francisco Hernandez (1517–1580) led the first scientific expedition to New Spain under the auspices of Phillip II to investigate the medicinal herbs of Mexico. In his report on 3000 plants, he mentioned five orchids, of which only two can be identified with certainty: *Vanilla planifolia* and *Stanhopea hernandezii*.

Illustrations of the remaining three derived from the 1651 *Thesaurus* distilled from the *Protomedico of Francisco Hernandez* and published by the Accademia dei Lincei in Rome

Fig. 9.5 *Trichocentrum cavendishianum* [as *Oncidium cavendishianum*]. From: Bateman J, *The Orchids of Mexico and Guatemala*, pl. 3 (1842) [M. Drake]

Fig. 9.6 *Myrmecophila tibicinis* (syn. *Schomburgkia tibicinis*) From: Bateman J, Drake SA, *The Orchidaceae of Mexico and Guatemala* (1843), pl. 30 [SA Drake]

are not sufficiently detailed, but there have been suggestions that:

Amazauhtli is a *Trichocentrum* (syn. *Oncidium*), possibly *Trichocentrum luridum* (syn. *Oncidium luridum*) or *Trichocentrum cavendisianum* (syn. *O. cavendishianum*);

Chichiltic tepelauhxochitl is either a *Laelia* or a *Schomburgkia*; and

Tzauxochitl is either a *Laelia* or an *Encyclia* (Ossenbach 2009) (Figs. 9.5 and 9.6).

Usage of orchids in Central America was well reviewed by Lawler (1984) and Ossenbach (2009), the latter providing data on pre-Hispanic usage derived from the *History of the Things of New Spain* authored by friar Bernardino de Sahagun (?1499–1590) who arrived in Santa Cruz de Tlatelolco in 1529 (Figs. 9.7 and 9.8).

Fig. 9.7 Bernardino de Sahagun (c. 1499–1590) wrote *La Historia Universal de las Casas de Nueue Espana* (now known as the *Florentine Codex*) in Nahuatl during the second half of the sixteenth century

Fig. 9.8 Page from de Sahagun's *Florentine Codex* illustrating the cultivation of corn in Nueue Espana (Mexico)

Many orchids were employed to treat dysentery, predominantly *Arpophyllum spicatum* (Nahuatl, *tzauhxilotl*), *Bletia campanulata* (Nahuatl, *tzacuxochitl xiuitl*), *Cranichis speciosa* (*Nahuatl, atzautli*) and *Encyclia pastoris* (Nahuatl, *tzacutli*) (Fig. 9.9). Pseudobulbs of these orchids were used as adhesives for paints and feathered ornaments, so it was reasonable to use them for stopping diarrhoea since they were binders. However, not every orchid employed as adhesive was used to treat diarrhoea: for instance, there was no report of *Bletia coccinea* (Nahuatl, *tonalxochitl*),

Fig. 9.10 *Catasetum maculatum*. From: *Edward's Botanical Register* vol. 26, pl. 62 (1840) [Miss Drake]

Fig. 9.9 *Arpophyllum spicatum* From: *Curtis Botanical Magazine*: t. 6009–6073, vol. 99 [ser. 3 vol. 29] t 6022 (1873) [WH Fitch]

Cranichis tubularis (Nahuatl, *acatzauhtli*), *Govenia liliacea* (Nahuatl, *iztactepetzacuxochitl*) and *G. superba* (Nahuatl, *cozticzacatzacuxochitl*) being employed medicinally in this manner. In the *Codex de la Cruz-Badianus* (1552) mention is made of the glue flower (Nahuatl, *Tzacouhxochitl: tascouth* meaning 'glue') being used as an adhesive as well as, in concoction, to treat shyness (*Timoris vel microspsychiae remedium*). It has been suggested that this may be an orchid, either *Bletia campanulata* or *Catasetum maculatum* (Mariano Ospina 1997) (Figs. 9.8, 9.9 and 9.10).

Catasetum maculatum (Mayan *ch'it cuc*) was employed by the Mayans in the Yucatan to treat

sores and tumours, whereas Aztecs used *Enchile citrina* to treat infected wounds (Fig. 9.10). The fruit was cooked in hearth ash for an hour and then applied to the sores. Two applications were required. The cooked fruit was grounded before application (Kunow 2003). *Catasetum intergerrinum* was employed to treat burns and other wounds (Garcia et al. 2014) (Figs. 9.11 and 9.12). Mayans employed pseudobulbs of *Myrmecophila tibicinis* [Maya: (medicinally) *dac kisin*; (commonly) *hom-ikim*] to facilitate childbirth. These pseudobulbs were more commonly employed as flutes or trumpets (Ossenbach 2009).

Stanhopea hernandezii (Nahuatl: *coatzonte-coxochitl*) was a tonic to relieve tiredness, and pods of *Vanilla planifolia* (Nahuatl, *tlilxochitl*, *mecaxochitl*; Maya, *sisbic*; Totonac, *zacanatum shanat*) provided an appetizer, digestive, diuretic and aphrodisiac. Powdered roots of *Stanhopea tigrina* was employed to treat sunstroke and

Fig. 9.11 *Catasetum integerrinum* [as *Catasetum wailesii*] From: *Curtis Botanical Magazine* vol. 68, t. 3937 (1842) [WH Fitch]

Fig. 9.12 *Catasetum integerrimum* From: *Curtis Botanical Magazine*, t. 3795–3879, vol. 67 [ser. 2, vol. 14]: t. 3823 (1841) [W.H. Fitch]

weakness (Lawler 1984). They were also used in rituals (see chapter on *Vanilla*) (Fig. 9.13).

Laelia autumnalis was used by the Aztecs to treat coughs. After the arrival of the Spaniards, candies made from pseudobulbs of *Laelia autumnalis* and *Laelia speciosa* were hawked on All Souls Day (Figs. 9.14 and 9.15).

The use of peyote (not an orchid) as a hallucinogen in Mexican Indian rituals is well known, but other plants were also employed. Cabecar used *Trichocentrum cebolleta* (syn. *Oncidium cebolleta*, Cabecar, *sulegli*; Bribri, *suler kili* meaning 'spear symbol') as a hallucinogen or to treat abdominal colic (Fig. 9.16). For the latter usage, a juice prepared by placing crushed stems in cold water is consumed intermittently until the patient is relieved. It would

appear that the medicinal compound is heat labile because the instruction was that the juice must not be heated. Bribri of Costa Rica and Panama use *suler kili* (*Trichocentrum cebolleta*) to treat heartache. Stems were cut into small pieces and boiled to prepare a drink which was to be taken three times a day, half a glass for an adult, 'spoonful' for young children. In this case the medicinal compound, if present, must be heat stable.

Some authors have pointed out that pre-Colombian golden jewellery dating from the eighth century at the Museum in Costa Rica which are thought to portray eagles or vultures actually show a remarkable resemblance to the labellum of *Trichocentrum cebolleta* (Alfaro

Fig. 9.13 *Stanhopea tigrina*. From: *Edward's Botanical Register* vol. 25: t. 1 (1839) [SA Drake]

Fig. 9.15 *Laelia speciosa*. From: *Edward's Botanical Register* vol. 30 (NS7) pl. 30 (1844) [SA Drake]

Fig. 9.14 *Laelia autumnalis*. From: *Edward's Botanical Register* vol. 25 (NS2) pl. 27 (1839) [SA Drake]

Fig. 9.16 *Trichocentrum cebolleta* [as *Oncidium cebolleta*] From: *Edward's Botanical Register* vol. 23: t. 1994 (1837)

Fig. 9.17 *Calanthe calanthoides* [as *Calanthe phajoides*] From: Reichenbach HG, *Xenia Orchidaceae* vol. 1, t. 79 (1858)

1, 1, 2, 3. Calanthe mexicana Rchb. fil. II, 4, 5. C. phajoides Rchb. fil.

1935, Atwood and de Retana, quoted by Ossenbach 2009), the orchid whose hallucinogenic alkaloid endowed it with sacred role among the ancient tribes. It was reportedly employed as a substitute for peyote (Ratsch 1998).

Post-Hispanic Period

Mexicans in the post-Hispanic period used powdered flowers of *Calanthe calanthoides* (syn. *Calanthe mexicana*) to stop epistaxis (Fig. 9.17), powdered roots of *Prosthechea citrina* (syn. *Cattleya citrina*) to relieve pain resulting from trauma (Fig. 9.18) and *Stanhopea tigrina* to treat sunstroke and weakness (Lawler 1984). A recent, in-depth investigation of *Prosthechea karwinskii* usage by the Mixtec community of Southern Mexico revealed that pseudobulbs, leaves and flowers of the plant are employed as infusion to treat cough (Fig. 9.19). Prepared with the addition of a gold coin or jewel, the infusion is imbibed for up to a month to prevent miscarriage and assist in childbirth. *Prosthechea karwinskii* is used as poultice for wounds and burns and chewed or consumed in the form of tea to manage diabetes. The flowers are fragrant, and standing them in oil

Fig. 9.18 *Prosthechea citrina* [as *Cattleya citrina*] From: Sander F, *Reichenbachia* vol. 1, 1858–1894, pl. 20 (1888)

for 15 days is a method for obtaining fragrant hair gel. At Easter, they are used to decorate icons and festive boats that carry *agua de gloria* (glory water) which is believed to confer protection. Flowers are also eaten, alone or added to food to which they impart a sweet-sour taste (Garcia et al. 2014).

Of the 433 orchid species native in Veracruz, 14 not employed elsewhere are employed locally to treat various conditions. *Catasetum integerrimum*, *Myrmecophila christinae* and *Ryncholaelia digbyana* are employed for wounds, and *Epidendrum anisatum* is employed to treat diarrhoea. A recent study uncovered an additional 11 species of which 3 are terrestrial and 8 are epiphytic. Epiphytic *Epidendrum chlorocorymbos* (local name *siempre viva*, forever alive) is used to 'lower cholesterol, stimulate dreaming and cure spots'. It is also employed as an eardrop. Heavy menstruation is treated with a leaf infusion of *Habenaria floribunda* (local name *clavocochinillo*, nail piglet) (Fig. 9.20). *Isochilus latibracteatus* is employed to treat abdominal colic ('pain in the mouth of the stomach and .. ..

colitis'), whereas leaves of *Isochilus majus* are made into a poultice for pain relief caused by trauma or inflammation. In the Zongolica region, people use pseudobulbs of the yellow flowered *Mormodes maculata* (*flor de mayo*) to make the poultice (Fig. 9.21). Poultice made with pseudobulbs of *Oestlundia luteorosea* (local name *topixcamite*) in alcohol is applied to the head to relieve pain caused by bumps. *Trichocentrum ascendens* (syn. *Oncidium ascendens*; local name in Sierra de Santa Martha *orquidea*) is used to treat injuries caused by sharp objects (Fig. 9.22). *Scaphyglottis fasciculata* is used to prevent miscarriage (Fig. 9.23). Fever is treated with *Sobralia macrantha* (local names *cabolin, Candelaria, lilio de San Antonio*), asthma with *Spiranthes eriophora* (local name *cecetzi y margaretilla*) and abdominal pain in women with an infusion prepared by boiling crushed pseudobulbs of *Stanhopea oculata* (local name *Tehuanxochitl*) (Fig. 9.24). Natives also practice *limpias*, a body and soul cleansing procedure, by sweeping across the body with *Trichocentrum ascendens* (syn. *Oncidium ascendens*) (Asseleih et al. 2015). Choco Indian fisherman living between Panama and Colombia mixes the flowers of *Cycnoches tonduzii* with (*Genipa americana*, not an orchid) to produce a rub which they apply on their hands to increase their catch (Ossenbach 2009) (Figs. 9.19, 9.20, 9.21, 9.22, 9.23 and 9.24).

Some orchids play a medicinal role in the Caribbean. Alcoholic tincture of *Brassia caudata* was used in the West Indies as an antispasmodic and to treat epilepsy and other nerve disorders; dried corms of *Bletia purpurea* (syn. *Bletia verecunda*) as a tonic. In Jamaica and Cuba, *Bletia purpurea* (*B. verecunda*) tubers were also used to treat stomach disorders. In the Bahamas, boiled tubers are eaten for relief after eating poisoned fish; sliced fresh bulbs are applied to wounds (Lawler 1984). *Scaphyglottis livida* was employed to eliminate parasites, relieve stomach-ache, relax muscles and prevent miscarriage (Meisel et al. 2014). Candy made with pseudobulbs of *Bletia coccinea* or *Cranichis*

Fig. 9.19 *Prosthechea karwinski* [as *Cattleya kawinski*] From: Martius CFP von, *Amoeritates botanicae monacenses/ Auswahl merkwurdiger Pflanzen*, t. 10 (1829–1831)

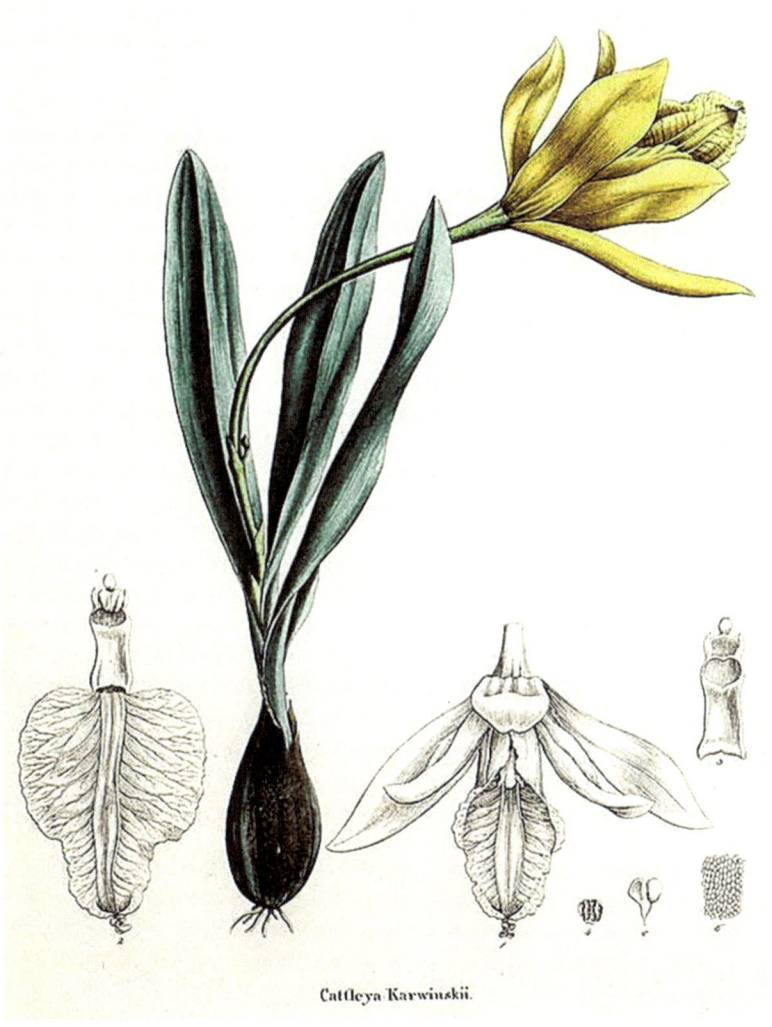

Fig. 9.20 *Habenaria floribunda* [as *Habenaria autumnalis*] Poeppig E, *Nova genera ac species plantarum* vol. 1, t. 75 (1835) [E Poeppig]

Fig. 9.21 *Mormodes maculata* var. *unicolor* [as *Mormodes pardina* var. *unicolor*] From: *l'Illustration horticole* vol. 1, t. 25 (1854)

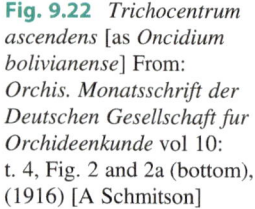

Fig. 9.22 *Trichocentrum ascendens* [as *Oncidium bolivianense*] From: *Orchis. Monatsschrift der Deutschen Gesellschaft fur Orchideenkunde* vol 10: t. 4, Fig. 2 and 2a (bottom), (1916) [A Schmitson]

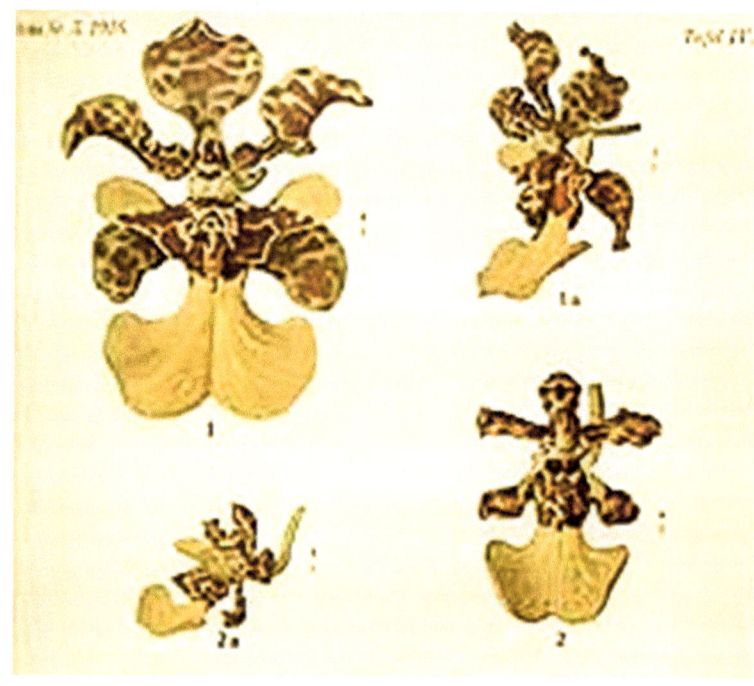

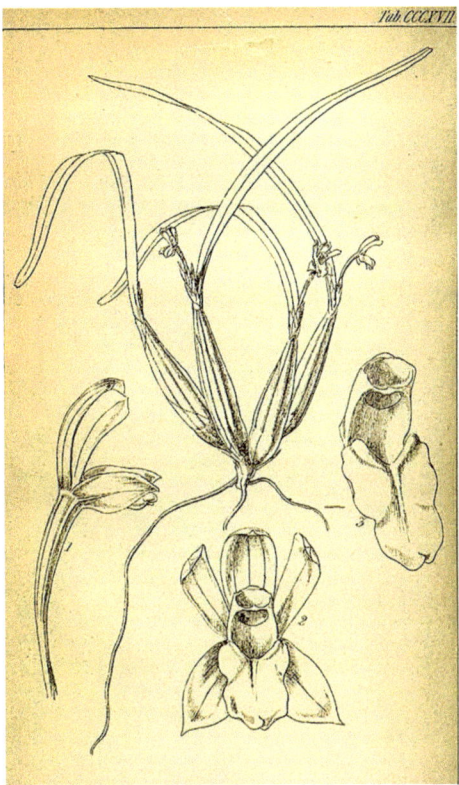

Fig. 9.23 *Scaphyglottis fasciculata*. From: Hooker WJ, Hooker JD, *Icones Plantarum* vol. 4, t. 317 (1841)

Fig. 9.24 *Stanhopea oculata*. From: *Gartenflora* [E von Regel] vol. 39: t.1335 Fig. 1 (1890)

Fig. 9.25 *Brassia caudata*. From: Hooker WJ, *Exotic Flora*, vol. 3: t. 179 (1827)

Fig. 9.27 *Govenia superba*. From: *Edward's Botanical Register* vol. 21: t. 1795 (1836) [SA Drake]

Fig. 9.26 *Bletia purpurea* [as *Limodorum purpureum*] From: Redoute PJ, *Les Lilacees* vol. 2: t. 83 (1805–1816)

Fig. 9.28 *Govenia lilacea*. From: *Edward's Botanical Register* vol 24: t. 13 (1838) [SA Drake]

Fig. 9.29 *Encyclia phoenicea* [as *Epidendrum phoeniceum*] From: Houtte L van, *Flore des terres et des jardins de l'Europe* vol. 4: pl. 306 (1848)

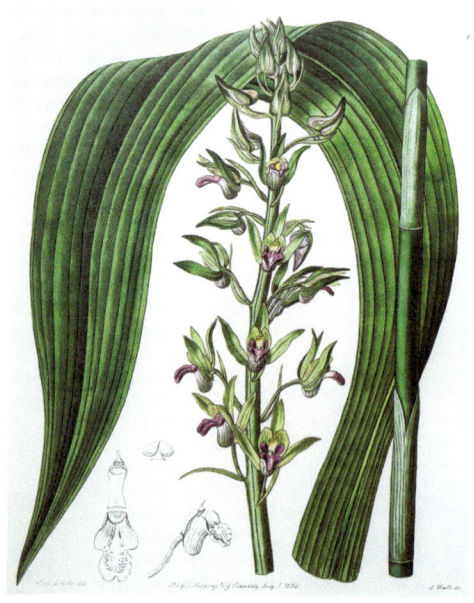

Fig. 9.30 *Eulophia alta* [as *Cyrtopodium woodfordii*] From: *Edward's Botanical Register* vol. 18: t. 1508 (1832) [SA Drake]

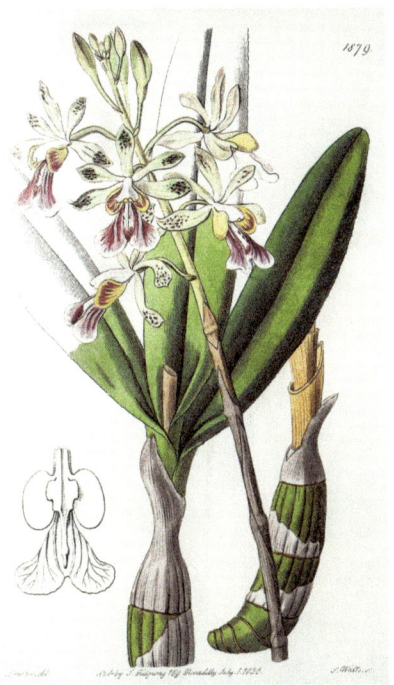

Fig. 9.31 *Psychilis bifida* [as *Epidendrum bifidum*] From: *Edward's Botanical Register* vol. 22: pl. 1879 (18336) [MIss Drake]

Fig. 9.32 *Ponthieva racemosa* [as *Ponthieva glandulosa*] From: *Curtis Botanical Magazine* vol. 22: p. 842 (1862)

Fig. 9.33 *Cyrtopodium punctatum* From Warner R, Williams BS, *The Orchid Album* vol. 1: t. 202 (1886) [JN Fitch]

speciosa, C. tubularis, Govenia liliacea, G. superba, Laelia autumnalis and *L. speciosa* became fashionable during the festival of All Souls Day following the arrival of the Spaniards (Ossenbach 2009) (Figs. 9.25, 9.26, 9.27 and 9.28).

Cubans had several different uses for *Cyrtopodium punctatum*: in decoction or as syrup for bronchial disease, as emollient for wounds and dislocations and an infusion for skin disease. In the past, Cubans employed *Encyclia phonicea* (syn. *Epidendrum phoniceum*) to induce abortion or menstrual flow, and relieve running nose. Mucilage of *Eulophia alta* (syn. *Cyrtopodium woodfordii*) was used in El Salvador as a remedy for disorders of the chest. In Martinique and Guadeloupe, juice of *Psychilis bifida* (syn. *Epidendrum bifidum*) was used to cause diuresis, treat worms and relieve constipation. *Ponthieva gracemorsa* (syn. *Ponthieva glandulosa*) was a Costa Rican emetic (Lawler 1984) (Figs. 9.29, 9.30, 9.31 and 9.32).

In the Dominican Republic, three endemic and one introduced orchid species are employed in herbal medicine. Roots of *Cyrtopodium punctatum*, an endemic species, are employed to treat coughs, fractures and kidney disease; *Vanilla dilloniana*, another endemic, is employed for anorexia, whereas pseudobulbs of *Broughtonia domingensis*, the third endemic species, are used for infections and kidney disorders. Pseudobulbs and roots of the naturalized, tropical species *Oeceoclades maculata* are employed to treat stomach disorders. All four species possess weak antibacterial activity: *Cyrtopodium punctatum*, *Vanilla dilloniana* and *Oeceoclades maculata* additionally possess weak antifungal activity (Bond et al. 2014) (Figs. 9.33, 9.34 and 9.35) (Table 9.1).

Fig. 9.35 *Oeceoclades maculata* [as *Angraecum maculatum*] From: *Edward's Botanical Register* vol. 8: pl. 618 [M Hart]

References

Anonymous (2019) Diego de Landa. Encyclopaedia Britannica (Google)

Asseleih LMC, Garcia R, Cruz J (2015) Ethnobotany, pharmacology and chemistry of medicinal orchids from Veracruz. J Agricul Sci Technol A5(2015):745–754

Bond MO, Aregullin MA, Laux MT (2014) Antimicrobial, cytotoxic and antiproliferative activities of native and invasive orchids in Dominican Republic. Ethnobotany Pennscience J 2014:23–28

Garcia GC, Gomez RS, Rivera LL (2014) Documentation of the medicinal knowledge of *Prosthechea karwinskii* in a Mixtec community in Mexico. Rev Bras Farmacogn 24(2):153–158

Kunow MA (2003) Maya medicine. Traditional healing in Yucatan. University of New Mexico Press, Albuquerque

Table 9.1 Medicinal orchids of Central America

Arpophyllum spicatum (*Nahuatl: tzauhxilotl*)

Bletia campanulata (*Nahuatl: tzacuxochitl xiuitl*)

Bletia coccinea

Bletia purpurea)

Brassia caudata

Broughtonia domingensis

Calanthe calanthoides

Catasetum maculatum

Cranichis speciosa

Cranichis tubularis

Cyrtopodium punctatum

Encyclia pastoris

Encyclia phonicea (syn. *Epidendrum phoniceum*)

Epidendrum chlorocorymbos

Eulophia alta (syn. *Cyrtopodium woodfordii*)

Govenia liliacea

G. superba

Habenaria floribunda

Isochilus latibracteatus

Isochilus majus

Laelia autumnalis

Mormodes maculata

Myrmecophila tibicinis

Oeceoclades maculata

Oestlundia luteorosea

Oncidium ascendens

Oncidium cebolleta

Ponthieva gracemorsa (syn. *Ponthieva glandulosa*)

Prosthechea citrina (syn. *Cattleya citrina*)

Prosthechea karwinskii

Psychilis bifida (syn. *Epidendrum bifidum*)

Scaphyglottis fasciculata

Scaphyglottis livida

Sobralia macrantha

Spiranthes eriophora

Stanhopea hernandezii

Stanhopea oculata

Stanhopea tigrina

Vanilla dilloniana

Vanilla planifolia

Lawler LJ (1984) Ethnobotany of the Orchidaceae. In: Arditti J (ed) Orchid biology reviews and perspectives, vol III. Cornell University Press, Ithaca, NY

Mariano Ospina H (1997) Orchids and the Aztecs. An herbal documented the use of orchids in the Americas. Orchids 66:1160–1163

Meisel JE, Kaufmann RS, Pupulin F (2014) Orchids of Tropical America: an introduction and guide. In:

VandebroekI JB, Calewaert S, De jonckheere SS; Lucio S, Van Damme P, Van Puyvelde L, De Kimpe N: Use of medicinal plants and pharmaceuticals by indigenous communities in the Bolivian Andes and Amazon. Bull World Health Organ vol. 82 n. 4 Geneva Apr. 2004

Ossenbach C (2009) Orchids and Orchidology in Central America. 500 years of history. Lankesteriana 9 (1–2):1–268

Ratsch C (1998) The encyclopedia of psychoactive plants: ethnopharmacology and its applications. Park Street Press, Rochester

Medicinal Orchids of South America

10

South America has a rich source of beautiful orchids, but not much has been written about the use of orchids as medicine in this continent in the English language. A recent study concluded that the Orchidaceae (the family of orchids) was one of the two most under-used plant families in Brazil, the other family being Poaceae. However, whereas Poaceae has few medicinal attributes from a chemical perspective, Brazilian orchids are not lacking medicinal properties. Rather, because orchids are uncommon and difficult to collect, they are being ignored because there is an abundance of other plants with proven efficacy (Medeiros et al. 2013). This team of scientists found only a single orchid species mentioned in their literature search, whereas popular plant families (*Asteraceae, Euphorbiaceae, Fabaceae, Myrtaceae, Malvaceae, Rubiaceae*) saw between 15 and 63 species being employed.

In the Bolivian Andes and Amazon, knowledge of medicinal plants lessened as distances increased between a respondent's residence and the nearest bus stop, the nearest town and the nearest public health centre (Meisel et al. 2014). Lack of contact with people naturally led to less knowledge. On the other hand, urbanization has contributed to disinterest and loss of knowledge regarding herbal remedies. Presently, investigators have difficulty in obtaining details of ethnic South American usage of orchids.

Orchids were not offered in the medicinal markets of La Paz and El Alto in Bolivia (Macia

et al. 2005), but this might be due to the fact that only single fresh leaf and other fresh parts were employed medicinally making it unprofitable for dealers to market medicinal orchids. However, two additional studies of medicinal plants in Bolivia which involved the participation of villagers also failed to unearth any medicinal usage for orchids in the Bolivian Andes (Bourdy et al. 2000; Quiroga et al. 2012), and only one study managed to identify four orchids (two *Maxillaria* species; two unidentified) being employed in Bolivian tribal medicine. Tubers and stem of *Maxillaria* were used and seeds of an unidentified orchid. Unfortunately, the report did not describe how the orchids were employed (Wilkin 2014).

Thus the report on the medicinal orchids of the Indians of the Columbian Amazon by Richard E Schultes which appeared in a 1990 issue of the American Orchid Society Bulletin (AOS Bulletin) is a rare ethnomedical work related to orchids in South America. Much of the data which follows are derived from this publication. Majority of 4000 orchid species native to Columbia are distributed in the highlands, and fewer species occur in the Amazonian lowlands. In the Columbian Amazon, a very small number of orchids are employed medicinally; furthermore, their usage appears to be solely for trivial albeit common complaints. Leaves of the terrestrial orchid *Cleistes rosea* (syn. *Pogonia rosea*) (Kubeo language, *po-te-moo*) which are astringent are

E. S. Teoh, *Orchids as Aphrodisiac, Medicine or Food*, https://doi.org/10.1007/978-3-030-18255-7_10

Fig. 10.1 *Cleistes rosea.* From: Poeppig E, *Nova genera ac species plantarum* vol. 1: t. 92 (1835) [E Poeppig]

Fig. 10.2 *Eriopsis sceptrum.* From: *Curtis Botanical Magazine* t. 8412–8471, vol. 138 [ser., vol. 8] t. 8642 [M Smith]

brewed to prepare a bitter tea for treating mouth sores. An alternative treatment used by Tanimukas and Yukanas for mouth sores is mucilage extracted from pseudobulbs of an *Epidendrum* sp. (in native Tanimukas, *wa-ta-ne-ke*; Yukanas, *wa-ka-roo*). Pseudobulbs of *Eriopsis sceptrum* (Makuna, *wan-oo-ma-ka*) are boiled to extract the mucilage which is applied to mouth sores for relief of pain by natives of Apaporis River. Its Makuna name translates as 'mouth medicine', *wan-oo-ma-ka* (Schultes 1990) (Figs. 10.1 and 10.2).

A wash prepared with the epiphytic plant of *Dichaea morrisii* (syn. *Dichaea muricata*) (Kofan, *sha-ha-se-he-pa*) is used to treat conjunctivitis. The second half of its native name means 'medicine' in Kofan. It is claimed that the fermented drink prepared from fruits (such as pineapple, apples and bananas) or tapioca root (*Manihot esculenta*) can be smoothened and made less intoxicating if large purple flowers of an *Elleanthus* sp. (Yakuna Indians of Rio Miritipartana, *too-re-wa-na-wee*; Tanimukas,

too-re-wa-na-ra) are crushed and added to it. *Chichi* is said to be good for the body and is usually consumed at festivals marked by dancing.

Infected cuts and wound are first cleaned with decoction prepared from the leaves of the large common orchid *Epistephium brevicristatum* before powder of its large purple flowers are applied on the wounds by the Kubeo in Columbia. This rank, weedy orchid which grow over low, scrubby vegetation on the xerophytic sandstone savanna of Yapoboda at the headwaters of the Kuduyari River in Columbian Vaupes is so profuse in some locations that 'one must cut one's way through its six-foot tall growths. Kubeo medicine men collect, dried and pulverize the flowers and store the powder thus prepared for emergency use' Schultes (1990).

Kofans, a small tribe living in the Ecuadorian and Colombian montana, boiled the entire diminutive plant of *Psygmorchis pusilla* (syn. *Oncidium pusillum*) (Kofan, *atti-pa-ka-shaikie-se-he-pa*) to prepare a wash for cleaning infected wounds and lacerations (Schultes 1990).

Fig. 10.3 *Sophronitis crispa* [as *Cattleya crispa*] From: *Edward's Botanical Register* vol. 14: t. 1172 (1828) [M Hart]

Valle-de-Sibundoy located in the foothills of the Columbian Andes is home to two ancient tribes, one of which is the Kamsa, a term that translated as 'men of here', i.e. indigenous people. Kamsa medicine men recommend a tea made with the entire plant of a *Masdevallia* (species unspecified) to promote urination in pregnant women and reduce bladder inflammation. The plant is possibly a strong diuretic. A different tea made with the terrestrial slipper orchid *Phragmipedium ecuadorense* (Kofan, *Shatifa-se-he-pa, To-pa-se-he-pa*) is consumed for relief when afflicted with stomach trouble (Schultes 1990).

Trichocentrum cebolleta (syn. *Oncidium cebolleta*) is employed to heal wounds in & 9.4Venezuela (Lawler 1984). In Brazil, *Cattleya crispa* (syn. *Cymbidium crispatum*) was used as a demulcent, whereas *Catasetum* and *Cyrtopodium* species were used to treat boils and abscesses, and a poultice made with pseudobulbs of *Eulophia alta* (syn. *C. woodfordii*) was applied to reduce swellings (Figs. 10.3 and 10.4). Pseudobulbs of *Catasetum fimbriatum* and *Rodriguezia lanceolata* (syn. *R. secunda*) were used as contraceptives in Paraguay and Brazil, respectively, the former in decoction and the latter in the form of a poultice applied over the woman's body. *Brachystele unilateralis* commonly known

Fig. 10.4 *Eulophia alta* [as *Cyrtopodium woodfordii*]. From: *Curtis Botanical Magazine* t. 1771–1859, vol. 43: t. 1814 (1816) [n.a.]

by its synonym (Figs. 10.5 and 10.6) *Spiranthes diuretica* was employed as a diuretic in Chile. In the northern part of South America, fever was treated with *Cyrtochilum aureum* (syn. *Maxillaria bicolor*) (Lawler 1984) (Figs. 10.7 and 10.8).

Mors and his colleagues are more specific regarding the use of *Cyrtopodium* species in

Fig. 10.5 *Catasetum fimbriatum* From: *Curtis Botanical Magazine* vol. 117 [ser. 3, vol. 47] t. 7158 (1881) [M Smith]

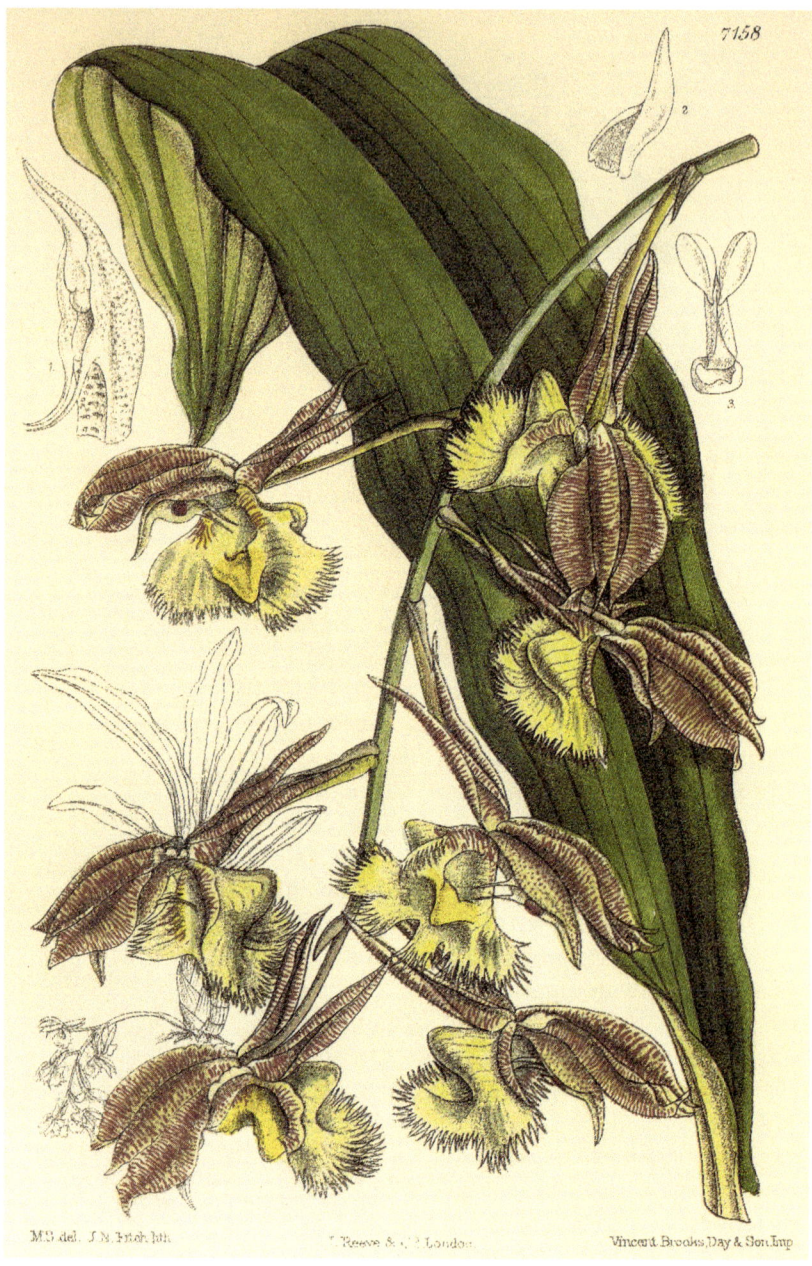

Brazil. According to the authors, pseudobulbs of *Cyrtopodium andersonii* (vernacular names *bisturi-do-mato*, *bisturi vegetal*, *colade sapateiro*, *rabo de tatu*, *sumare*, *sumare da pedra*) are employed to manage haemoptysis, to prepare an anti-catarrhal syrup and to heal infected wounds. Juice of *Cyrtopodium punctatum* (vernacular names *bisturi-do-mato*, *bisturi vegetal*, *sumare da mata*, *sumare do pau*) is used to promote cicatrization (Mors et al. 2000) (Figs. 10.9 and 10.10). Charote Indians in Argentina employ pseudobulbs of *Cyrtopodium*

Fig. 10.6 *Rodriguezia lanceolata.* From: *Curtis Botanical Magazine* t. 3458–3541, vol. 63 [ser. 2, vol. 10]: t. 3524 (18836) [n.a.]

punctatum to prepare a decoction for use as emetic and to control high blood pressure (Bond et al. 2014).

A *Scaphyglottis* species was employed for pain relief. Two compounds present in *Scaphyglottis livida* and *Camaridium densum* (syn. *Maxillaria densa*), namely, 5-alpha-lanosta −24, 24dimethyl-9(11)25dien3beta-ol, cyclobalanone, gigantol and 3, 4, dihydroxy3, 4, 5trimethoxybibenzyl, elicit anti-inflammatory and anti-nociceptive effects, partially by activation of opioid receptors in mice and rats (Deciga-Campos et al. 2007). This effect may explain why *Scaphyglottis* works for pain relief (Figs. 10.11 and 10.12).

Orchids were not listed in the medicinal plants traded in Suriname (van Andel et al. 2007). In Trinidad and Tobago, *Sarcoglottis acaulis* (syn. *Spiranthes acaulis*) is 1 of the 14 plants (the remaining 13 are not orchids) employed to treat diabetes (Lans 2006). It was the only orchid listed among the medicinal plants in the country (Barclay 2012) (Fig. 10.13).

Some countries in the northern region, like Guyana and Tortola, use *Psychilis bifida* (syn. *Epidendrum bifidum*) in a similar manner as people in Martinique and Guadeloupe, i.e. it was used as a diuretic and purgative to get rid of intestinal worms in the body (Lawler 1984).

Magical Cures

In South American native culture medicinal preparations are used as first-line cures for symptoms such as fever, cough, pain, diarrhoea and arthritis and for cuts and wounds. These are sold at herbal markets. Parts of trees and common herbs make up the bulk of such traditional remedies. Orchids are seldom, if ever, offered in herbal markets (Bourdy et al. 2000; Macia et al. 2005; Quiroga et al. 2012). When the symptoms persist and are recognized as forming an illness, the patient resorts to shamanistic healers because illness is generally attributed to banal supernatural influences. Possibly on account of their unusual character or relative rarity, some orchids are included in the preparation of a magical drink, ritual bath and talisman (*seguro*, a bottle with herbs and perfumes that serves as a protective charm handed to patients for continued protection).

There are 212 genera and 2020 species of orchids in Peru, of which 775 species in 137 genera are endemic (Gamarra and Leon 2006) of which only 6 species have been found to have medicinal usage (Bussmann and Sharon 2006a). The paper by Rainer W Bussman from the University of Hawaii and Douglas Sharon from the San Diego Museum of Man reported

Fig. 10.7 *Brachystele unilateralis.* From: Feuille LE, *Journal des observations physiques, mathematiques et botaniques, Faites sur les Cotes Orientales de l'Amerique meridionale et dans Indes Occidentalis* vol. 2: t. 17 (1714)

their in-depth study of traditional plant usage in Northern Peru where traditional herbal medicine has been practiced for almost 2000 years. In this ancient culture, treatment is usually performed at the *mesa* (a healing altar) in the healer's home or at sacred sites in the countryside or by lakes high up in the mountains. Healing involves spraying the patient with herbal extracts believed to possess magical power to ward off evil, followed by nasal application of tobacco, perfumes or hallucinogen. Purification baths are an important component of the treatment. The patient or supplicant may also be supplied with a *seguro*. Magical/ritual ailments were managed by the use of 207 plant species. This was followed by respiratory disorders (95 species), urinary disorders (85), female genital tract infection (66), liver disorders (61), inflammation (59), stomach complaints (51) and rheumatism (45). Among the 510 plant species that Bussmann and Sharon identified, there were six orchids. A single fresh leaf of *Aa paleacea* (*Herba de la Soledad, Hierba Sola*) was boiled in a cup of water which was drunk once a year to combat loneliness or depression and prevent unwanted pregnancy. Alternatively, the leaf was mixed with *Tapa tapa*, and one litre of it was

Fig. 10.8 *Cyrtochilum aureum* [as *Genus novum*] From: Ruiz H, Pavon J, *Drawings of the Royal Botanical Expedition to the Viceroyalty of Peru* (1777–1816)

Fig. 10.9 *Cyrtopodium andersonii.* From: *Curtis Botanical Magazine* t. 1771–1859, vol. 43: t. 1880 (1816) [n.a.]

consumed every week. A cold drink prepared by boiling 50 g of dried leaves and stems of *Epidendrum calanthum* (local names, *Semora Negra*, *Semora Curandera*) was drunk daily to cope with miasma, bad air (*mal de susto*). To treat kidney infection, the advice was to prepare a drink by boiling 10 g fresh stems of *Sudamerlycaste gigantea* (syn. *Lycaste gigantea*; *Ida gigantea*; local name *Cana Cana*) with 10 g of *Linanza*, *Berro*, *Peta de Perro*, *Papa Madre* and *Espiga de Maiz* in 500 ml of water for 5 minutes. When the mixture cooled, half a cup was to be drunk twice daily for 8 days; it was also provided as *seguro* (Fig. 10.14). Fresh leaves and stems of *Pachyphyllum pastii* (*Guaimi guaimi*,

huami huami, *huaime*, *huaime*) boiled with other strong herbs was employed 'topically and as *seguro*' for protection against everything (Fig. 10.15). A single fresh stem of *Stelis eublepharis* (*Hierba del Oro*, *Boton de Oro*) was combined with *Hierba de la Plata*, *Hierba de la Justica*, *Hierba del Dominio*, *Encanto*, *Sigueme Sigueme* and other plants that confer strength and good luck to prepare a lotion after boiling for 20 minutes (Fig. 10.16). This was used

Fig. 10.10 *Cyrtopodium punctatum*. From: *Lindenia, Iconography of Orchids* (E. Lindemann) vol. 4: t. 344 (1892)

Fig. 10.11 *Scaphyglottis livida* [as *Leaoa monophylla*]. From: *Achivos do Jardin Botanico do Rio de Janeiro*, vol. 3: t. 27 (1922) [onleesbaar]

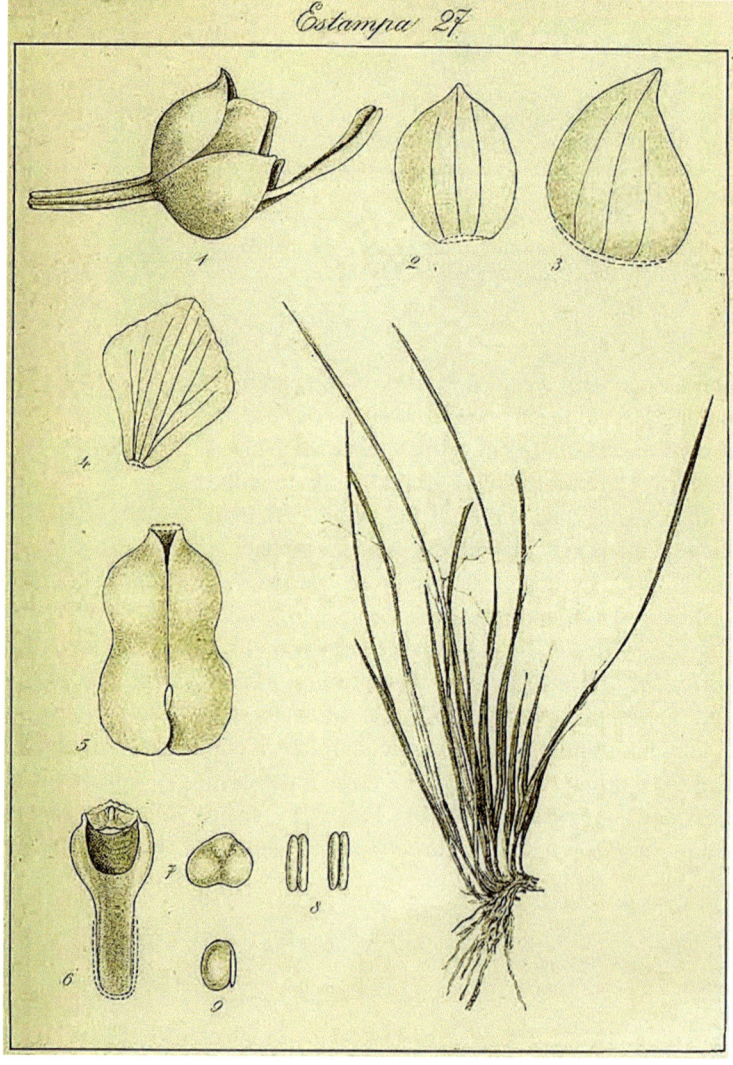

Fig. 10.13 *Sarcoglottis acaulis* [as *Spiranthes picta*] From: *Annales de flore et de pomone ou journal des jardins et des champs*, ser. 2, vol. 4: t. 48 (1817) [n.a.]

Fig. 10.12 *Maxillaria densa*. From: *Edward's Botanical Register* vol 21: t. 1804 (1836) [SA Drake]

as a bath and topically three times a week for 1–6 months to confer fragrance, remove nerves and bring good fortune in love, work, business and travel. A paste made with the entire plant of *Stelis* species (*Huaime huaime*, *Cucharilla*) and *Agua de Florida* was applied to treat facial paralysis and inflammation of the womb and ovaries (Bussmann and Sharon 2006a). Spiritual wellbeing is particularly important to the people of African descent (van Andel et al. 2007).

Three orchid species were employed by native healers in Loja Province of Southern Ecuador who shared a similar belief that magical practice is essential for the treatment of disease. Healers are shamans. Three orchid species were identified among 215 medicinal plants collected during a study of this population by Bussmann and Sharon 2006b). Flowers of *Epidendrum acrorhodum*

Fig. 10.14 *Sudamerlycaste gigantea* (syn. *Ida gigantean*, *Lycaste gigantea*) Photo: Henry Oakeley

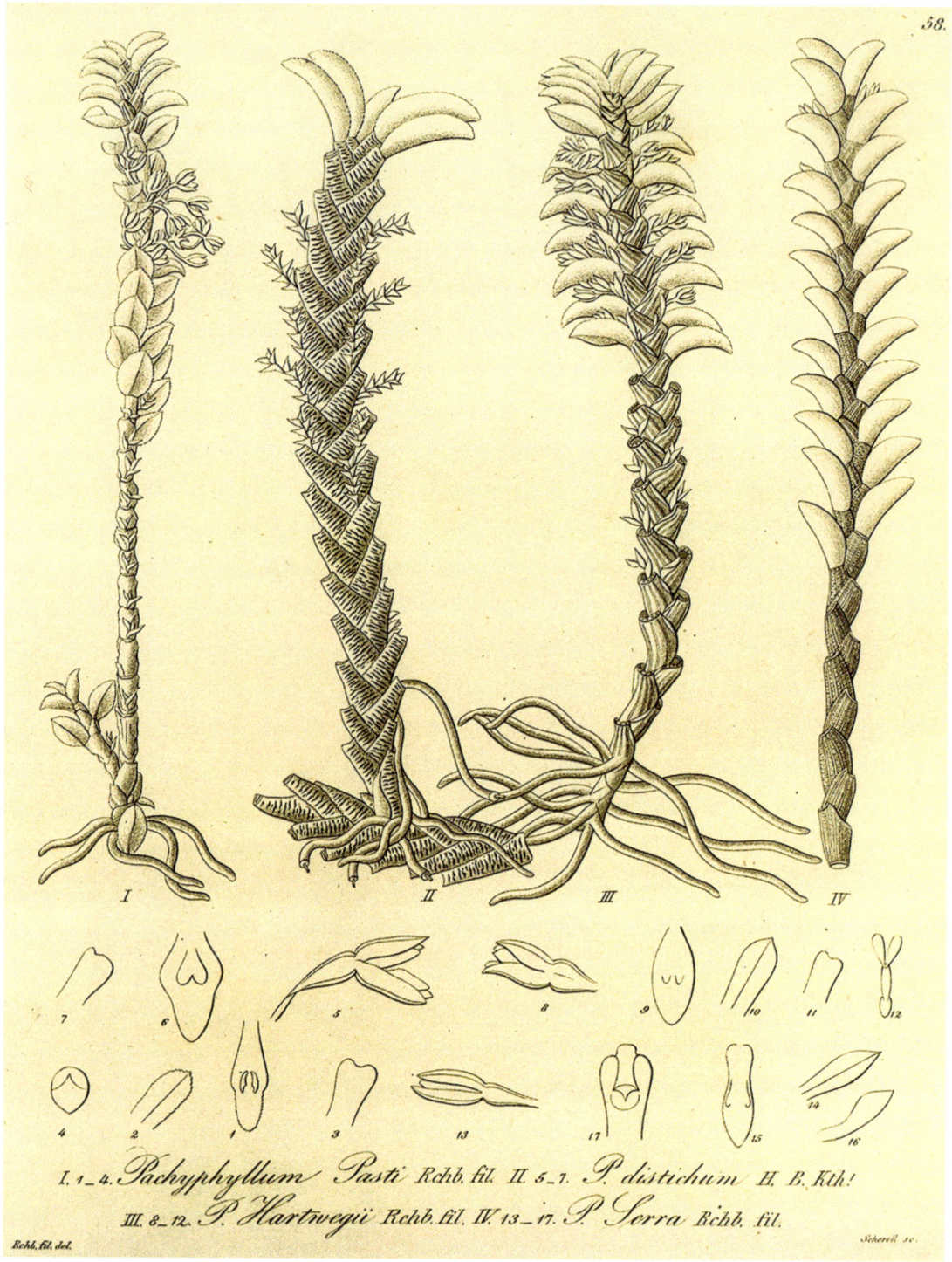

Fig. 10.15 *Pachyphyllum pastii* [as *Pachyphyllum distichum*] Reichenbach HG, *Xenia Orchidaceae* vol. 1: t. 58 (1858)

Fig. 10.16 *Stelis eublepharis*. From: *Curtis Botanical Magazine* t. 6469–6533, vol. 106 [ser. 3, vol. 36] t. 6521 (1880) [M Smith]

Fig. 10.17 *Epidendrum cochlidium* (*Flor del Christo*) (ⓒTeoh Eng Soon 2019. All Rights Reserved)

(local name, *Hierba del Caballero*) were added to cold water containing sugar, *chirimoya*, orange and lime flowers, *toronjil*, *congona*, white roses and a few drops of *aguardiente* to prepare a fragrant drink that conferred good luck. A similar preparation is made by adding fresh flowers of *Epidendrum cochlidium* (*Flor de Christo*) to cold water containing sugar, *chirimoya*, orange and lime flowers, *toronjil*, *congona*, white roses and some drops of *aguardiente* to treat patients affected by nerve disorders (Fig. 10.17). Fresh flowers of *Lycaste gigantea* are boiled with Cola de Caballo, Llanten, Matico Culantrillo and Cadillo for a drink to treat kidney disease. Fresh fruits of *Sudamerlycaste gigantea* (*simayuca*) are boiled with *preñadilla* and *guayusa* to a decoction for correcting male infertility (Begar et al. 2001; Bussmann and Sharon 2006b).

Table 10.1 lists the orchid species employed medicinally in South America.

Table 10.1 Orchid species employed as medicine in South America

Aa paleacea (Kunth.) Rchb.f.

Brachystele unilateralis (Poir.) Schltr.

Camaridium densum (syn. *Maxillaria densa*)

Catasetum fimbriatum (C. Morren) Lindl.

Catasetum sp.

Cattleya crispa Lindl. (syn. *Cymbidium crispatum*)

Cleistes rosea Lindl., syn. *Pogonia rosea* [(LIndl.) Rchb f.]

Cyrtochilum aureum (Lindl.) Senghas (syn. *Maxillaria bicolor*)

Cyrtopodium andersoni (Lamb. Ex Andrews) R.Br

Cyrtopodium punctatum (L.) Lindl.

Cyrtopodium sp.

Dichaea morrisii Fawc. & Rendle [syn. *Dichaea muricata* (Sw.) Lindl.]

Epidendrum acrorhodum Hagster & Dodson

Epidendrum calanthum Rchb. f. & Warsz.

Epidendrum cochlidium Lindl.

Epidendrum sp.

Epistephium brevicristatum R.E. Schult.

Eriopsis sceptrum Rchb. f. et Warsecwicz

Eulophia alta (syn. *C. woodfordii*)

Sudamerlycaste gigantea Lindl.

Masdevallia sp.

Trichocentrum cebolleta (Jacq.) M.W.Chase & N.H. Williams [syn. *Oncidium cebolleta* (Jacq.) Sw.]

Pachyphyllum pastii Krenzl. Ex. Weberb.

Phragmipedium ecuadorense Garay, Harling et Spaare

Psychilis bifida [(Aubl.) Sauleda] (syn. *Epidendrum bifidum*)

Psygmorchis pusilla (L) Dodson et Dressler [syn. *Oncidium pusillum* (L) Rchb.f.]

Rodriguezia lanceolata Ruiz. & Pav. (syn. *R. secunda* Kunth.)

Scaphyglottis livida (Lindl.) Schltr.

Sarcoglottis acaulis (Sm.) Schltr. [syn. *Spiranthes acaulis* (Sm.) Cogn.]

Brachystele unilateralis (Poir.) Schltr. [syn. *Spiranthes diuretica* (Willd.) Lindl.]

Stelis eublepharis Rchb. f.

Stelis sp.

References

Barclay G (2012) Medicinal plants of Trinidad and Tobago. In: Reid B (ed) Caribbean heritage, Chapter: 16. The University of the West Indies Press, Kingston, pp 221–235

Begar E, Bussmann R, Roq C, Sharon D (2001) Herbs of Southern Ecuador. L.H.Press, Spring Valley

Bond MO, Aregullin MA, Laux MT (2014) Antimicrobial, cytotoxic and antiproliferative activities of native and invasive orchids in Dominican Republic. Ethnobotany. Pennscience J 2014:23–28

Bourdy G, DeWalt SJ, Chávez de Michel LR, Roca A, Deharo E, Muñoz, Balderrama VL, Quenevo C, Gimenez A (2000) Medicinal plants uses of the Tacana, an Amazonian Bolivian ethnic group. J Ethnopharmacol 70:87–109

Bussmann RW, Sharon D (2006a) Traditional medicinal plant use in Northern Peru: tracking two thousand years of healing culture. J Ethnobiol Ethnomed 2:47–50

Bussmann RW, Sharon D (2006b) Traditional medicinal plant use in Loja province, Southern Ecuador. J Ethnobiol Ethnomed 2:44–46

Deciga-Campos M, Palacios-Espinosa JF, Reyes-Ramirez A, Meta R (2007) Antinociceptive and anti-inflammatory effects of compounds isolated from Scaphyglottis livida and Maxillaria densa. J Pharmacol 114(2):161–168

Gamarra JR, Leon B (2006) Orchidaceae endemicas del Peru. Rev Peru Boil 13(2):759s–878s

Lans CA (2006) Ethnomedicines used in Trinidad and Tobago for urinary problems and diabetes mellitus. J Ethnobiol Ethnomed 2(45):2006

Lawler LJ (1984) Ethnobotany of the Orchidaceae. In: Arditti JA (ed) Orchid biology. Reviews and perspectives, vol III. Cornell University Press, Ithaca, NY

Macia MJ, Garcia C, Vidurra PJ (2005) An ethnobotanical survey of medicinal plants commercialized in themarkets of La Paz and El Alto, Bolivia. J Ethnopharmacol 97:337–350

Medeiros PM, Ladio AM, Albuquerqua UP (2013) Patterns of medicinal plant use by inhabitants of Brazilian urban and rural areas: a macroscale investigation based on available literature. J Pharmacol 150 (2):729–746

Meisel JE, Kaufmann RS, Pupulin F (2014) Orchids of tropical America: an introduction and guide. In: Vandebroek I, JB Calewaert De Jonckheere SS, et al. Use of medicinal plants and pharmaceuticals by indigenous communities in the Bolivian Andes and Amazon. Bull World Health Organ vol. 82 n. 4 Geneva Apr. 2004

Mors W, Rizzini CT, Pereira NA (2000) Medicinal plants of Brazil. Reference Publications, Algonac

Quiroga R, Lidia Meneses L, Bussmann RW (2012) Medicinal ethnobotany in Huacareta (Chuquisaca, Bolivia). J Ethnobiol Ethnomed 8:29

Schultes RE (1990) Medicinal orchids of the Indians of the Colombian Amazon. Am Orchid Soc Bull 59 (2):159–161

van Andel T, Behari-Ramdas J, Reinout Havinga R, Groenendijk S (2007) The medicinal plant trade in Suriname. Ethnobot Res Appl 5:351–372

Wilkin PF (2014) Transmission and commoditization of medicinal plant knowledge in the marketplaces of Oruro, Bolivia. Ph.D. Thesis, University of Kent

Cherokee was a great nation when the earliest European settlers arrived in the United States. Their peoples were spread over Southeastern United States (principally Georgia, Tennessee, North Carolina and South Carolina) living in towns and villages made up of clans and families, with White and Red leaderships. Reds were warriors, young men led by a leader and a council with representations from each of seven clans. They took charge during emergencies and in time of war. Whites were tribal elders. They had their own headman and council-administered tribal affairs in times of peace. Healers were shamans, intermediaries between human and spirits. They were consulted if the home remedy did not produce the desired effect, when an illness was serious or prolonged and for emotional, sometimes even for social, problems tormenting individuals.

Cherokees were close to nature and cherished the sanctity of all living things. They maintained the concept that one should not take from nature that which cannot be used, nor more than can be used (Hamel and Chiltoskey 1975). This applied equally to plants and animals; they needed no advice on the importance of conservation.

There are many charming Cherokee legends about the origin of woman, forests, edible plants, disease and medicinal plants. In their creation legend, the first woman appeared in a stalk of corn. To remind herself of her source, she planted the crop. Corn became the staple diet of the Cherokees. It is seldom missing from a meal.

Cherokees believed that diseases were caused by animals in retaliation to human encroachment on their domains and their wanton slaughter. Fortunately, plants were always friendly towards humans. When plants learnt about the havoc caused by animals, they decided to help by volunteering to turn themselves into cures for disease. Trees, grasses, mosses, fungi and beautiful orchids all responded. Should the herb be inadequate, the code of Cherokee medicine (*nunwati* also spelled *nvwoti*) specified that an appeal must be made to spirits for their help. This is when the healer transforms into a shaman. Knowledge of *nunwati* was traditionally transmitted vertically within a family. It was a closely guarded secret (Fig. 11.1).

The Cherokee healer was usually conversant with the medicinal usage of 400–600 plants, but collective knowledge might reach 800. In 1975, Hamel and Mary U Chiltoskey managed to identify 497 medicinal plants by collating data from the twelfth-, eighteenth- and twentieth-century records. In the list there were eight orchids [*Spiranthes lucida, Cypripedium acaule, Cypripedium parviflorum* (incorrectly identified as *Cypripedium calceolus* which is a Eurasian species, whereas *Cypripedium calceolus* var. *parviflorum* is the pan-American *Cypripediium parviflorum*), *Platanthera ciliaris* (syn.

© Springer Nature Switzerland AG 2019
E. S. Teoh, *Orchids as Aphrodisiac, Medicine or Food*, https://doi.org/10.1007/978-3-030-18255-7_11

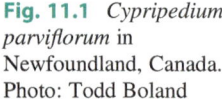

Fig. 11.1 *Cypripedium parviflorum* in Newfoundland, Canada. Photo: Todd Boland

Habenaria ciliaris), *Goodyera pubescens*, *Goodyera repens*, *Aplectrum hyemale* and *Liparis loeselii*] (Hamel and Chiltoskey 1975; Mergen 2006).

Many early white settlers, just like those who went to the Far East, adopted American Indian remedies because they could not obtain adequate medical supplies from Europe. Several orchid remedies remained official in the *United States Pharmacopeia* well into the twentieth century. It is interesting to note that American Indian medicinal orchid species with universal subarboreal distribution (e.g. *Cypripedium*, *Epipactis helleborine*, *Goodyera pubescens* and *Goodyera repens*) are also employed in Chinese herbal medicine (Figs. 11.2 and 11.3, Table 11.1).

Moccasin flower, Partridge Moccasin or *Cypripedium acaule* and *Cypripedium parviflorum* were the two attractive slipper orchid species employed by Cherokee for medicinal purposes. Depending on which species were prevalent where they lived, other Indian tribes substituted with different *Cypripedium* species to treat the same conditions. Roots were gathered in spring or autumn and brewed to treat various nervous disorder such as hysteria, fits, spasms and nerve pain. Diabetics also drank tea made from the roots, and hot tea was recommended for flu and neuralgia. Additional uses include treatment for worms, stomachache and pain arising from hernias. To treat stomach cramps, the Cherokee prescription called for the addition of *Aristolochia* sp. (Black snakeroot); for kidney problems,

addition of bastard toadflax (*Comandra umbellata*. Sandalwood family, *Santalaceae*): and for worms, chickweed (*Stellaria media*, *Caryophyllaceae*) (Hamel and Chiltoskey 1975). Women used the orchid plant for pain relief during menstruation and childbirth. Menominis (Wild Rice People that originally occupied 40,000 km^2 of Wisconsin) employed *Cypripedium acaule* for male disorders and used *Cypripedium parviflorum* for female disorders (Smith 1923). Ojibwas who lived around the Great Lakes used root of *Cypripedium pubescens* to treat

Fig. 11.2 *Cypripedium parviflorum* [as *Cypripedium calceolus*] on the United States postage stamp issued in 1991

1·65

SVERIGE

GUCKUSKO

O.S.del   Cypripedium calceolus   1982 ZJ·sc

Fig. 11.3 *Cypripedium calceolus* featured on Swedish postage stamp

resembling valerian in its action, but it was less potent. It was used as an antispasmodic and recommended for neuralgia, hysteria and hypochondriasis. The drug was administered in powder (1 g three times daily), in the form of infusion or fluid extract. The taste was sweetish, bitter and somewhat pungent. It contained a volatile aromatic oil, resinous matter, tannic acid, sugar, starch, etc. *Cypripedin* was the term applied to a resinoid mixture obtained by precipitating a strong tincture of the drug with water (*British Pharmaceutical Codex* 1911). British medicine spread its usage far and wide: Ayurvedic medicine recommended administering 2–4 ml of a 1:1 45% alcoholic extract of *Cypripedium pubescens* to treat anxiety state and insomnia (Karnick 1994). However, by 1909 there were already calls for *Cypripedium* to be deleted from the *United States Pharmacopeia* (Russell 1909; Capps 1909).

Cypripedium parviflorum and *C. pubescens* were employed by Cherokee to treat worms. Their women also employed *C. parviflorum* for pain relief in labour, hysteria and insomnia (Lawler 1984). Perhaps to the Cherokee, all *Cypripedium* species shared the same properties. In retrospect, the old *United States Pharmacopeia* and the *British Pharmaceutical Codex* were also imprecise in their designation of *Cypripedium* species. Even botanists in the past have been confused with *Cypripedium*, a genus with multiple synonyms for numerous species.

Unlike modern botanists, Cherokee also do not distinguish between *Goodyera pubescens* (common names: downy rattlesnake plantain, downy rattlesnake orchid, scrofula weed) and *Goodyera repens* (common names: creeping goodyera, lesser rattlesnake plantain, northern rattlesnake plantain). Indeed when the plants are not in bloom, it is difficult to distinguish between the two species because their mottled leaves look fairly similar. They are referred to as rattlesnake or plantain. Most species have been used as medicine (Correll 1950). Leaves were used to prepare a cold tea for treating colds. The tea is held in the mouth to relieve toothache. Leaves were also employed to treat kidney problems, burns, sore eyes and to induce vomiting. A decoction of the

female problems. Cherokee women relied on *Cypripedium parviflorum* to help them sleep. It was also used for hysteria. In Appalachia, natives used it to relieve headaches. *Cypripedium reginae* was a popular herb among North American Indians who employed an aqueous extract of the roots as a sedative and antispasmodic to treat a variety of nerve disorders that included hysteria and chorea (Vogel 1970; Hand 1976; Lawler 1984) (Figs. 11.4 and 11.5).

Fluidextractum Cypripedii, USP or extract of *Cypripedium parvifolium* and *C. pubescens* were official in the *United States Pharmacopeia*, and it was included in the *British Pharmaceutical Codex* just a hundred years ago. To prepare the product in England, *Cypripedium*, in No. 60 powder, was exhausted with alcohol (49%), and the strength of the final product was adjusted by evaporation, etc. so that the strength of the fluid extract shall be 1 in 1. *Average dose* is 1 ml (15 minims). Known by its common name, American Valerian, *Cypripedium* was reported as

Table 11.1 Accepted names and synonyms of medicinal orchid species employed by North American Indians

Aplectrum hyemale (Muhl. ex Willd.) Nutt.

Arethusa bulbosa L.

Bletia purpurea (syn. *Bletia verecunda, B. acutipetala*)

Corallorhiza maculata Rafin.[a]

Corallorhiza odontorhiza (syn. *Corallorhiza wisteriana*)

Cypripedium acaule Alton

Cypripedium parviflorum Salisb. Canada and the United States (syn. *C. calceolus* var. *parviflorum*)[b]

Cypripedium pubescens Willd. [= *C. parviflorum* var. pubescens (Willd.) O.W. Knight]

Epipactis gigantea Douglas ex Hook.

Epipactis helleborine (L.) Crntz.

Goodyera oblongifolia Raf. (syn. *Goodyera menziesii*)

Goodyera pubescens (Willd.) R.Br (L.) R.Br.

Goodyera repens (L.) R.Br.

Habenaria ciliaris [= *Platanthera ciliaris* (L.) Lindl.]

Liparis loeselii (L.) Rich.

Malaxis unifolia Michx. (syn. *Microstylis ophioglossoides*)

Platanthera ciliaris (L.) Lindl. (syn. *Habenaria ciliaris*)

Platanthera dilatata (Pursh) Lindl. ex L.C.Beck (syn. *Habenaria dilatata*)

Platanthera dilatata var. *leucostachya* (Lindl.) Hulten (syn. *Habenaria leucostachya*)

Platanthera grandiflora (Bigelow) Lindl. (syn. *Habenaria psycodes* var. *grandiflora, Habenaria fimbriata*)

Platanthera hookeri (Torr.) Lindl. (syn. *Habenaria hookeriana*)

Platanthera orbiculata (Pursh) Lindl.

Spiranthes lucida (H.H.Eaton) Ames

[a]*Corallorhiza maculata* (Rafin is distributed in Canada and the United States, whereas *C. maculata* Schltr. only occurs in Mexico and Guatemala.)

[b]*Cypripedium calceolus* referred to in several articles on North American medicinal orchids probably refers to *Cypripedium calceolus* var. *parviflorum* which is *Cypripedium parviflorum* Salisb. because *Cypripedium calceolus* L. is only distributed in Europe and Northern Asia to Japan

plant with wild cherry, wild ginger and *Xanthorhiza simplicissima* (yellowroot, a herb, not an orchid) was prescribed as a blood tonic (Hamel and Chiltoskey 1975).

Goodyera species were also employed for medicinal purposes by other Indians and early white settlers. Delawares who occupied the Atlantic seaboard and coastal woodlands of Canada and New Jersey in the United States before being forced to migrate to Oklahoma used root of *Goodyera pubescens* to treat lung infection, aching bones and as a prophylactic following childbirth. Delaware joined the Cherokee nation in Oklahoma, and the tribes could have influenced one another in the use of medicinal herbs. Mohegans, 'People of the Wolf', an Indian tribe living in Connecticut, used *Goodyera pubescens* and *G. repens* to prevent infections in

the mouth (probably thrush) in babies. Indians and European settlers ate *Goodyera pubescens* and made fresh leaf poultice to treat scrofula. A decoction was used for cleaning sores (Lawler 1984). Once known as 'cancer weed', the young leaves were employed to treat cancers, lupus and ulcers. In 1814, Federick Pursh in his *Flora Americae Septentrionalis* stated that the plant was administered to people who caught rabies; macerated leaves were applied on bites by rattlesnakes (Correll 1978): there were no comment about results, but presumably they all died, regardless. Potawatomi Indians, a Native American people occupying the Upper Mississippi River area, Michigan, Illinois, Indiana and Ontario, used *Goodyera repens* for female disorders, stomach ailments and bladder diseases and its leaves for snake bites. European

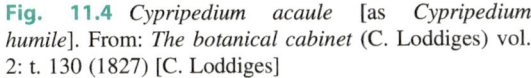

Fig. 11.4 *Cypripedium acaule* [as *Cypripedium humile*]. From: *The botanical cabinet* (C. Loddiges) vol. 2: t. 130 (1827) [C. Loddiges]

Fig. 11.5 *Cypripedium reginae* [as *Cypripedium spectabile*]. From: *The botanical cabinet* (C. Loddiges) vol. 7: t. 697 (11827) [G. Cook]

settlers used *Goodyera repens* for scrofula, eye infections and as a demulcent (Figs. 11.6, 11.7 and 11.8).

Goodyera oblongifolia Raf. (syn. *Goodyera menziesii*; common name: Menzies' rattlesnake plantain) is widespread in Western Canada and the United States, occurring in dry, moist coniferous or mixed forests, commonly at high elevations, up to 3300 m. Its flowering season is last week of June to first week of September, depending on location (Correll 1978). Coast Salish Indians of Vancouver Island boiled leaves of *Goodyera oblongifolia* to prepare a liniment for treating sore and stiff muscles. Split leaves

were applied, split side downwards, on bruises and sores (Turner and Bell 1971).

Aplectrum hyemale (putty-root, Adam and Eve) is a terrestrial orchid which thrives in deciduous woodlands, commonly under sugar maple or beech, throughout the eastern half of North America excluding Florida and Mexico. The distinctive features of the plant are the silvery pin stripes running longitudinally along the dark green, elliptical and undulating leaves which arise singly from white underground corms that are 2.5 cm in diameter. Corms are paired, two arising from the rhizome, thus the common name, Adam and Eve (Richburg 2003). Inflorescence arises in the spring after the leaf has withered. It is 30–60 cm

Fig. 11.6 *Goodyera repens*. From: Mosclef A, *Atlas des plantes de France* vol. 3: t. 336, Fig. A (1893)

Pl. 336.

A. *Spiranthe d'automne.*   Spiranthes autumnalis Rich.

B. *Goodyera rampante.*   Goodyera repens R.Br.

Fig. 11.7 *Goodyera oblongifolia* [as *Peramium decipiens*]. From: Walcott MV, *North American Wild Flowers* vol. 5: t. 350 (1925–1927)

tall with several small flowers of pale green, yellow or white merging into purple at tips of segments: sepals long and prominent, petals small and not spread outwards. Lip is white. Flowering season is May to early June, but flowers rarely appear on the plants; in nature only two or three out of a hundred will be in bloom (Correll 1950) (Figs. 11.9 and 11.10).

Cherokee made an emollient with the roots of *Aplectrum hyemale* for use against headache and boils. Root tea was used for bronchial disease, and corms were fed to children to fatten them and

confer eloquence (Hamel and Chiltoskey 1975). Catawba Indians in Southeastern United States apply macerated woodstock and corms to boils (Correll 1978). Corms were worn on the arm as amulets in Southern United States. When dropped into water, they were used for fortune telling (Richburg 2003).

In a different habitat, one finds *Spiranthes lucida* (common name: wide-leaved ladies tresses). This is another terrestrial orchid that grows in Eastern United States and Canada. It is found in fens at low elevations, 165–800 m, besides streams and river banks, particularly those with calcareous substrate. Only a small shrub with three to four leaves, its flower stalk is erect, bearing small white flowers arranged in two ranks towards the tip. Cherokee infants were given a warm bath made with the orchid to ensure they grew strong and tall. Adult Cherokee employed the orchid to treat urinary problems.

Another inconspicuous bog orchid, *Liparis loeselii* (yellow wide-lip orchid), was also used by Cherokees to treat urinary problems, but this fact is not well known, and such usage does not constitute a threat to the species. Hydrologic factors, ditching and drainage of fens for hay production or to obtain water for cattle are the main threats to the existence of *Liparis loeselii* in North America today (Hamel and Chiltoskey 1975; Rolfsmeier 2007) (Figs. 11.11 and 11.12).

In contrast, *Platanthera ciliaris* (syn. *Habenaria ciliaris*; common names: yellow fringed orchid, orange fringe, orange plume, rattlesnake master) is a showy species that enjoys a wide longitudinal distribution from Ontario along the Eastern American states to Texas and Florida. It grows in wide range of habitats, in full sun or partial shade, with a preference for seepage hillside bogs and is dependent on periodic fires to create a favourable, well-lit habitat. It occurs from sea level at coastal plains to 1800 m in North Carolina and Tennessee (Luer 1975; Correll 1978). It is rare in Ontario and in the Northeastern American states but is globally secure (Sharp 2004). Plant is stout, 0.25–1 m tall, with lanceolate leaves, 7–30 by 0.6–6 cm, ensheathing the stem. A showy species, inflorescence is upright, raceme 3.5–20 cm with many flowers of golden

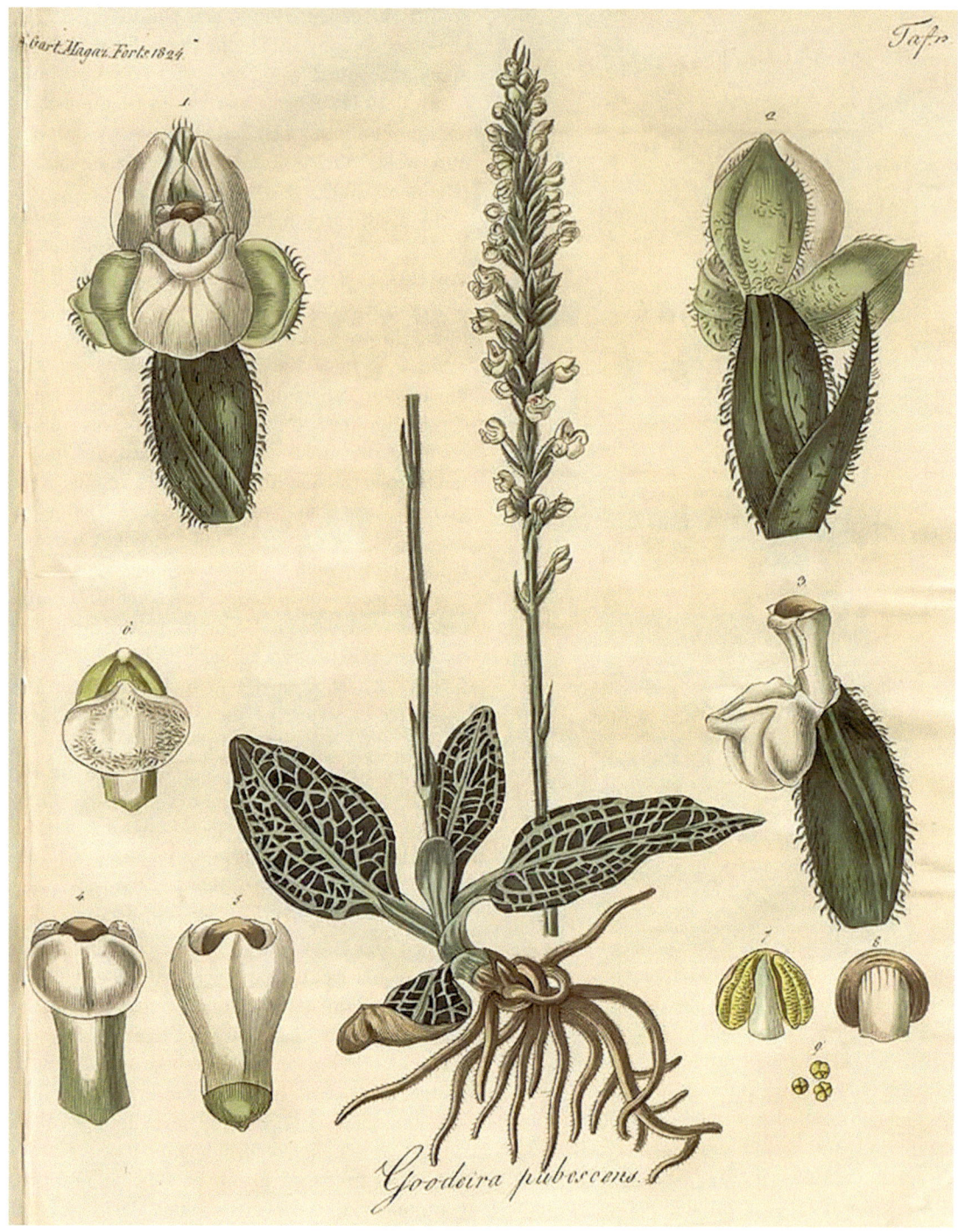

Fig. 11.8 *Goodyera pubescens*. From: *Allgemeinen teutschen* Garten-Magazin (JF Bertuch) vol. 8: t. 13 (1824)

Fig. 11.9 *Aplectrum hyemale*. From: Sharp H, *Watercolor sketches of North American plants especially New England* (1898) [H. Sharp]

Fig. 11.10 *Spiranthes lucida*. From: Sharp H, *Watercolor sketches of North American plants especially New England* (1888–1919) [H. Sharp]

yellow or orange appearing from late June to late September and staying good on the plant for 2 weeks. Henry Baldwin in1884 described it as a 'flaming orchids, a fit symbol of the wealth and glow of August' (Luer 1975). Roots are fleshy, thickened into tuberoids proximately, up to 4 long and 1.5 cm in diameter (Corell 1978). Cherokee remedies included cool root infusion of *Platanthera ciliaris* (yellow fringed orchid) for headache, and warm tea, drunk on the hour, for diarrhoea. Root of *Platanthera ciliaris* was a convenient bait for fishing (Hamel and Chiltoskey). Picking of its pretty flowers is a potential threat to the species (Sharp 2004) (Fig. 11.13).

When he named *Platanthera orbiculata* in 1813, F T Pursh stated that 'It is known in the mountains by the name of '*Heal-all*' (Correll 1978) without providing details. Plants carry a pair of large, shiny, succulent and green leaves that hug the ground. The species is distributed across Canada from Newfoundland to British Colombia and in Eastern United States in shady beech woods from Pennsylvania and Virginia (Luer 1975). Juice from the roots of *Platanthera dilatata* (syn. *Habenaria dilatata*; common names: bog candle, white bog orchid, tall white bog orchid; fragrant orchid, boreal bog orchid) was employed by the Newfoundland Micmac-Montagnais (Mik'maq) to treat 'gravel' or urinary stones (Lawler 1984). The Mik'maq are an ancient people who lived in Northeastern Canada and Maine. They were hunters and gatherers, and migratory, moving with the seasons but always living close to water. The tall white bog orchid whose inflorescence is up to 1 m tall would have been a familiar sight to the Mik'maq and an easy plant for them to locate. Plant is erect, to 1.2 m tall, sometimes taller with linear to lanceolate leaves, 30 m long and 5.5 cm wide, which ensheath the stem. Flowers are white, occasionally tinged with yellow or green and emit the fragrance of clove or vanilla. Flowering season is April to September depending on area. *Platanthera dilatata* is widely distributed across Canada south to the Great Lakes and also occurs in Alaska and Washington and Oregon (Luer 1975; Correll 1978). Another species of *Platanthera*

Fig. 11.11 *Liparis loeselii*. From: Thomas OW, *Flora van Duetschland Oesterreich und der Schweiz Taflon*, vol. 1: t. 135, Fig. A (1885)

Fig. 11.12 *Platanthera ciliaris.* From: Andrews HC, *The botanical repository* vol. 1: t. 42 (1797–1798)

(*P. hookeri*) is employed by the Mik'maq to treat wounds and cracks on their feet (Lawler 1984). *Platanthera hookeri* (syn. *Habenaria hookeriana*; common name: Hooker's orchid) is distributed in the American northeast and is considered threatened or extinct in some states of the United States but it is widespread on the Canadian side, occurring in woods of beech and maple, in semi-shade or at the edge of wet areas where few under-shade plants grow. Hooker's orchid has the distinguishing feature that its flowers resemble

upturned 'hooks' when the lips are seen in profile. The flower has also been likened to a gargoyle on account of the hood formed by the dorsal sepal and the long, upcurved lip which extends forwards like an elongated chin. Flowers appear from May to August (Correll 1978). A modest herb, in Canada, is only threatened by forest succession, deer grazing and acid rain (Reddoch and Reddoch 2007). *Platanthera dilatata* var. *leucostachya* (syn. *Habenaria leucostachya*, white rein orchid, Sierra orchid, Sierra bog

Fig. 11.13 *Platanthera orbiculata*. From: Hooker WJ, *Flora boreali Americana, or, the botany of the northern parts of British America* vol. 2: t. 200 (1840) [n.a.]

orchid) is distributed along the coastal states of Canada and the United States, occurring in wet meadows, seeping slopes, river banks and roadsides from sea level to 3400 m. It is distinguished from the other two varieties in that the spur is longer than the lip (Anonymous 2012). Nlaka'pamux, more commonly known as the Thompson River Indians, who are native Americans of British Columbia and Washington State, placed leaves of *Platanthera dilatata* var. *leucostachya* on stones in a hot bath to treat

rheumatism. They used *Goodyera oblongifolia* Raf. (syn. *Goodyera menziesii*) to assist childbirth (Lawler 1984) (Figs. 11.14 and 11.15).

As its name suggest, *Platanthera grandiflora* (syn. *Habenaria psycodes* var. *grandiflora*, *Habenaria fimbriata*; common names: large purple-fringed orchid, large butterfly orchid, greater purple-fringed orchid, plume royal) has the largest flowers in the genus. In an attractive plant, its light purple flowers emit the fragrance of lavender. Plants vary in height from 27 to 120 cm

Fig. 11.14 *Platanthera dilatata* [as *Habenaria dilatata*]. From: Hooker WJ, *Exotic Flora* vol. 2: t. 95 (1825)

Fig. 11.15 *Platanthera hookeri* [as *Habenaria hookeri*]. From: Sharp H, *Watercolor sketches of North American plants especially New England* (1888–1910) (H. Sharp)

and bear two to six narrow, pointed leaves, 13–24 by 2.5–6 cm along its stem. Flowering season is June to August. It occurs in wet areas from sea level to 2000 m in the northeastern part of North America from Newfoundland to the Great Lakes (Luer 1975). During the nineteenth century, powdered roots were used to treat worms (Lawler 1984) (Fig. 11.16).

Arethusa bulbosa (common names: dragon's mouth, bog rose, wild pink) is a rare, spectacular bog orchid distributed in temperate, Northeastern United States and Canada. It is in the endangered list compiled by the US Department of Agriculture. Corm is 0.5–1.5 cm in diameter. Plant is 15 cm tall. Inflorescence is upright bearing a single, terminal and pink flower, up to 6 cm tall with a

Fig. 11.16 *Platanthera grandiflora*. From: Mechan T, *The native flowers and ferns of the United States in their botanical, horticultural, and popular aspects*, vol. 1: t. 23 (1878–1879)

Fig. 11.18 *Arethusa bulbosa* on the United States postage stamp

Fig. 11.17 *Arethusa bulbosa*. From: Sprague I, Harvey AB, *Flowers of the field and forest*, p. 43, t. 4 (1880) [I. Sprague]

showy lip marked by white and yellow crests and purple spots. Flowering season is early May in its southern locations to early August in the Far North. A single grasslike leaf up to 12 cm long develops after the plant finishes flowering. Early settlers in North America used corm of *Arethusa bulbosa* to treat toothache (Luer 1975; Correll 1978). They were also employed for boils and tumours (Lawler 1984) (Figs. 11.17 and 11.18).

Bletia purpurea (syn. *Bletia verecunda, B. acutipetala*; common names: pine pink, purple Bletia, sharp-petaled Bletia) is found only in Florida in the United States and is under threat there, but it is widely distributed in Central and South America. A pretty pink species is also cultivated. Plant is terrestrial, sometimes saxicolus, with round corm 2–3 cm in diameter bearing four lanceolate, plicate leaves 20–90 cm long and a lateral inflorescence. Flowers are pink to deep purple. Tea made with dried corm of *Bletia purpurea* was used as a tonic or to treat stomachache and fish poisoning. Fresh corms, known as 'wild ginger' because they contained a bitter and irritating juice, were used for cuts and skin abrasions (Correll 1978) (Fig. 11.19).

Corallorhiza maculata (common names: spotted coralroot, large coralroot, many flowered coralroot) is a mycoheterotrophic orchid. It has no green leaves and is totally dependent on mycorrhiza (associated fungi) for its carbon supplies. In summer, inflorescences arise from coral-shaped rhizomes, breaking through the ground to a height of 15–45 cm, and each carries numerous, small brown flowers with white lips speckled with crimson spots. Colour of flowers and fruit is variable, and there is a deep purple form. *Corallorhiza maculata* is widely distributed in the United States and Canada occurring in acid soils in coniferous or deciduous forests. *Corallorhiza maculata* is the commonest coralroot; nevertheless, it is not common. When they have finished flowering, rootstocks may stay dormant for years resulting in plants disappearing from an area only to reappear several years later (Luer 1975) (Figs. 11.20 and 11.21).

Rhizome of *Corallorhiza maculata* was used by the Paiute and Shoshone Indians of Nevada as a diaphoretic, febrifuge and sedative. Tea made with dried stems was administered to patients with pneumonia to build up their blood (Correll 1978). *Corallorhiza maculata* was also used to treat worms (Lawler 1984).

Corallorhiza odontorhiza (syn. *Corallorhiza wisteriana*; common names: fall coralroot, small-flowered coralroot, dragon's claw, chicken toes) is also mycoheterotrophic. Occurring in

Fig. 11.19 *Bletia purpurea* [as *Limodorum tuberosum*]. From: Jacquin NJ von, *Icones plantarum rariorum* vol. 3: t. 602 (1786–1793)

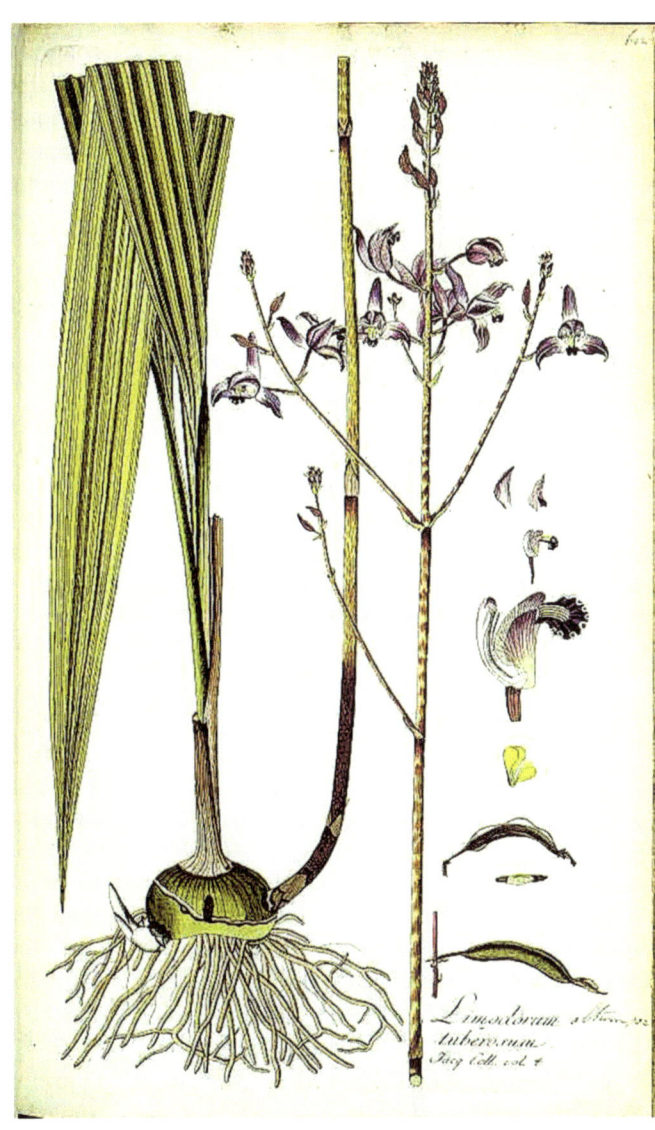

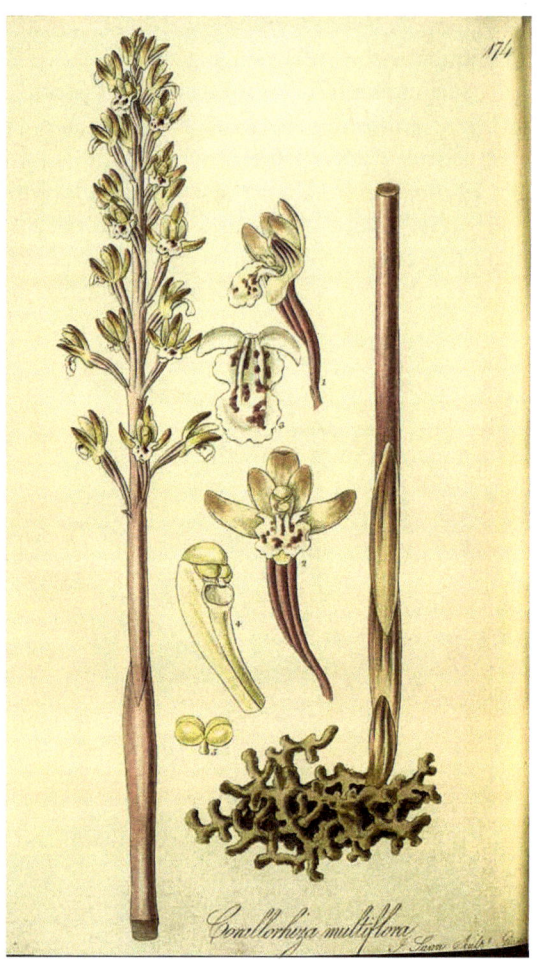

Fig. 11.20 *Corallorhiza maculata* [as *Corallorhiza multiflora*]. From: Hooker WJ, Exotic Flora vol. 3: t. 174 (1827)

Eastern United States excluding Florida, plant is small and frail, inconspicuous and the ugly duckling of the genus (Luer 1975). Inflorescence is short and sparsely flowered; the flowers are small, brown suffused with green. Pendent fruit are commonly found on newly emergent inflorescences, probably flowers self-fertilize. It flowers from August to October. Roots have a strong, peculiar odour. Roots and rhizome of *Corallorhiza odontorhiza* was also used as a diaphoretic to promote sweating and reduce fever, as sedative (Correll 1978) and to treat tumours, boils, skin disease, dysmenorrhea, scurvy, varicose veins, typhus and night sweats (Lawler 1984).

The Giant Helleborine (*Epipactis gigantea*) also known as stream orchid, chatterbox (on account of a distal-oscillating segment to its protruding lip), giant orchid and false lady slipper is a native of Western North America. It is abundant in limestone regions on the West Coast. It is found from sea level in California to 650 m in British Columbia to almost 3000 m in Colorado. It requires a damp habitat and becomes dormant during drought. Nevertheless, it is usually locally abundant, persistent and continues to flourish for decades in suitable habitats (Luer 1975). Plant is 30–100 cm tall, with numerous lanceolate leaves. Inflorescence is upright and carries three large, showy and green to pale yellow or brown flowers with purple veins on the sepals. Flowering season is March to August depending on location. California Indians living in Mendocino County used to boil the roots to treat mania and any serious illness that makes a person bedridden (Correll 1978) (Figs. 11.22 and 11.23).

Epipactis helleborine is a naturalized orchid which was introduced from Europe. It is a tough plant that survives in diverse habitats, appearing unexpectedly near homes, farms and ranches, earning it the common ignoble name, weed orchid. Inflorescence bears numerous nodding, green flowers. Early American immigrants used it to treat gout.

Malaxis unifolia (syn. *Microstylis ophioglossoides*; common names: green adder's mouth, tenderwort, Adder's tongue tenderwort) is a small, modest herb that is widely distributed in the eastern half of North America and the Caribbean, occurring in dense woods. Plant is 10–20 cm tall, the upright stem bearing a single ovate leaf, 3–10 by 2–7 cm, above the stem but ensheathing it up to the base. Stem is swollen near the base forming a globular pseudobulb, 2 cm in diameter. Inflorescence carries a dense cluster of 20–70, tiny, pale green flowers at the tip. Flowering may occur at any time of the year (Correll 1978). The plant is found in several habitats but has a preference for acidic and sandy soils (Ames 2006). Ojibwas, a seminomadic tribe which originally occupied Ontario, Michigan, Wisconsin, Minnesota and North

Fig. 11.21 *Corallorhiza odontorhiza*. From: Torrey J, *A flora of the state of New York* vol. 2: t. 126, Fig. 2 (1843)

Fig. 11.23 *Epipactis helleborine* (©Teoh Eng Soon 2019. All Rights Reserved)

Fig. 11.22 *Epipactis gigantea*. From: *Curtis Botanical Magazine* vol. 125 [ser. 3, vol. 55] t. 7690 (1899) [M. Smith]

Dakota, employed the root of *Malaxis unifolia* as a diuretic (Lawler 1984) (Fig. 11.24).

In this chapter I have only discussed medicinal orchid usage in the United States (excluding Hawaii) and Canada because the two countries form a distinct botanical biosphere. Their flora and ancient, indigenous culture is distinct from that of Meso-America (Mexico, Belize, Guatemala, El Salvador, Honduras, Nicaragua and Costa Rica) and South America. Therefore, Central and South American medicinal orchids are discussed in separate chapters.

Fig. 11.24 *Malaxis unifolia* [as *Malaxis ophioglossoides*]. From: *Edward's Botanical Register* vol. 15: t. 1290 (1829) [M. Hart]

References

Ames D (2006) Green-adder's-mouth (*Malaxis unifolia*). Native Orchid News 8(4): December 2006. Native Orchid Conservation Inc.

Anonymous (2012) Flora of North America (online). www.eFloras.org

Capps P (1909) McCrae and Halsey. JAMA 53:792

Correll DS (1978) Native orchids of North America, north of Mexico. Stanford University Press, Stanford

Hamel PB, Chiltoskey MU (1975) Cherokee plants their uses – a 400 year history. (Self published, address not stated)

Hand WD (1976) American folk medicine. University of California Center for Study of Comparative Folklore and Mythology, Pub. No. 4. University of California Press, Los Angeles

Karnick CR (1994) Pharmacopoeial standards of herbal plants, vol 2. India Sri Satguru Publications, New Delhi

Lawler LJ (1984) Ethnobotany of the Orchidaceae. In: Arditti J (ed) Orchid biology. Reviews and perspectives III. Cornell University Press, Ithaca, NY, pp 27–149

Luer CA (1975) The native orchids of the United States and Canada. New York Botanical Gardens, New York

Mergen DE (2006) *Cypripedium parviflorum* Salisb. (Lesser yellow lady's slipper): a technical conservation assessment. USDA Forest Service, Rocky Mountain Region. http://www.fs.fed.us/r2/projects/scp/assessments/cypripediumparviorum.pdf

Reddoch JM, Reddoch AH (2007) Population ecology of *Platanthera hookeri* (Orchidaceae) in southwestern Quebec, Canada. J Torrey Bot Soc 134(3):369–378

Richburg JA (2003) *Aplectrum hyemale* (Muhl. ex Willd.) Nutt. Puttyroot. Conservation and research plan for New England. New England Wild Flower Society, Framingham, MA. www.newfs.org

Rolfsmeier SB (2007) *Liparis loeselii* (L) Rich (yellow widelip orchid): a technical conservation assessment.

USDA Forest Service, Rocky Mountain Region, Lakewood, CO

Russell MH (1909) Trans Am M Assn, Sect: Pharm Therap, 204

Sharp PC (2004) *Platanthera ciliaris* (L.) Lindl. Yellow fringed orchid. (Online). www.newfs.org

Smith HH (1923) Ethnobotany of the Menomini Indians. Bull Public Mus City Milwaukee 4:1–174

Turner NC, Bell MAM (1971) The ethnobotany of the Coast Salish Indians of Vancouver Island. Econ Bot 25:63–99

Vogel VJ (1970) American Indian medicine. Univesity of Oklahoma Press, Norman

Economic benefit was the prime motivation behind European exploration of new lands during the fifteenth to seventeenth centuries. Spices being a commodity employed by both rich and the less well to do, the spice trade attracted much competition among traders and countries. Consequently, this generated a general interest in plants as an economic resource. It was matched by scholarly works resulting in the publication of herbals and renewed efforts at the classification of plants.

Unique plants received special attention. Later, the revolutionary ideas of Wallace and Darwin, and, in particular, the discovery of beautiful, exotic orchid flowers in the tropics and Darwin's work on orchid pollination, brought orchids into focus during the nineteenth century. Many important contributions on orchids made during this century are well known, but much of the work produced two to four centuries earlier laid hidden in private libraries until they were discovered in the late twentieth century.

The first important botanical work to reach Europe from the East was the *Hortus Indicus Malabaricus*, a massive encyclopaedia of 740 medicinal and other economic plants of the Western Ghats commissioned by Hendrik Adriaan van Rheede tot Drakenstein and published between 1678 and 1703. van Rheede (1636–1691) was an aristocratic military man who served as Commander of Dutch Malabar from 1669 to 1676, 8 years after the Dutch East India Company wrested Malabar from the Portuguese in order to have monopoly over pepper and cinnamon. During his tenure as Commander, van Rheede embarked on the historic botanical publication for which he is best remembered. Malabar botanists regard him with much fondness, and recently they have attached Rheede's name to several orchid species (Fig. 12.1).

van Rheede had no formal botanical training, only a great love for plants. The meticulously detailed accounts of trees and their epiphytes in his dairies attest to his love for nature. His interest as a naturalist spurred him to employ a staff of 25 people in 1674 to compile a compendium of useful plants in the Malabar region, and in 1675 the first draft of *Hortus Indicus Malabaricus* was completed. Rendered in Dutch, and then translated into Latin, Volume I was published in 1678. Latin was employed in the publication being the norm for scholarly works at that time. The undertaking did not end there, but it continued for three decades and involved nearly a hundred participants which included scholars, botanists, physicians, native healers, professors of medicine, clergymen, translators, illustrators and engravers, both Indian and foreign.

At the start, three Brahmins, Ranga Botto, Vinaique Pandito and Apu Botto, were requested

Fig. 12.1 Henrik Adriaan van Rheede tot Drakenstein (1636–1691)

word *maravara* appears to describe epiphytes, referring to both epiphytic orchids and ferns; however, this term was also employed for terrestrial orchids. Although 20 orchids are illustrated and described, the actual number of species is considerably lower according to contemporary classification.

The foxtail orchid, *Rhycostylis retusa*, is described twice, first as *Ansjeli maravara* (local name: *Ponoffou kely*). The description is accompanied by a line drawing showing a vandaceous orchid that bears a long inflorescence with numerous (over 40) spotted flowers, densely arranged around its rachis. Next it is labelled as *Biti maram maravara* (local name: *Giriy*) together with a drawing of a similar looking plant but with sparsely arranged flowers also in

to assemble all the plants mentioned in the ancient manuscript *Manhaningattnam* and to contribute whatever additional information they had on these plants. Palm-leaf manuscripts in the possession of Itty Achuden, a native Pegu physician, were made available to the scholars. Achuden himself contributed many native plant names, and he explained the reputed medicinal properties of several species. A Portuguese medical text based on a Malabarese tome compiled by Achudem's teacher, Coladda, was consulted by van Douet, the physician who prepared the translation of the *Hortus* into Latin (van Steenis 1948). *Hortus Indicus Malabaricus* was eventually published in 12 volumes in four languages— Latin, Sanskrit, Arabic and Malayalam. K S Manilal's recent translation of the text into English should now give this monumental, original contribution the recognition that it deserves.

In the final volume of the *Hortus*, 20 species of orchids and their medicinal usage are described. They include both epiphytic and terrestrial orchids. Beautiful, detailed line drawings accompany the descriptions, albeit, on occasion, the flowers were not drawn in sufficient botanical perspective to make for easy identification. The

Fig. 12.2 *Hortus Indicus Malabaricus* by Rheede tot Drakenstein, Hendrik van. Title page

good numbers around a longer inflorescence. The *Hortus* stated that along the Malabar Coast during the late seventeenth century, juice obtained from the whole plant was used to treat fits in infants, tics and spasms. When mixed with sugar, the extract also served as a remedy for headache, giddiness and fever. Palpitation was treated with ash prepared by incinerating the plant. The larger leaves were employed to treat urinary stones or to induce menstruation (van Rheede 1703) (Figs. 12.2 and 12.3).

The earliest mention in European literature of *Taprobanea spathulata* (syn. *Vanda spathulata*) being employed as a medicinal herb is in van Rheede's *Hortus Indicus Malabaricus*. *Taprobanea spathulata*, a lowland, monopodial orchid is commonly found in scrub scrambling over rocks and bushes. Stems are leafy but generally leafless at the bottom portion. Inflorescences are borne at leaf axils, upright, bearing golden, butter-coloured flowers. The species is native to Sri Lanka and Southern India. It has become rare. Rheede reported that the entire plant of this orchid, *Ponnampu maravara* (local name: *Suanna puspa*) was pounded, boiled with rice and coconut juice and then mixed with honey and administered to stop diarrhoea and dysentery. It also corrected biliary disorders. Pulverized flowers were used to treat tuberculosis, asthma and mania (van Rheede 1703). Later writers reported similar usages, but it is unclear whether they were merely quoting van Rheede or the orchid had continued to be used in such manner in the Malabar region. For instance, it was reported that powder prepared from the dried flowers was administered for consumption, asthma and psychiatric disorders (Dymock et al. 1893; Trivedi et al. 1980; Singh and Duggal 2009), whereas juice from the plant tempered bile and abated frenzy (Dymock et al. 1893; Caius 1936; Yoganarasimhan and Chelladurai 2000). Leaves were employed to treat consumption, asthma and mania (Nadkarni 1954; Duggal 1972) (Fig. 12.4).

Acampe praemorsa (syn. *Acampe papillosa*) is the commonest lowland orchid in the Western Ghats. Huge colonies of this robust, lowland, monopodial orchid thrive on roadside trees throughout Kerala. Hence, it is not surprising that it would come to the attention of van Rheede. *Thalia maravara*, the name mentioned in the *Hortus*, is derived from its original name, *Thaliyamaravara* in the Malayalam dialect, the spoken language of the Malabar region. *Acampe praemorsa* is also common on the eastern seaboard of the Indian subcontinent as well as in Sri Lanka. The plant was employed to treat bone ache and painful joints (van Rheede 1703). Its root was

Fig. 12.3 *Rhynchostylis retusa* [as *Ansjeli maravara*; local name *Ponooffou kely*] From: van Rheede, *Hortus Indicus Malabaricus* vol. 12: t. 1 (1703)

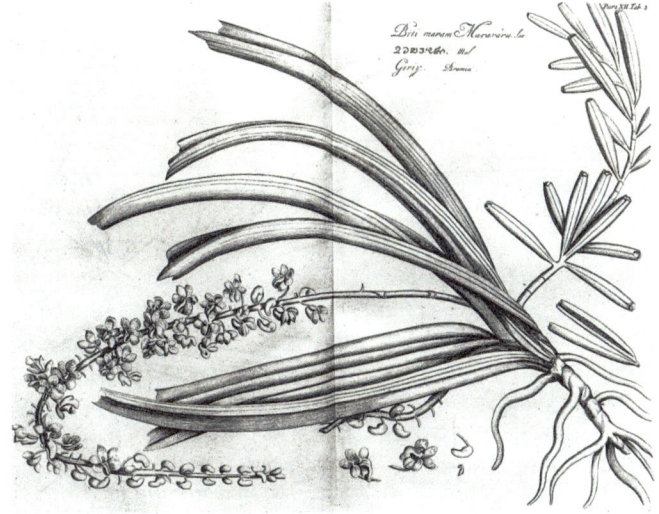

used interchangeably with root of *Vanda tessellata* (syn. *Vanda roxburghii*) by practitioners of Traditional Indian Medicine who had not bothered to examine the plants and flowers and did not know that the two species were different. They referred to both as *Rasna* (Dutt 1900) (Fig. 12.5).

Acampe praemorsa is still employed all across India in a similar manner (Rao and Sridhar 2007).

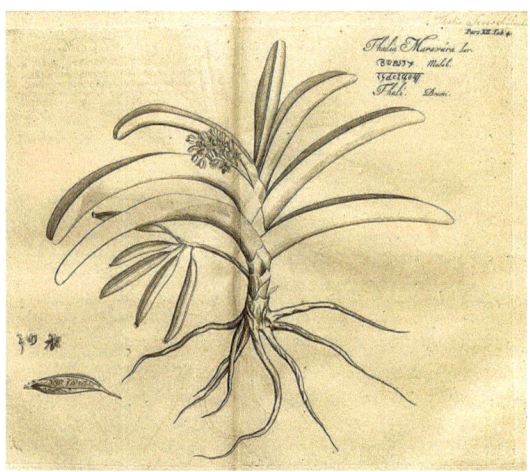

It is also used as a tonic in India and Bangladesh (Trivedi et al. 1980; Musharof Hossain 2009). In Andhra Pradesh on the Eastern Ghats, the Koya tribe prepares a paste made from pulverized whole plant of *Acampe praemorsa*, egg white and lime (calcium) to apply on fractured limbs in order to promote healing (Akarsh 2004). When they suffer from arthritis, members of the Dongria Kandh, a primitive tribe in Southeast Orissa, consume a tablespoon of a paste prepared from roots of the *Acampe praemorsa* and *Asparagus racemosus* (not an orchid) on an empty stomach, twice daily for 15 days to obtain relief (Dash et al. 2008). *Acampe praemorsa* is also used in Indian folk medicine to treat typhoid (Gupta and Tandon 2004).

Mav Tsjerou Mavarava or *Ambo keli* of the *Hortus* is *Cleisostoma tenuifolium*, a small epiphytic orchid which occurs in Southern India, Sri Lanka and Thailand from sea level to 300 m. Stem is short, erect and completely sheathed by the bases of long, linear, pointed leaves. Inflorescence is 5 cm long, bearing up to a dozen, small brownish flowers with a white lip. Although the plant is rare, it is known by a dozen names with many Indian authors calling it *Sarcanthus pauciflora* or *Sarcanthus peninsularis*. The drawing of *Kolli Tsjerou Mav-maravara* or *Ambo-tia* in the *Hortus* does not show any flowers, but it illustrates a diminutive, monopodial orchid which, in the local context, should be a *Cleisostoma*. It may be confidently identified as *Cleisostoma tenuifolium* because this is the only species of *Cleisostoma* that occurs on the Malabar Coast. The entire plant was made into a poultice to reduce pain and swelling of abscesses by promoting their rupture. 'Reduced to dust' (blended) with vinegar, it was administered to expel kidney stones or to treat dysuria, gonorrhoea, leucorrhoea and heavy menstrual loss (van Rheede 1703) (Fig. 12.6).

Rheede's *Anantali Maravara* is easier to identify. It is *Dendrobium ovatum*, a common, attractive, deciduous, white *Dendrobium* that is endemic in India. It occurs in open, deciduous forests in Bengal and southern India from sea level to 900 m (Santapau and Kapadia 1966;

Fig. 12.6 *Cleisostoma tenuifolium* [as *Koli Tsjerou Maravara*; local name *Ambolia*] From: van Rheede, *Hortus Indicus Malabaricus* vol. 12: t. 6 (1703)

Abraham and Vatsala 1981). First described as a medicinal plant employed for treating all sorts of aches and pain, especially tummy ache, by van Rheede in 1703, the plant was still used as an emollient to relieve stomachache more than 200 years later. Juice from a living plant was employed for this purpose. It acts as a laxative (Caius 1936). In the northeastern state of Uttar Pradesh, it is used as a tonic or antiphlogistic and employed to treat disorders of the chest and rheumatism (Trivedi et al. 1980) (Fig. 12.7).

Cymbidium aloifolium (Rheede's *Kaus Jiram Maravara*) is another common epiphyte flourishing as large clumps that cling to palm trees and exposed branches of numerous tree species in lowland forests throughout India and Southeast Asia. *Cymbidium bicolor* is generally considered a variety of this variable species. The plant, pounded with ginger (*Zinzibere*) and diluted, was used to induce vomiting and diarrhoea. *Cymbidium aloifolium* was employed to treat darkened vision, vertigo, paralysis and other diseases of old age (van Rheede 1703). The tubers are sometimes traded as salep and alleged to be aphrodisiac (Puri 1970) (Figs. 12.8 and 12.9).

van Rheede next described *Wellia-theka-maravara* or *Tolassi* which is *Pholidota imbricata*, sometimes referred to by its synonym, *Pholidota pallida*, or as the necklace orchid. *Valiya* (*wellia*) means 'big', and *theka* refers to the host tree (*Tectona grandis*, teak). An alternative name was *Mau maravara*, in this case referring to an alternative host tree, the mango (*Mangifera indica*, local name *mau*). *Pholidota imbricata* is a montane epiphyte with creeping, tightly spaced rhizomes and large, solitary, leathery, erect, plicate, greyish-green leaves. Long, pendent scapes arise from the tips of younger pseudobulbs, and each carries two lateral rows of small, white, poorly expanded flowers. Towards the close of the dry season, leaves arise in the axils of large bracts on the rhizomes, near the base of old pseudobulbs. An inflorescence emerges near the base of a leaf with the arrival of the rains. Gradually, as the season advances, the base of the leaf begins to swell, forming a pseudobulb below the attachment of the rachis of

Fig. 12.7 *Dendrobium ovatu*m [as *Anantali Matravara*] From: van Rheede, *Hortus Indicus Malabaricus* vol. 12: t. 7 (1703)

the fruiting inflorescence. Young pseudobulbs are ovoid and smooth, but with advancing age grooves appear. Mature pseudobulbs are conspicuously angulated (Santapau and Kapadia 1966) (Fig. 12.10).

On the Malabar Coast of India, crushed roots were applied on the shaven head or the entire plant on the feet to relieve fever. A poultice made from the entire plant was applied at the

waist to facilitate childbirth, induce menstrual flow and diuresis. Fruit eliminated chronic or malignant ulcers, was invigorating and got rid of one's biliousness (van Rheede 1703) (Fig. 12.11).

In contemporary traditional Indian medicine, pseudobulbs of *Pholidota imbricata* are finely macerated in mustard oil and applied on joints to relieve rheumatic pain (Rao 2004). In Nepal, pseudobulb juice is applied to boils (Manandhar

Fig. 12.8 *Cymbidium aloifolium* [as *Kansjiram Maravara*] From: van Rheede, *Hortus Indicus Malabaricus* vol. 12: t. 8 (1703)

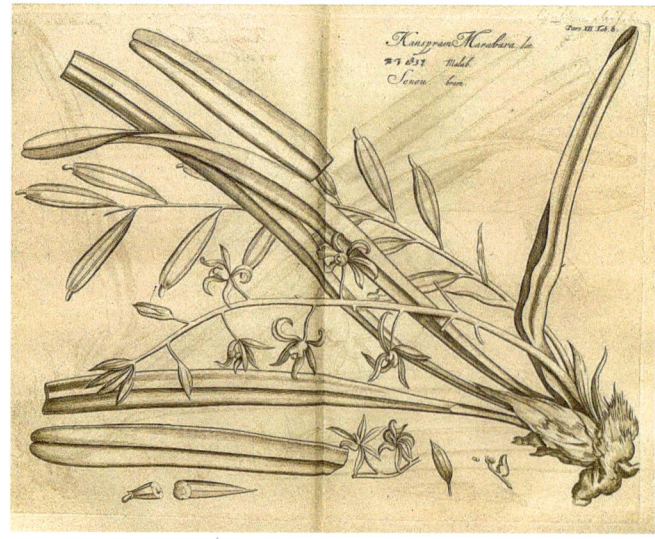

and Manandhar 2002), or to relieve abdominal and rheumatic pain (Pant and Raskoti 2013). Leaves and roots are made into a paste for treating fractures in Bangladesh (Musharof Hossain 2009). Juice of *Pholidota imbricata* may have some action on boils because it contains a 9,10-dihydrophenanthre derivative, imbricatin (Majumder and Sarkar 1982), a phytoalexin that is bacteriostatic (Arditti et al. 1975).

In his English translation of the *Hortus Indicus Malabaricus*, K.S. Manilal identified *Katou theka maravara* as *Eulopha graminea*, a widely distributed species which, nevertheless, was apparently very rare in Malabar during the sixteenth century because the authors reported that 'it had never been seen by local physicians'. The illustrated specimen was a gift from the King of Calicut to van Rheede. The plant grew in dense forests on old oak trees. Hence the name *Kathou* (wild) *theka* (teak) *maravara* (epiphyte). The alternative name *Maravarazhakka* appears in the Malayalam manuscript (Manilal 2003).

Whereas traditional physicians had not seen the plant, they seem to hold it in high regard when van Rheede showed it to them. It was claimed that its dried roots repelled serpents. Roots were ground into powder to treat poisoning

Fig. 12.9 *Cymbidium aloifolium* (©Teoh Eng Soon 2019. All Rights Reserved.)

Fig. 12.10 *Pholidota imbric*ata [as *Wella Theka Maravara*; local name *Tolassi*] From: van Rheede, *Hortus Indicus Malabaricus* vol. 12: t. 24 (1703)

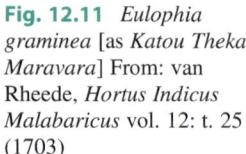

Fig. 12.11 *Eulophia graminea* [as *Katou Theka Maravara*] From: van Rheede, *Hortus Indicus Malabaricus* vol. 12: t. 25 (1703)

from dog bites and severe wounds. Pseudobulbs dried and roasted resolved abscesses. When cattle were fed by the plant, they became enraged (van Rheede 1703; Manilal 2003).

Hortus Indicus Malabaricus describes two species of *Eulophia* which some scholars consider to be the same species, but Manilal (2003) identified them as separate species. *Katou kayda maravara* (local name *Boin cadeki*) is *Eulophia epidendraea*, an Indian species found only in the southern India (Maharashtra, Kerala and the Eastern Ghats from Andhra Pradesh to Tamil Naidu) and Sri Lanka. Plant has large ovoid, green pseudobulbs, 7 by 2.5 cm which are above ground and numerous linear, grass-like leaves that measure 30–60 cm. Inflorescence is erect, simple or branched and carries many fragrant, white, pale green or yellow-green flowers with a white lip that is marked by pink veins and red crests. It occurs from sea level to 1000 m. It is the only terrestrial orchid found at sea level in southern India growing among rocks in scrub and grassland or dry deciduous forests (Santapau and Kapadia 1966). Widely differing flowering seasons are reported by different authors even from the same state. It flowers in April and May

in India (Bose and Bhattacharjee 1980; Reddy et al. 2005), 'November to December in the wild' possibly referring to Kerala (Abraham and Vatsala 1981), August to October in Kerala, November to January in Mumbai (Santapau and Kapadia 1966) and in February, June, August and October in Sri Lanka (Jayaweera 1981) (Fig. 12.12).

The *Hortus* stated that juice from the roots of Katou kayda maravara (bitter if the plant was growing on Kanjiram) promoted flow of bile and relieved abdominal discomfort. Juice of pseudobulb and leaves when mixed with canine blood healed burns and scalded skin. The powder neutralized poisons. Plants harvested from a Java tree had the ability to kill worms, relieve fever, strengthen the heart and expel flatus. Plant juice was blended to make an emollient to treat abscesses on the back. When applied it lessened pain and caused the abscesses to rupture. The plant was expensive in the seventeenth century (van Rheede 1703).

More recent publications report that the pseudobulb of *Eulophia epidendraea* is sometimes used as a vermifuge (Chopra 1933; Uphof 1968), demulcent and analgesic. It is applied externally

Fig. 12.12   *Eulophia nuda*
[as *Katou kaida maravara*;
local name *Boin cadeki*]
From: van Rheede, *Hortus
Indicus Malabaricus* vol.
12: t. 26 (1703)

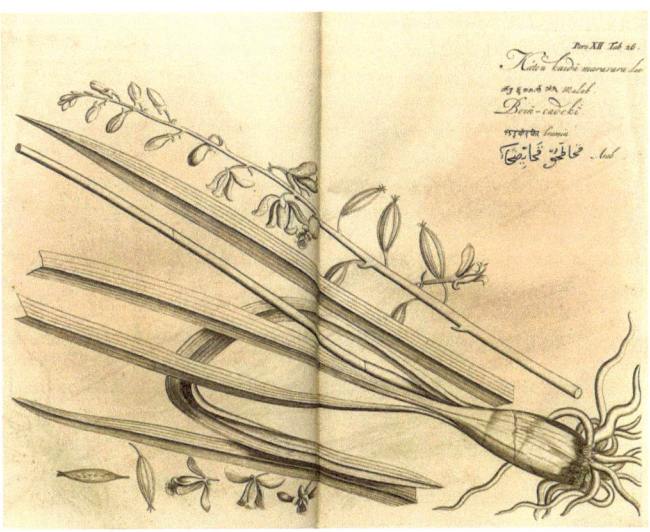

for muscular pain (Trivedi et al. 1980). Tubers of the orchid and two non-orchidaceous plants, *Withania somnifera* (local name: *Penneru gaddalu*) and *Curculigo orchioides* (local name: *Nela Taadi)* in the ratio of 2:1:1, are crushed with 'a sufficient amount of pepper and garlic', and the extract is administered once a day for a week to restore appetite in anorexic subjects in Anantapur district of southern India. An anthrax remedy from the same district employs 100 g of the tubers, with 50 g each of the fruit of *Terminalia bellirica*, *Terminalia chebula* and *Emblica officinalis* and 'a sufficient amount of pepper'. These items are crushed, and the mixture is administered orally, once a day for 15 days (Reddy et al. 2005). *Eulophia epidendraea* is a traditional remedy for tumours, abscesses, wounds and diarrhoea among the Yadav community of Tamil Nadu.

Beta-sitosterol, beta-sitosterol glucoside, beta-amyrin and luperol have been isolated from tubers of *Eulophia epidendraea*, and four flavonoids (apigenin, luteolin, kaempferol and quercetin) were isolated from the leaves. Beta-sitosterol and its glucoside are common in plants. They exhibit anti-inflammatory activity in vitro (Maridass and Ramesh 2010), but there is no data on animals or humans.

In a separate volume, Rheede describes and illustrates *Ela Pola* which grew in sandy soil. It is terrestrial in contrast to the other two *Eulophia* species which grew on trees. Therefore the term maravara was not employed. The name *Ilapola* (Malayalam *ila* leaf; *pola* sheath) employed in the Malayalam manuscript is not in current usage, but no new name has been given to the species. The orchid is *Eulophia nuda* (commonly referred to by its pseudonym, *Eulophia spectabilis*) (Manilal 2003). This widespread terrestrial orchid is distributed throughout southern China, India, Myanmar, Thailand, Indochina, Malaysia, Sumatra, Borneo and the Philippines. It thrives in the lowlands to an altitude of some 900 m, in open, grassy or lightly shaded areas. It often occurs in open, disturbed habitats (Wood et al. 2011). Inflorescence is erect, 30–80 cm tall and bears a dozen, well-spaced, attractive flowers which open about five at a time. Comb-like appendages are present on the lip (var. *andersonii*). Flower colour varies from region to region. Flowers of *Eulophia nuda* are purple in the Western Ghats (Abraham and Vatsala 1981). Floral and leafy shoots appear simultaneously. Leaves are still young when the plant is in bloom (Fig. 12.13).

Fig. 12.13 *Eulophia nuda*
(syn. *Eulophia spectabilis*)
[as *Katou kaida maravara*]
From: van Rheede, *Hortus
Indicus Malabaricus* vol.
12: t. 36 (1703)

An ointment prepared with the roots and leaves of *Eulophia nuda* together with fresh turmeric (*Curcuma*) is applied to the temples to relieve 'cephala' (van Rheede 1703; Manilal 2003).

Ayurvedic practitioners claimed that consuming pseudobulbs of *Eulophia nuda* (syn. *Eulophia spectabilis*) improved one's appetite. They employed the tubers to treat tuberculous glands in the neck, other tumours, bronchitis and diseases of the blood. It was an antidote for poisoning. A fresh poultice made from a living plant was applied to boils and abscesses to get them to point and drain. A powder made from the tubers was given for intestinal worms (Caius 1936). Seeds were used to treat worms and scrofula (Nadkarni 1954).

The orchid is used to treat stomachache and related complaints in the Nicobar Islands (Dagar and Dagar 2003). Today, the tubers serve as an appetizer for tribal dwellers at Kudremukh National Park in Karnataka who additionally employ them to treat tumours and bronchitis (Rao 2007).

Tubers of many *Eulophia* species, in particular *Eulophia nuda* (syn. *Eulophia spectabilis*), are regarded as an aphrodisiac in India (Dymock et al. 1893; Kumar 2003). They are boiled and eaten with flowers of *Madura longifolia* var. *latifolia* (which is not an orchid) (Kumar 2003). Paradoxically, the Dongria Tribe in Orissa believes that a mixture of dried pseudobulbs of *Eulophia nuda* when suitably combined with other herbs (10 g dried tuber and 5 g dried leaves of *Withania somnifera*, 5 g dried leaves of *Curculigo orchioides* and 5 g black pepper, pounded and mixed with water, to be consumed daily for 20 days) is effective as an antiaphrodisiac (Dash et al. 2008). *Manya* and *Goruma* were used to treat bronchitis, tumours and worms (Duggal 1972). *Manya* refers to 'neck', and the orchid is so-named because of its alleged resemblance to the scrofulous glands of the neck for which it was considered an appropriate treatment (Dymock et al. 1893). It is also employed as a blood purifier (Trivedi et al. 1980). Rao and Sridhar (2007) working in Karnataka in Western Peninsular India reported that the decoction prepared from the tubers was called *amarcana* (Sanskrit), and it was employed not only as an appetizer but also to treat tumours and bronchitis. It was incorporated into many medicinal formulations. The Dongria hill tribe in Orissa uses its leaves in decoction as a vermifuge

(Dash et al. 2008). Tubers are employed in Bangladesh as vermifuge, tonic, an antidote for poisoning, and to treat bronchitis (Musharof Hossain 2009).

Eulophia nuda (syn. *Eulophia spectabilis*) contains at least nine phenanthrenes including eulophiol and nudol. One of the phenanthrenes showed some anticancer activity but only in laboratory tests (Shriram et al. 2010). It is unlikely to have practical application.

Bela Pola is *Geodorum densiflorum*, a terrestrial, moisture-loving species with characteristic nodding inflorescence bearing a cluster of small, pink or white flowers near the apex. In the later seventeenth century, it was known in the Malabar region as *Vellapola* (*vella* water; *pola* sheath), an allusion to its damp habitat. But this Malayalam name is no longer in use (Manilal 2003). *Geodorum densiflorum* is widely distributed from India and Sri Lanka eastwards to the Ryukyu Islands, the Philippines and Queensland, Australia. Its pseudobulbs were made into a paste to relieve phlegm or to treat abscesses and other swellings (van Rheede 1703; Manilal 2003) (Figs. 12.14 and 12.15).

There is some difficulty in identifying the next two species in the *Hortus* which are rare plants with confusing names. Presently, *Baasala poulou maravara* or *Keli* has been tentatively identified

Fig. 12.15 *Geodorum densiflorum* (©Teoh Eng Soon 2019. All Rights Reserved.)

as *Liparis rheedei* (syn. *Malaxis rheedii*), formerly named *Malaxis plicata* by William Roxburgh who authored *Flora of India* in 1820. John Lindley in 1830 identified it as *Malaxis*

Fig. 12.14 *Geodorum densiflorum* [as *Bela Pola*] From: van Rheede, *Hortus Indicus Malabaricus* vol. 12: t. 35 (1703)

vesicolor. It was also known as *Malaxis odorata* Willd. Dennstedt (1818) identified *Katau parnam maravara* of van Rheede as *Malaxis odorata*, whereas he considered *Baasala poulou maravara* to be *Malaxis rheedii.* It is evident that this genus complex needs to be studied and revised.

Already in the *Hortus*, it was commented that plants belonging to this genus were very similar. Indeed there is still much confusion over individual identities in the entire *Liparis-Malaxis* complex (Chase et al. 2015). Dash, Sahon and Bal (2008) refer to the species as *Seidenfia rheedei.* *Seidenfia* is an endemic Indian terrestrial genus named by Szlachetko, but the term is not listed by the Royal Horticultural Society (RHS) nor in the International Plant Nomenclature Index (IPNI). There are nine species within this newly proposed genus which has been separated from other genera in the *Liparis* Tribe on account of the shape of the lip which lacks the auricles protruding backwards on both sides of the column and the fringed apex (Seidenfaden 1999). Currently, the correct name for *Seidenfia rheedei* should be *Crepidium resupinatum* (RHS World Checklist of Plant Names; www.theplantlist.org).

Crepidium resupinatum (*Liparis rheedei*) is a terrestrial herb with pseudobulbs close to one another on a 10–16 cm long rhizome, each bearing 3–5 ovate, acute, plicate green leaves 15–25 cm by 4–12 cm and an ensheathing petiole of 3 cm length. A non-flowering plant resembles *Malaxis latifolia.* Inflorescence is 20–45 cm tall, erect bearing numerous flowers that open gradually from the base. Flowers are green or red, or bicolored, usually changing from green to red as they age. They are 1 cm across. Petals and sepals are filiform. Lip is dark purple, oblong with a central groove and recurved at its midpoint (Fig. 12.16).

The colour of the plant and its flowers is dependent on the amount of light that it receives. Plants growing in the shade are deep purple whereas those growing in the light are a pure green. On an inflorescence leaning towards light on the roadside, the lowermost flowers still in shade are purple, the middle ones are yellow, and the uppermost flowers receiving full sunlight are green. All flowers on plants that grow in the deep shade of the forest floor are a deep purple (Santapau and Kapadia 1966).

Crepidium resupinatum (*Liparis rheedei*) occurs in mountain forests from southern India, across Thailand, Malaysia and Indonesia to New Guinea at 450–1500 m (Comber 2001), growing

Fig. 12.16 *Crepidium resupinatum* [as *Baasala poulou maravara*] From: van Rheede, *Hortus Indicus Malabaricus* vol. 12: t. 12 (1703)

at the forest floor, in humus, in deep shade. It flowers in July to August at Karnataka and the Western Ghats (Santapau and Kapadia 1966), July at Nilgiris (Joseph 1982) and from August to September in Tamil Nadu (Matthew 1995; Seidenfaden 1999).

van Rheede suggested that this orchid should be cultivated in medicinal gardens because it was difficult to locate in the forest. It was employed to clear bile, relieve fever, control infantile fits, treat measles and promote sweating (van Rheede 1703). Today, it is used as a tonic in Karnataka (Rao 2007). It is one of the eight ingredients of the much admired Ayurvedic tonic known as *Ashtavarga*, another being the orchid, *Malaxis muscifera* (LIndl.) Kuntze (syn. *Microtis muscifera* Ridl.) (Rao 2004). Hill tribes in Orissa use the root to treat cholera. For this purpose,

approximately 250 g of root is decocted in 1000 ml of water until the volume is reduced to a third of its original. After cooling, 5 ml of the decoction is mixed with 2 ml of honey and orally administered twice a day on an empty stomach for 15–21 days (Dash et al. 2008) (Fig. 12.17).

Katou Ponnam Maravara has been frequently identified as *Liparis nervosa*. Sathish Kumar (personal communication) opined that the correct species is *Liparis odorata* (syn. *Liparis paradoxa*), the only species occurring on the Malabar Coast. *Liparis nervosa* is not native to this region.

In illustrations in *The Native Orchids of Japan*, *Liparis odorata* appears to prefer a more exposed habitat than *Liparis nervosa*: its raceme is longer with numerous flowers, whereas *L. nervosa* is few flowered. Tepals are a light purple. Lip is a dull yellow. The species is distributed in forests and grassy slopes at 600–3100 m in Taiwan, Fujian, Guangdong, Hong Kong, Guangxi, Guizhou, Jiangxi, Hubei, Hunan, Zhejiang, Yunnan, Sichuan and in India, Nepal, Bhutan, Myanmar, Thailand, Vietnam, Japan and the Pacific Islands. It flowers from April to July or August in China (Chen and Wood 2009; Jin et al. 2009) and July to September in India (Bose and Bhattacharjee 1980).

Liparis odorata was employed to treat elephantiasis in Malabar (van Rheede 1703; Rao and Sridhar 2007). This disease is characterized by gross swelling of the lower limbs due to lymphatic blockage caused by a filarial worm which is transmitted by mosquito bites. (This should be easy to test by using Sydney Brenner's roundworm, *Caenorhabditis elegans*.) Tribals in Karnataka offer pseudobulbs of *Liparis odorata* for sale among their tribal medicines (Rao 2007). Juice extracted from the leaves was used to treat fever and oedema and juice from the roots to treat burns, inflammation, gangrene and tumours (Dalgado 1896, 1898, both quoted by Lawler 1984). This practice still exists in northwestern India (Medhi and Chakrabarti 2009).

Liparis odorata appears to be a new addition to the Chinese pharmacopoeia because its usage is described only in the 2000 edition of *Zhonghua bencao* (Hu et al. 2000). The herb removes 'wind' and dispels 'dampness'. It is employed to treat

Fig. 12.17 *Liparis nervosa* [as *Katou Ponnam Maravara*; local name Pon kely] From: van Rheede, *Hortus Indicus Malabaricus* vol. 12: t. 28 (1703)

Fig. 12.18 *Bulbophyllum
sterile* [as *Theka Maravara,
Rou kely*] From: van
Rheede, *Hortus Indicus
Malabaricus* vol. 12: t. 22
(1703)

Fig. 12.19 A miniature
Bulbophyllum [as *Tsjerou
Theka Maravara*] From:
van Rheede, *Hortus Indicus
Malabaricus* vol. 12: t. 23
(1703)

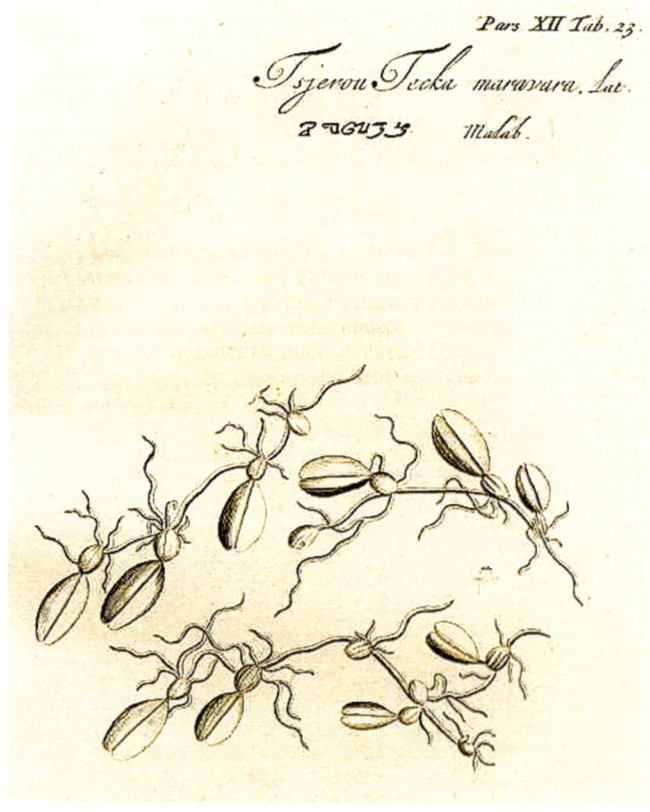

flu-like symptoms, peripheral neuritis, leucorrhoea, discomfort at the waist, ulcers and swellings. The herb is prepared by boiling 6–15 g of sliced, dried pseudobulbs (Hu et al. 2000).

Two species of *Bulbophyllum* were described in the *Hortus. Theka maravara, Zuka keli, Rou kakely* is *Bulbophyllum sterile* (Lam.) Suresh, based on the accompanying illustration and the natural occurrence of this epiphytic-saxicolous orchid. The commonest *Bulbophyllum* species at low elevations in Peninsular India grows abundantly on trees and moist rocks. Its appearance resembles that of *Bulbophyllum careyanum*, and flowers are foul-smelling. It is distributed in Kerala, Tamil Nadu, Karnataka, Goa, Maharashtra and Northwest India (Figs. 12.18 and 12.19).

The plant was washed and boiled to treat all forms of cold. It was pounded and mixed with leaves of *Cassia fistula* (Golden Shower tree; *Raja Kayu*), ginger and other ingredients and then dried and toasted over a fire, and the resultant mix was used to treat herpes, shingles, smallpox and plague. Powdered fruit was mixed with the oil of a nut (*Nucis indica*) to promote the flow of urine (perhaps meaning to overcome urinary obstruction rather than anuria). Juice was administered to improve hearing and deafness caused by trauma (van Rheede 1703). Currently, chopped pseudobulbs are boiled in coconut oil to prepare an embrocation to treat rheumatism. Plant paste is applied on swellings (Shanavaskhan et al. 2012).

Tsjerou thecka maravara, *Theka maravara minor, bonka kely*, is a smaller *Bulbophyllum* species, recently identified as *Rhytionanthos rheedei* (Manilal & Sathish Kumar) Garay. A small, creeping epiphyte, the species is found only in the states of Kerala and Karnataka at 800 m. It is uncommon. Inflorescence is an umbel with 2–4, shoe-shaped flowers, 7 mm in length. Flowers are not fragrant. Flowering season is July. Medicinally, it is a substitute for *Bulbophyllum sterile* (van Rheede 1703).

Jean Ferdinand Caius (1936)

After a lapse of over 200 years, a Jesuit monk, Rev Fr. Jean Ferdinand Caius (1877–1944), published a new report on the medicinal and poisonous plants of India, this time in English. Father Caius was a professor of Chemistry at St. Joseph's College, Trichy (1911–1922), director of the Chemistry Department at St. Xavier's College, Bombay, and subsequently biochemist in charge of the Pharmacological Laboratory of the Haffkine Institute (1924–1932). He was particularly interested in snake bites on which he carried out extensive testing with more than 300 - Indian remedies, including two orchid species, *Flickingeria fimbriata* (sic) and *Vanda tessellata*. His conclusion was that all local remedies were ineffective against cobra and viper bites (Caius 1936) (Fig. 12.20).

The medicinal and poisonous orchids of India that Caius described belonged to 12 genera: *Acampe, Cymbidium Dendrobium* (including *Desmotrichum* and *Flickingeria*), *Eulophia, Habenaria, Hetaeria, Luisia, Oberonia, Orchis, Rhynchostylis, Saccolabium (Acampe), Vanda, Vanilla* and *Zeuxine*. In the present discussion, orchids already discussed by van Rheede (1703) are omitted unless there was new information.

Of salep, Caius mentioned that Indian salep came from the hills of Afghanistan, Baluchistan, Persia and Bokhara, also, in part, from the Nilgiri Hills and Ceylon. *Eulophia nuda* (syn. *E. spectabilis*), *Eulophia dabia* (syn. *E. campestris*), *E. pratensis* (endemic in the Western Peninsular India) and *E. herbacea* (distributed from Western Himalayas down to the western peninsula) were sold as salep in the Indian bazaars. They were much prized by Indians as a tonic and aphrodisiac. *Habenaria commelinifolia, Orchis latifolia* and *Zeuxine stratoumatica* were also presented as salep. Although scientific opinion concurred that salep was devoid of medicinal value, in the mid-1930s local faith in its efficacy was 'unshaken and the

Fig. 12.20 *Vanda tessellata*. From Warner R, Williams BS, *The orchid album* vol. 1883: t. 59 (1883) [JN Fitch]

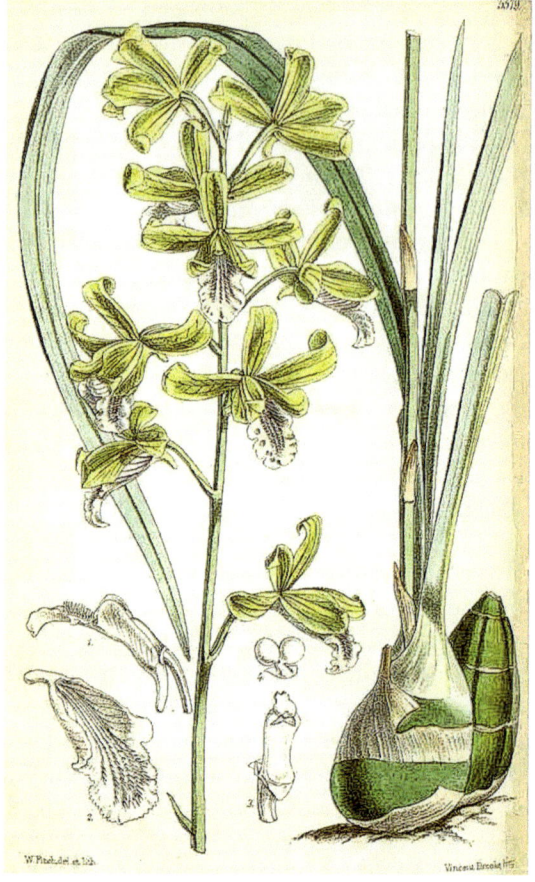

Fig. 12.21 *Eulophia pratensis* [as *Eulophia virens*] From: *Curtis Botanical Magazine* vol. 92 [ser.3, vol. 22] t. 5579 (1866) [WH Fitch]

dry tuber has an immense reputation as an aphrodisiac, restorative and fattener. It is much used by the *vaids* and *hakims* in conjunction with other nervine tonics'. Substitution was prevalent in the herb markets. 'Wealthy Orientals are known to have paid a handsome price for pounded potatoes and gum. 'Royal Saḷep', said to be much used as food in Afghanistan, has been identified as the product of a liliaceous plant, *Allium macleanii* Baker' (Caius 1936) (Figs. 12.21 and 12.22).

A decoction of *Dendrobium macraei* together with other herbs was recommended by Sanskrit writers as a remedy for disorder of bile, blood and phlegm. Plant was also used a stimulant and tonic for debility due to seminal loss. The fruit was an aphrodisiac. *Luisia tenuifolia* which occurs in the Western Peninsula and Sri Lanka formed the basis of an emollient applied to boils, abscesses and tumours. Malays used *Hetaeria obliqua* to prepare a poultice for sores (Caius 1936) (Fig. 12.23).

Caius commented that orchids were remarkable for being generally harmless as a group, and only a handful were found to possess toxic chemicals. Leaves and stem of the Malayan *Vanilla griffithii* contain a latex that caused skin irritation. Dermatitis has been reported by workers handling beans of *Vanilla planifolia* (Teoh 2016).

After Indian Independence

Even today, many tribal communities in India have poor access to modern medicine, so they are ideal populations for studies on native knowledge of medicinal herbs. Being isolated, when

Fig. 12.22 *Habenaria commelinifolia* From: *Annals of the Royal Botanic Gardens, Calcutta* vol. 9: t. 134 (1888–1921) [R Pantling]

they fall sick, tribals have to depend on herbalists, witch doctors and medicine men to treat their illnesses. Since many ethnic groups, and particularly the nomads, believe that illness is caused by the wrath of gods, evil spirits, witchcraft or sorcery, medicine men often assume the role of shamans. The use of botanicals is not inconsistent with such belief.

Contact with the west has not completely removed such beliefs. On the other hand, the availability of modern medicine in the form of pills, cream, ointments and plasters have diminished the need for a total reliance on remedies provided by the forests. Coupled with the fact that healers are neither well-paid nor well-respected, young tribals are not keen to study herbal medicine. During the past 30–40 years, several ethnobotanical studies were undertaken among tribals with a view to gathering their ethnobotanical knowledge before it disappeared completely, as

Fig. 12.23 *Hetaeria obliqua*. From Blume CL, *Collection des Orchidees les plus remarkquables de l'archipel Indien et du Japon*, t. 34, Fig. 1 (18858) [AJ Wendel]

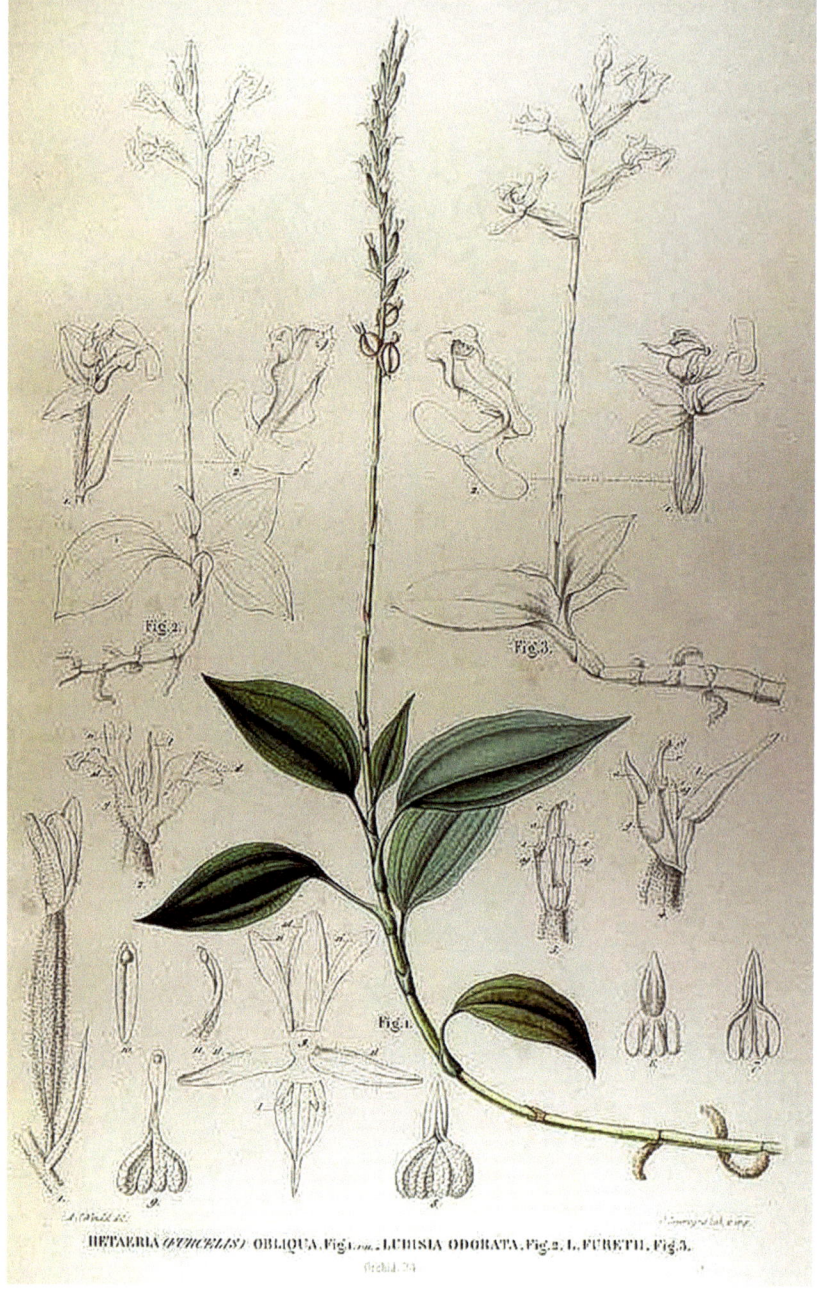

well as to conserve the orchids, many of which are under serious threat. Many investigators found that some tribes are already ignorant about the use of medicinal orchids.

In Indian Himalaya, *Ashtavarga*, a concoction of eight herbs that include four orchids is a popular item of folk medicine. It is employed to manage sexual dysfunction, pain, fever, urinary problems

Fig. 12.24 *Crepidium acuminatum* [as *Microstylis acuminata*] From: *Annals of the Royal Botanic Gardens, Calcutta* vol. 8 (2): t. 18 (1891) [R Pantling]

and the slow the progress of aging. It is claimed that this usage can be traced to ancient medical texts dating as far back as the seventh to eleventh century (Indian Medieval Period), but such texts are vague and mysterious in their description of many medicinal plants. Nevertheless, Indian ethnobotanists agree that the four orchid species in *Ashtavarga* are probably *Crepidium acuminatum* (syn. *Malaxis acuminata*), *Malaxis muscifera* (syn. *Microstylis muscifera*), *Platanthera edgeworthii* (syn. *Habenaria edgeworthii*) and *Habenaria intermedia* (syn. *Habenaria arietina*). All four species are rare terrestrial orchids found only at elevations of 1200–4000 m in Himalaya (Teoh 2016) (Figs. 12.24, 12.25, 12.26, 12.27 and 12.28). The four non-orchidaceous plants in *Asthavarga* are *Roscoea procera*, *Fritilaria royeli*, *Polygonum cirrifolium* and *Polygonum verticilliatum* (Mohanty et al. 2015).

The four orchid species employed in *Asthavarga* grow in the moist shady habitats in temperate Nagdev forests in Pauri Garhwal, Uttarakhand. They generally appear during the rainy season and are in bloom in July and August. Three more medicinal species that have a more extended flowering season also occur in the forests, namely, *Satyrium nepalense* whose dried tuber is used as tonic and to treat malaria and dysentery, *Herminium lanceum* whose tuber is employed to relieve difficult urination and *Goodyera repens* which is employed in decoction for a variety of complains—toothache, urine infection, insect bites, injuries, menstrual irregularity and poor appetite (Khajuria et al. 2017) (Figs. 12.29 and 12.30). Given the possibility

Fig. 12.25 *Malaxis muscifera* [as *Microstylis muscifera*] From: *Annals of the Royal Botanic Gardens, Calcutta* vol. 8 (2): t. 25 (1891) [R Pantling]

that a *Flickingeria* (*Dendrobium*) species might be the mythical *Sangeevani* (the Himalayan herb which restored life to mortally wounded Lakshmana in the *Ramayana*), a hundred truckloads of the orchid was once imported from Nepal to prepare an aphrodisiac, by analogy, restore life to a flaccid body part (Teoh 2016).

Pseudobulbs, roots and rhizomes of *Cephalanthera ensifolia*, *Habenaria acuminata*, *Habenaria susannae*, *Orchis latifolia*, *Microstylis wallichii* and *Pholidota articulata* are boiled, curried or rendered into a drink for consumption by ethnic tribes in northeastern India (Medhi and Chakrabarti 2009; Mohanty et al. 2015).

Tribals in Northeast India have extensive knowledge of traditional herbal medicine, and they are familiar with orchids. In this region, 15 native orchids are employed in traditional medicine (Medhi and Chakrabarti 2009). *Dendrobium fimbriatum* (local name *Mokya tu*) is employed to treat liver disorders and nerves. The entire plant of *Conchidium musicola* (syn. *Eria muscicola*) is used for diseases of the chest, eye, ear and nerves. Leaf paste is applied to cuts and wounds. Kamti tribals consume approximately 5 g of freshly grounded leaves and fruit of *Cymbidium aloifolium* (local name *Mok Hang Meew*) twice daily for 1–6 months to enhance their memory or to treat epilepsy, and if the condition did not respond, they would continue the treatment for a year. Alternatively epilepsy may be treated by administration of 2 g *Cymbidium bicolor* (local name *Mok Hang Meew*) root powder twice daily for 1–6 months. This treatment is believed to be also suitable for treating depression. Paste made with flowers of the orchid is

Fig. 12.26 *Platanthera edgeworthii* [as *Habenaria edgeworthii*]. From: *Annals of the Royal Botanic Gardens, Calcutta* vol. 9 (2): t. 139 (1906) [H Hormusji]

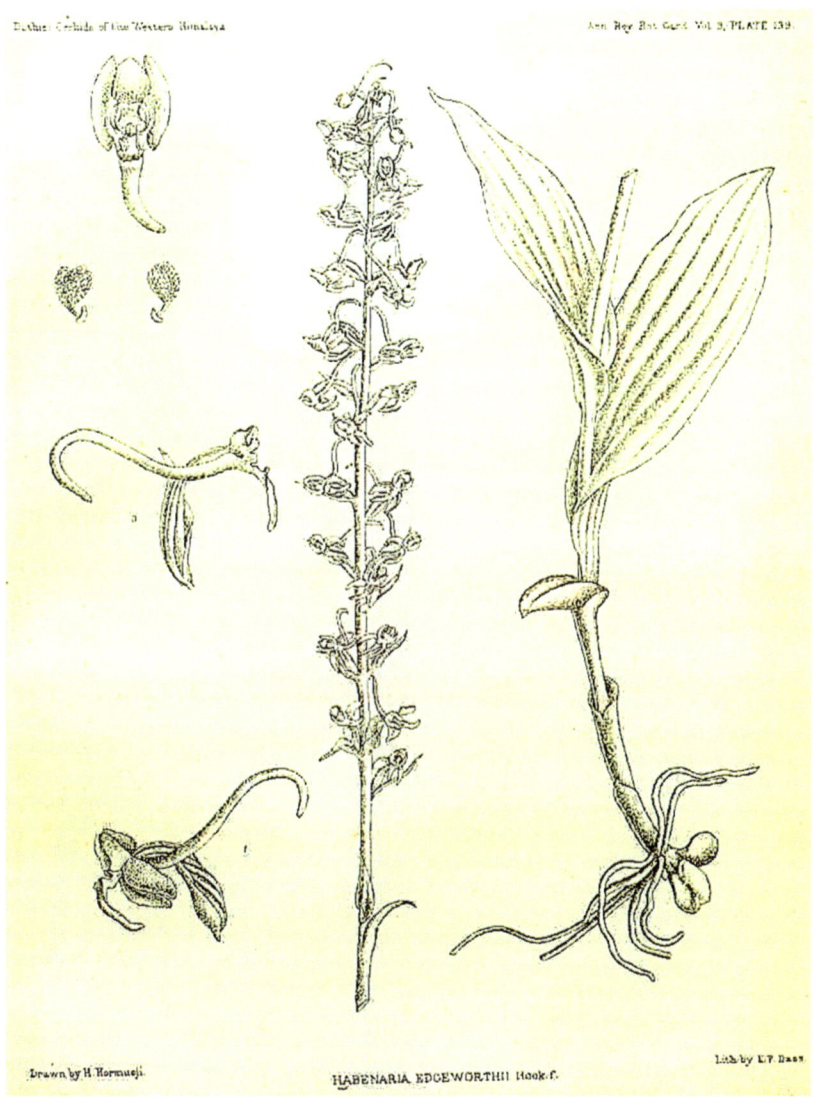

applied to burns on the face and to lighten skin pigmentation. Leaf paste of *Cymbidium bicolor* is applied to painful joints and inflamed skin. Leaves of *Mycaranthes pannea* (syn. *Eria pannea* local name *khadla*) is rendered into a paste which is applied to fractured bones for 2–6 months and to wounds and inflamed skin for a shorter period of 7–10 days (Chowlu et al. 2017). Leaf paste made with *Papilionanthe teres* is applied to the forehead to relieve fever; a stem is fastened to the loin to protect against cold and cough (Medhi and Chakrabarti 2009) (Figs. 12.31 and 12.32).

Tubers of *Eulophia nuda* (syn. *E. spectabile*) is a valued as a tonic and aphrodisiac, but it is also a remedy for cough, stomachache and paralysis. Tuber of *Orchis latifolia* is also a nerve tonic and aphrodisiac. Juice of *Liparis odorata* is applied to burns, ulcers and gangrenous parts. Pseudobulbs of *Phaius tankervilliae* are used in dysentery and fractures. *Vanda coerulea* is used to treat diseases of the eye and leaf extract for skin disease and diarrhoea. Leaves of *Vanda cristata* provide a tonic. *Vanda tessellata* has many uses: poultice for fever, ear drops to relieve earache and roots

Fig. 12.27 *Habenaria intermedia.* From: *Annals of the Royal Botanic Gardens, Calcutt*a vol. 9 (2): t. 131 (1906) [H Hormusji]

for pain relief, arthritis, nerve disorders, syphilis and poisoning. In Meghalaya, juice of *Dendrobium moschatum* leaf is used as eardrops. The Minpa tribe employs stem of *Cleisostoma williamsonii* for fracture (Medhi and Chakrabarti 2009). In the Khasi Hills of northeast India, juice extracted from the leaves of *Cymbidium iridioides* D. Don (syn. *C. giganteum* Wall ex Lind.) is used to promote blood clotting when a person is wounded. It is also used to treat diarrhoea (Jalal et al. 2010). Among the list of tribal cures, two may have a plausible basis: the use of powdery seeds *of Dendrobium nobile, Cymbidium aloifolium* and *C. giganteum* to treat fresh cuts. Bleeding would stop through platelet aggregation (Figs. 12.33 and 12.34).

Nagaland, one of the eight NE states of India, is a floristic hotspot, and it is estimated that 360 species of orchids belonging to 87 genera occur in the province. Local knowledge of medicinal usage of

Fig. 12.28 *Habenaria intermedia*. Photo: Bhaktar B Raskoti

orchids is scarce. However, by reviewing the Indian literature, Nongdam (2014) was able to identify 30 species with potential medicinal usage. Among them is the CITIES I critically threatened *Renanthera imschootiana* whose leaf paste is employed for treating skin disorders (Deorani and Sharma 2007; Nongdam 2014). Pseudobulbs of *Bulbophyllum neilgherrensis* are used as a tonic and for 'the restoration of youthfulness' (Deb and Imchen 2008), i.e. as aphrodisiac (Fig. 12.35).

The Vallaiyans of Vellimalai Hills of Tamil Nadu use *Cymbidium aloifolium* (*panaipulluruvi*) to relieve earache (Ganesan and Kesavan 2003). It was the only orchid employed in their medicinal repertoire. On the Eastern Ghats, stems of *Dendrobium nobile* are used to prepare an aphrodisiac (Jonathan and Raju 2005). Poliyars of Anaimalai Hills in Tamil Nadu at the tip of Peninsular India treat skin diseases with a paste prepared with crushed leaves of *Nervilia biflora* (Sivakumar et al. 2003). Tribals of Anaimalai Hills enjoy juice from pseudobulbs of *Bulbophyllum fuscopurpureum* to which they add jaggary (Sivakumar and Murugesan 2005). Tribals of the Kolli Hills ate fresh tubers of *Habenaria longicorniculata* to reduce scrotal swelling (Subramani and Goraya 2003). Ayurveda

practitioners in Tamil Nadu value *Habenaria* species (*H. ovalifolia*, *H. rariflora* and *H. roxburghii*) as *Riddi* and employ their tubers to treat wasting diseases, haemorrhage, other blood disorders and fainting *Crepidium acuminatum* (syn. *Malaxis rheedei*) is *Rshabba* with similar uses as *Habenaria,* which is also an aphrodisiac (Yoganarasimhan and Chelladurai 2000).

Data from all the recent studies are summarized in Tables 12.1, 12.2 and 12.3. Differences in the latitudinal or latitudinal distribution of orchids and from province to province influence the choice of species for ethnomedicinal usage. Many small terrestrial species are only found in the highlands of Himalaya; hence, more species are employed medicinally in northern India (Table 12.1). Comparison of three publications on the use of medicinal orchids on the western coast of Peninsular India shows significant differences in the range of species employed medicinally (Table 12.4). Perhaps this might be expected after the passage of three centuries, there were also differences between two reports made only 66 years apart. Extent of the surveys and selection of interviewees could influence the result: another would be the loss or scarcity of some species.

The flora in the Nicobar Islands located between east of Tamil Nadu and Sri Lanka is closer to that of Southeast Asia. Here, five orchid species are employed, but tribes also differ in their usage of particular species (Dagar and Dagar 2003).

Four Nicobarese tribes use *Dendrobium crumenatum* to treat body ache and chest pain whereas one tribe employs it to treat fever and another to treat headache and giddiness. *Pinalia bractescens* (syn. Eria *bractescens*) is employed by two tribes to treat fever, whereas one tribe uses it particularly for malaria and another to treat body ache. Only a single tribe employs *Eulophia nuda*, for stomachache, and *Pomatocalpa wendlandorum* for fits. A single and different tribe uses common *Spathoglottis plicata* to treat earache (Dagar and Dagar 2003) (Fig. 12.36).

In herbal practice, different tribes may find different uses for a particular herb, this being evident in the reports from Nepal, Thailand and Malaysia. Also, the manner by which a herb is prepared may also determine its usage. More than

Fig. 12.29 *Herminium lanceum*, From Wight R, *Icones Plantarum Indiae Orientalis* vol. 5(1): t. 1691 (1846) [Govindoo]

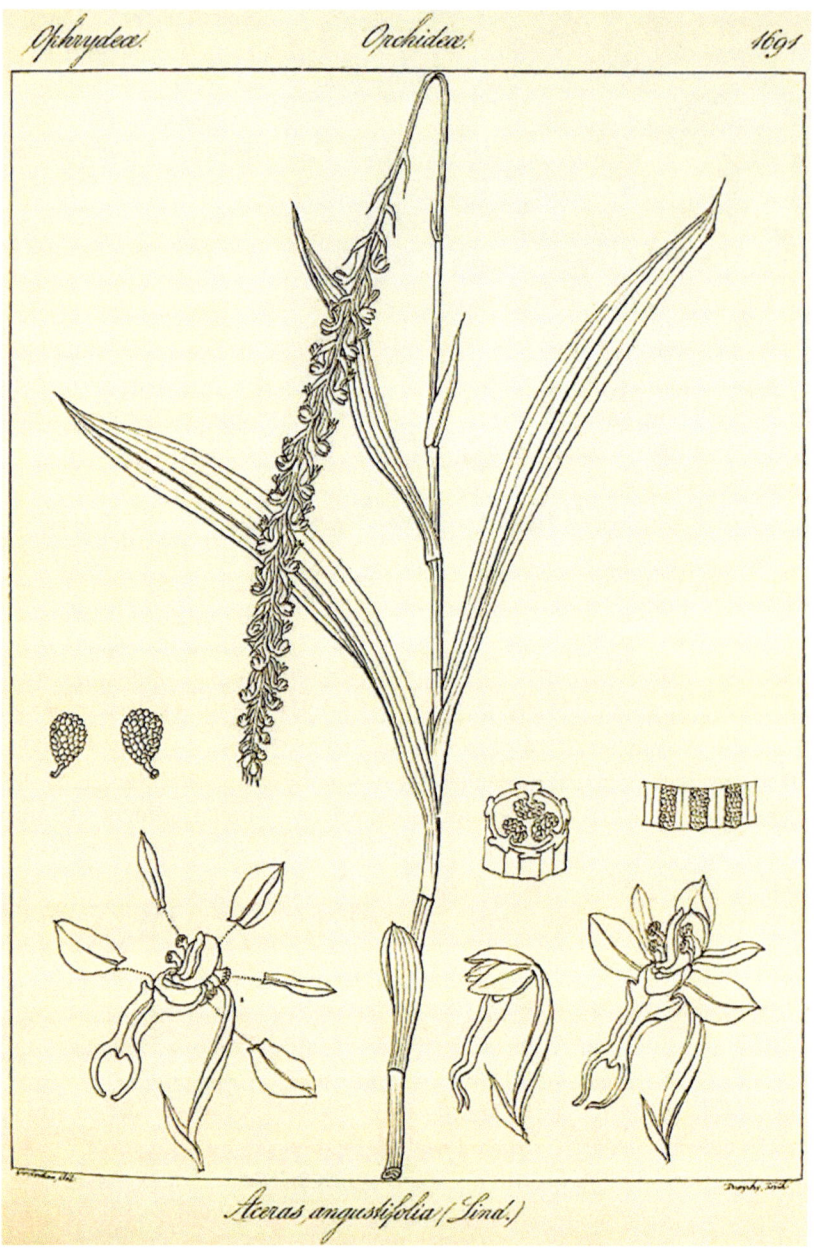

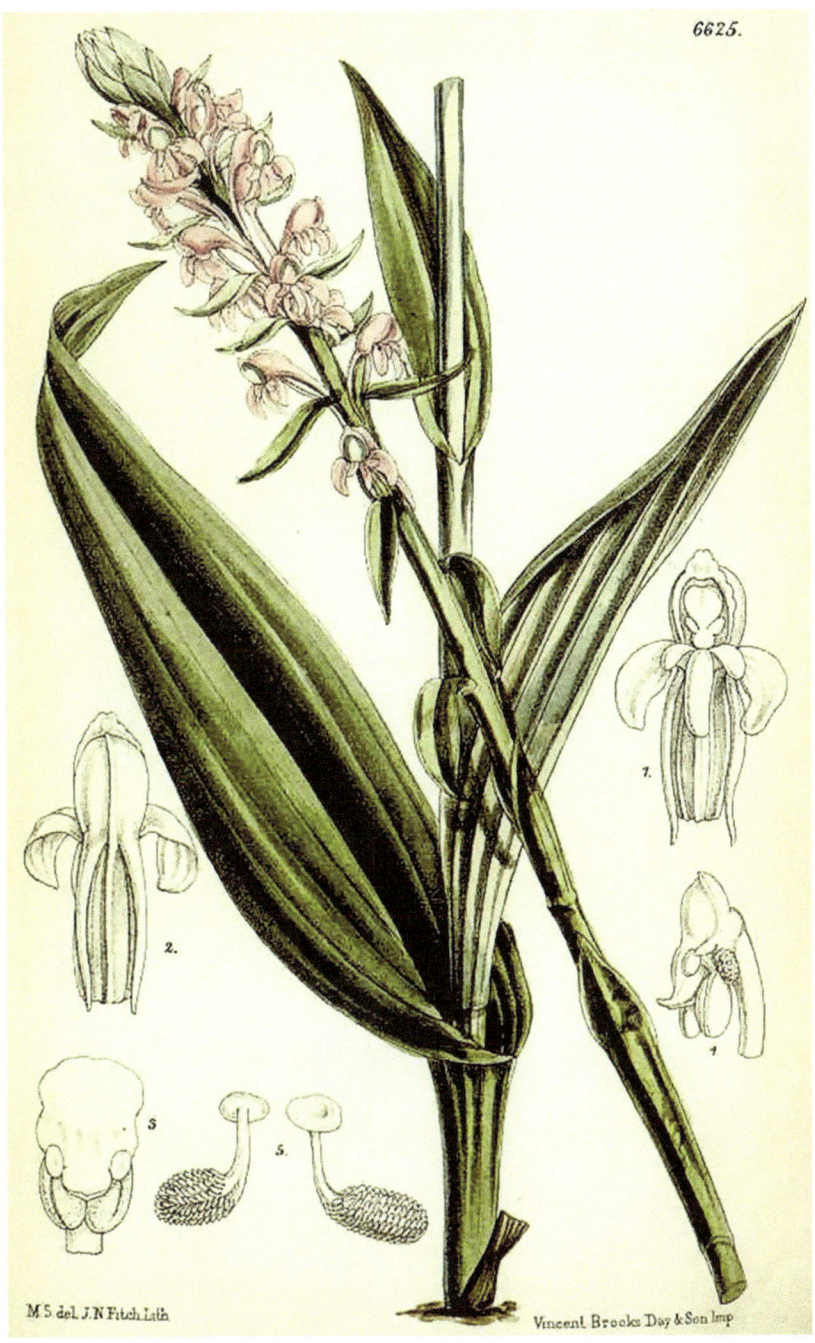

Fig. 12.30 *Satyrium nepalense.* From: *Curtis Botanical Magazine* t.6000–6664 vol. 108 [ser.3 vol 38] t. 6625 (1882) [M Smith]

Fig. 12.31 *Cymbidium bicol*or. From: *Edward's Botanical Register* vol. 27: t. 38 (1841) [SA Drake]

Fig. 12.33 *Cymbidium iridioides* [as *Cymbidium giganteum*] From Warner R, Williams BS, *The orchid album* (1887) vol. 1: t. 284 (1887) [JN Fitch]

Fig. 12.32 *Papilionanthe teres* (©Teoh Eng Soon 2019. All Rights Reserved.)

Fig. 12.34 *Phaius tankervilleae* [as *Phaius grandifolia*] Warner R, Williams BS, *The orchid album* vol. 189de7: t. 502 (1897)

Fig. 12.35 *Bulbophyllum neilgherensis* Wight. From: *Curtis Botanical Magazine t. 5035–5090* vol. 84 [ser. 3, vol.14]: t.5050 (1858) {WH Fitch} 812

a century ago, Ridley (1906) reported that *Senna* which is the dried leaf of *Cassia angustifolia* could be employed for divers conditions depending on how it was prepared. Religious belief added weight to such claims since the herb was imported from Arabia. Thus: '(1) If taken with rose water it will cure disease in the chest. (2) If taken with sugar it will expel cold from the body and act as an aphrodisiac. (3) If taken with sugar candy it will strengthen bones and abdomen. (4) If taken with ghee and sugar for three mornings it will remove venereal disease. (5) If taken with fresh butter it will cure headache and moisten the brains and remove offensive odours from the mouth. (6) If taken with milk it will expel poison and the same will not be dangerous. (7) If taken with goat's milk it will strengthen the body and add to virility . . .' and so on, 15 variations altogether.

In a recent article, T Ananda Rao (2007) reported on the usage of 18 orchid species in Karnataka, a province on the Malabar Coast. Three centuries ago, six of these species (*Acampe praemorsa, Cymbidium aloifolium, Dendrobium ovatum, Eulophia nuda, Liparis rheedii* and *Rhynchostylis retusa*) were described as

Table 12.1 Medicinal orchid genera from INDIA (total genera 49, species 112)

North	Peninsular	Tamil Nadu
Acampe (2)	*Acampe* (1)	*Acampe* (1)
Aerides (3)		
Agrostophyllum (2)		
Anacamptis (1)		
Anoectochilus (2)		
Arundina (1)		
Betillia (1)		
Bulbophyllum (1)	*Bulbophyllum* (1 + 2)	*Bulbophyllum* (2)
Calanthe (3)		
Cephalanthera (1)		
Cleisostoma (1)	*Cleisostoma* (1)	
Coelogyne (1)		
Corymborkis (1)		
Crepidium (2)	*Crepidium* (1)	
Cremastra (1)		
Cymbidium (2)	*Cymbidium* (1)	*Cymbidium* (1)
Cypripedium (1)		
Dactylorhiza (1)		
Dendrobium (9)	*Dendrobium* (3)	
Dienia (1)		
Eria (*Mycaranthes*) (2)		
	Eulophia (5)	*Eulophia* (3)
Flickingeria (1)	*Flickingeria* (1 + 1)	*Flickingeria* (1)
Geodorum (3)	*Geodorum* (1)	*Geodorum* (1)
Goodyera (2)		
Gymnadenia (2)		
Habenaria (5)	*Habenaria* (7 + 3)	*Habenaria* (7)
Herminium (2)		
	Hetaeria (1)	
	Liparis (2)	
Luisia (2)	*Luisia* (2)	
Malaxis (2)		*Malaxis* (1)
Nervilia (2)	*Nervilia* (2)	*Nervilia* (2)
	Oberonia (1)	
Orchis (1)		
Pachystoma (1)		
Pecteilis (1)		
Phaius (1)		
Pholidota (2)	*Pholidota* (1 + 1)	

(continued)

Table 12.1 (continued)

North	Peninsular	Tamil Nadu
Plantanthera (2)		
Pleione (1)		
Ponerorchis (1)		
Polystachya (1)		
Rhynchostylis (1)	*Rhynchostylis* (1)	*Rhynchostylis* (1)
Satyrium (1)	*Satyrium* (1)	*Satyrium* (1)
Seidenfadenia (1)		
Spathoglottis (1)		
Tropidia (1)		
Vanda (5)	*Vanda* (3)	
	Vanilla (1)	
Zeuxine (1)	*Zeuxine* (1 + 1)	*Zeuxine* (1)
Genera Total: 46	21	12
Species Total: 82	46	22

Number of species are indicated in brackets

medicinal by van Rheede (1703). Their current usage differs in four species. Only the employment of *Acampe praemorsa* to treat rheumatism and *Dendrobium ovatum* to treat stomachache has remained unchanged over the years. Six medicinal orchid species mentioned in the *Hortus* (van Rheede 1703) and eight species described by Caius (1936) were not included in the list drawn up by Rao (2007).

A survey of epiphytes and parasitic medicinal plants employed by tribals in Kerala reported on 28 species of which 12 were orchids (Shanavaskhan et al. 2012). Seven species were previously described in van Rheede's *Hortus Indicus Malabaricus* (1703) and seven by Caius (1936), but some of the species were different between the two lists. Two species that had not been previously described were *Luisia tristis* (syn. *Luisia zeylanica*) employed as emollient to treat boils abscesses and tumours [similar usage of this plant are described for Karnataka, Uttar Pradesh and Nepal (Teoh 2016)] and *Vanda testacea* whose leaves are employed to treat asthma, malaria, rheumatism and nervous conditions, these being unique uses only

Table 12.2 Medicinal orchid usage in Peninsular India (in the state of Orissa)

Orchid species	Part used	Indications
Acampe carinata	Root juice	Scorpion and snake bites
(*A papillosa*)	Leaf paste	Chest pain, stomachache
Acampe praemorsa	Root paste	Arthritis
Aerides odorata	Root and leaves	Joint pains and swellings
	Leaf juice	Tuberculosis
Bulbophyllum cariniflorum	Root	Abortifacient
Cymbidium aloifolium	Root	Paralysis
Dendrobium herbaceum	Pseudobulb	Anti-aphrodisiac
	Leaf	Worms
Eria bambusifolia	Whole plant	Stomachache
Flickingeria fimbriata	Root	Eczema
Geodorum recurvum	Root	Tumours
Geodorum densiflorum	Root	Menstrual irregularity
Habenaria commelinifolia	Root	Spermatorrhoea
H. longicorniculata	Tubers	Leucoderma
H. marginata	Tuber	Malignant ulcer
Luisia trichorhiza	Root	Muscle ache
Polystachya concreta	Tuber	Arthritis
Rhynchostylis retusa	Root	Dysentery
	Plant	Wounds
Seidenfia rheedii	Root	Cholera
Vanda testacea	Leaf and root	Fractures
	Root	Asthma
	Plant	Earache
Vanda tessellata	Root	Sexually transmitted diseases, rheumatism, nerve disorders

Reference: Dash et al. (2008)

employed by the Kanikkar tribe of Thiruvananthapuram in Kerala (Shanavaskhan et al. 2012). Another unique practice by the Kanikkar is to apply leaf juice on the nipple to relieve stomachache (Shanavaskhan et al. 2012) (Fig. 12.37).

Tribals also employ orchids to treat their domestic animals. Gonds of Karimnagar district in Andhra Pradesh apply root paste of *Vanda tessellata* (*Ippa vajrnika*) for 5–6 days to treat rheumatism in cattle (Reddy et al. 2003). Tribals at the Nilgiri Biosphere Reserve use a paste made with the stem of *Vanilla walkeriae* (vernacular names: *Kundu*, Pirandai) to treat cattle when they suffer from fever (Balasubramaniam and Prasad 1996). *Dendrobium* pseudobulbs are fed to cows to increase milk production and *Cymbidium* pseudobulbs to cattle to promote health (Medhi and Chakrabarti 2009).

Orchids are much admired by tribal people in NE India who tend to them with care. *Rhynchostylis retusa* (*Bihul* orchid; in Assamese *Kopou Phul*) is the state flower of Arunachal Pradesh. In Assam, it is regarded as the symbol of love, fertility and joy. An entire inflorescence is worn by maidens during the spring festival of *Rongali Bihu* and employed in marriage ceremonies. Flowers of *Papilionanthe teres* are included in offerings to Buddha. For the Monpas of Arunachal Pradesh, flowers of *Cymbidium grandiflorum* constitute an important item for worship. In the district of Kameng, *Dendrobium gibsonii*, *D. hookerianum* and *D. nobile* are symbols of sanctity and purity (Medhi and Chakrabarti 2009).

Much of the population in Sikkim rely on their forests to provide food and medicine, fodder for their livestock and fuel for their kitchens. A recent study conducted in alpine and tropical regions of Sikkim identified 36 orchid species being employed to promote wound healing (9 species), for pain relief (8 species), cough and cold (6), inevitably as aphrodisiac (5) or tonic, also to treat skin disorders and as haemostatic, emetic and anti-diarrhoeal, besides having other uses. Several species are employed to treat multiple, unrelated conditions: for instance, *Gymnadenia orchidis* Lindl. is employed as haemostatic albeit it major usage is as aphrodisiac and tonic (Fig. 12.38). Likewise *Dactylorhiza hatagirea*, popular as an aphrodisiac throughout Indian

Table 12.3 Medicinal orchid usage in Peninsular India (in the states of Andhra Pradesh and Tamil Nadu)

Orchid species	Part used	Indications for usage
Acampe papillosa	Root	Rheumatism, nerve pain, uterine disorders, syphilis
Aerides crispum	Plant	Earache, deafness
Aerides odorata	Leaf	Boils in nose and ear
	Fruit	Wounds
Anoectochilus regalis	Stem, leaf	Medicinal oils
Arundina bambusifolia	Stem	Cracks on heels
Bulb. neilgherrense	Pseudobulb, leaf	Leucoderma
Calanthe sylvatica	Flowers	Epistaxis
Calanthe triplicata	Roots	Swollen hands, diarrhoea
	Flowers	Toothache
	Pseudobulb	Indigestion
Cephalanthera longifolia	Root, rhizome	Tonic
Cleisostoma williamsonii	Leaf	Fracture, swelling of limbs
Corymborkis veratrifolia	Leaf	Emetic, childhood fever
Crepidium acuminatum	Stem	'Tonic', promotes spermatogenesis
Cremastra appendiculata	Root	Toothache, emollient
	Tuber	Abscess, scrofula, freckles, snakebite
Cymbidium aloifolium	Plant	Chronic illness, poor vision, vertigo, paralysis In medicated oil for benign and malignant tumours
	Root	Fractures
Cymbidium hookerianum	Seeds	Haemostatic for wounds
Dendrobium densiflorum	Leaf	Fracture
Dendrobium nobile	Pseudobulb	Aphrodisiac
Eria pannea	Root, leaf	Ague
Eulophia epidendraea	Rhizome	Abscess, engorged breast
Goodyera repens	Root, leaf	Female illnesses, stomach and bladder disorders; snake bites; infant rash
Goodyera schlectandaliana	Plant	Tonic for internal injuries

(continued)

Table 12.3   (continued)

Orchid species	Part used	Indications for usage
Gymnadenia orchidis	Tuber	Salep, aphrodisiac, tonic
Habenaria plantaginea	Rhizome	Chest pain, stomachache
H. roxburghii	Rhizome	Cooling effect
Malaxis muscifera	Tuber	Tonic for kidneys
Phaius tankerville	Pseudobulb	Swelling of limbs, abscess
Pholidota chinensis	Pseudobulb	Scrofula, stomachache, toothache, internal bleeding asthmatic cough, tuberculosis, dysentery
Pholidota imbricata	Pseudobulb	Rheumatic pain
Pholidota pallida Lindl.	Pseudobulb	Infection of fingers
Pleione maculata	Pseudobulb	Liver disorders, stomachache
Ponerorchis chusua	Tuber	Diarrhoea, dysentery, chronic fever
Rhynchostylis retusa	Root	Rheumatism
	Plant	Asthma, tuberculosis, tics, cramps infantile epilepsy, vertigo, palpitations, urinary stones, menstrual disorders
Satyrium nepalense	Tuber	aphrodisiac, malaria, dysentery
Spathoglottis plicata	Plant	Rheumatism
Tropiidia curculidiodes	Plant	Malaria
	Root	Diarrhoea
Vanda coerulea	Leaf	Diarrhoea, dysentery, skin disorders
Vanda tessellata	Root	Fractures, food
	Leaf	Earache, skin infection
Zeuxine longilabris	Plant	Whooping cough

References: Rao (2004) and Rajendran et al. (1997)
Note: Orchid species employed in Orissa and Andhra Pradesh are very different. Only four species are used in both states: *Acampe praemorsa, Aerides odorata, Cymbidium aloifolium* and *Rhynchostylis retusa*. The list by Rajendra et al. (1997) that covers Andhra Pradesh and Tamil Nadu contains eight species that are not listed by Rao (2004). Only one species is included in both lists, *Cymbidium aloifolium*

Table 12.4 Medicinal orchids employed in the Malabar States: a comparison of early and late periods

van Rheede (1703)	Caius (1936)	Rao (2007)
Acampe praemorsa	+	+
Bulbophyllum neigherense syn. *B. sterile*		
Bulbophyllum rheedei		
		Calanthe triplicata
Cleisostoma tenuifolium		
Cymbidium aloifolium	+	+
Dendrobium ovatum	+	+
	*Dendrobium pumilum**	
	Eulophia campestris	
Eulophia epidendraea		
	E. herbacea	
E. nuda		+
	E. pratensis	
		E. pulcra
	Flickingeria fugax	+
Geodorum densiflorum		+
		Habenaria acuminate
	Habenaria commenlinifolia	
		Habenaria crinifera
		H. heyneana
		H. longicorniculata
	*Hetaeria obliqua***	
	Luisia tenuifolia	
		Luisia zeylanica
Malaxis rheedii		+
Liparis odorata		+
		Nervilia aragoana

(continued)

Table 12.4 (continued)

van Rheede (1703)	Caius (1936)	Rao (2007)
	Oberonia anceps	
	Orchis latifolia	
Pholidota imbricate		
Rhynchostylis retusa	+	
		Satyrium nepalense
Vanda spathulata	+	
		V. testacea
	V. tessellata	+
	*Vanilla griffithii****	
	Zeuxine strateumatica	

*, ** and *** Caius included three medicinal orchids from Malaysia. Perhaps these were brought to Kerala by sailors

Himalaya, is used for treating cough and cold, tuberculosis, painful micturition and general debility. The significant number of orchids employed as aphrodisiac in Sikkim is unusual given that they have not been reported to have similar usage elsewhere in India, notably *Cephalanthera longifolia, Crepidium acuminatum, Dendrobium macraei, Dendrobium nobile, Gymnadenia orchidis* and *Satyrium nepalense*. Another uncommon usage is the employment of a decoction of root or leaves of *Mycaranthes pannea* to bathe the skin during chickenpox to reduce the rash and subsequent pigmentation (Panda and Mandal 2013) (Fig. 12.39).

Farming of Medicinal Orchids and Conservation

A study conducted by Nautiyal, Maikhuri, Rao and Saxena (2003) on inhabitants of buffer zone villages of Nanda Devi Biosphere Reserve (NDBR) of Garhwal Himalaya (east of Kashmir) found that local people depended entirely on the surrounding vegetation for their food, medicine,

Fig. 12.36 *Pinalia bractescens* (syn. *Eria bractescens*) (Photo: Teoh Eng Soon, Asian Orchids. Times Books International, Singapore, 1980)

fodder, fuel and timber. The community they studied was small, total population 2253. They cultivated 272.6 ha of land, devoting 11.4 ha to medicinal plants. Among the plants collected from the wild, *Calanthe triplicata* (local name: *Garur panja*) and *Dactylorhiza hatagirea* (local name: *Hathazari*) were used medicinally. Root of *Calanthe triplicata* was employed to treat jaundice and typhoid. A paste made with tubers of *Dactylorhiza hatagirea* was applied on cuts and wounds. Decoction of the tuber which was believed to be an excellent tonic is mixed with milk, sugar and spice and fed to patients to accelerate recovery from illness. With the designation of the area as a National Park in 1982 and Biosphere Reserve in 1988, restrictions were imposed on the collection of resources from the wild. *Dactylorhiza hatagirea* which is found at elevations of 2500–4500 m in Central Himalayas is now cultivated in NDBR. It grows well in shady locations and requires porous soil which is rich in humus. Manuring twice a year and weeding are recommended for better yield. Plants are propagated by seed and root cuttings. Ideally they should be harvested 5 years after planting,

but, in practice, villagers harvest the tubers after 2–3 years (Nautiyal et al. 2003).

Pandey, Joshi and Mudaiya et al. (2003) have also been concerned about the management and conservation of medicinal orchids of Kumaon and Garhwal in the western Himalayan state of Uttarakhand. They mentioned that 22 orchids are considered medicinal in Kumaon and Garhwal Himalaya, and all of them have become scarce and need protection. *Dactylorhiza hatagirea* (local name: *Munjataka*) provides a tonic and aphrodisiac. People suffering from tuberculosis, other debilitating illnesses, spermatorrhoea and impotence are the main consumers. Over-collection from the wild poses a threat to this orchid species, and steps are being taken to regulate the practice. In NE India sheep and cattle graze in pastures that harbour terrestrial orchids.

Root of *Vanda tessellata* (local name: *Rasna*) is used throughout India to treat rheumatism. The *Vanda* is disappearing due to deforestation and loss of host trees. The much admired *Eulophia* species (*E. nuda*, *E. dabia* and *E. herbacea*; local name, *Munjataka, Salep*) are easily grown in a pot or on the ground, but medicinal species have become rare due to environmental factors. *Acampe papillosa* (local name: *Rasna*) is a substitute for *Vanda tessellata*, and it faces similar threats. *Seidenfadeniella filiformis* (Rchb.f.) E.A. Christ & Ormerod [syn. *Saccolabium chrysanthum*, *Schoenorchis filliformis* (Wight) Schltr.] is used to treat rheumatism or as a demulcent. It too needs protection. *Cymbidium aloifolium* is attracting attention because its sap promotes blood coagulation. It is a hardy plant and widely distributed and is currently not under threat. Several terrestrial herbaceous orchids used medicinally also need protection and should be cultivated, namely, *Malaxis acuminata* (local name: *Jivak*) and *Malaxis muscifera* (local name: *Risvak*) which are employed as tonics: *Habenaria commelinifolia*, *H. latilabris* (local name: *Ridhi Vridhi*), *Plantanthera susannae* (local name: *Ridhi Vridhi*), *Pogonia gammiena* (local name: *Shankhaluka*), *Goodyera repens*,

Fig. 12.37 *Luisia tristis* [as *Luisia teretifolia*] From: *Annals of the Royal Botanic Gardens, Calcutta* vol. 8 (3): t. 271 (1891) [R Pantling]

Herminium monophyllum, Herminium lanceum, Satyrium nepalensis and *Zeuxine strateumatica.* The last four species are known locally as *salep.* They are collected for sale to *salep* traders. This term *salep* is also applied to *Dendrobium chrysanthum, D. lindleyi* and *D. fimbriatum,* but the first two are used as tonics, stimulants, stomachic, pectoral and antiphlogistic, whereas *Dendrobium fimbriatum* is used to treat liver upsets and nervous debility (Pandey et al. 2003) (Fig. 12.40).

In contrast to this apprehension over over-collection by traders, there is the reassurance that the local people in NE India treasure and care for orchids because of the important role enjoyed by these plants in their culture and religion. In the Kameng district of Arunachal Pradesh, *Dendrobium gibsonii, D. hookerianum*

Fig. 12.38 *Gymnadenia orchidis*. From: *Annals of the Royal Botanic Gardens, Calcutta* vol. 8 (3): t. 401 (1891) [R Pantling]

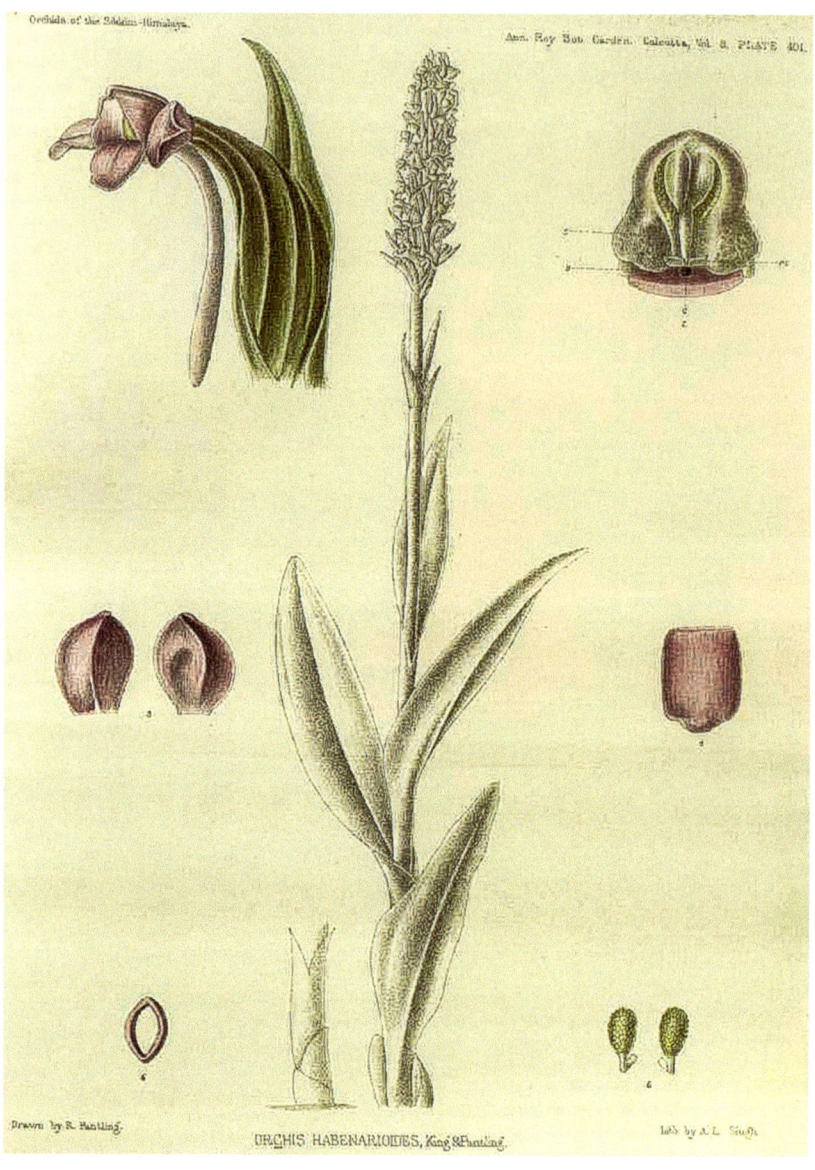

and *D. nobile* are symbols of purity. Among the youth of Assam, *Rhynchostylis retusa* (local name *Kopou Phul*) is a symbol of love, merriment and fertility. Young women adorn their hair with its flowers during the spring festival. It is employed during the wedding ceremony. So admired is this orchid in Assam that most families cultivate the orchid in their homes. Assamese women also wear flowers of *Coelogyne nitida* and *Vanda tessellata*. Young Naga women in Manipur wore golden flowers of *Dendrobium densiflorum* behind their ears. In autumn they place blue flowers of *Vanda coerulea* in their hair for the puja festival when they make offerings to Buddha, gods, monks and temples. *Cymbidium grandiflorum* flowers are an important item for worship in Kameng, and flowers of *Papilionanthe teres* are offered to Buddha and to

Fig. 12.39 *Cephalanthera longifolia* (©Teoh Eng Soon 2019. All Rights Reserved.)

spirits by Kamtis and other Tai ethnics in Assam and Arunachal Pradesh (Medhi and Chakrabarti 2009). Arunachal Pradesh is an important state for orchid conservation because it has the greatest number of orchid species among the Indian states. There are 183 species belonging to 55 orchid genera in the Sessa Orchid Sanctuary which covers an area of only 100 km^2. There are 112 medicinal orchid species in Arunachal Pradesh (Tsering et al. 2017).

Numerous approaches have been proposed to conserve orchids that are in danger of becoming rare or extinct (Teoh 2011, 2016; Luo 2016). Wild orchids require special habitats and the opportunity to establish symbiosis with specific fungi. Montane subtropical pine forest is possibly the most suitable habitat for medicinal orchids in the state of Mizoram, India (Vanlairuati et al. 2012). Appangala in Karnataka, Loleyangaon and Darjeeling areas in West Bengal are designated as orchid reserves by their state governments. Orchid sanctuaries have been set up in Deorali and Singtam in Sikkim, and Sessa in Arunachal Pradesh (Behera et al. 2013). Efforts have been made in India, Nepal, China, Korea, Thailand, Malaysia and elsewhere to conserve the gene pool of precious medicinal orchids. Nevertheless, desirable orchids already scarce are becoming rarer in the wild by the day.

Much effort has been put into micropropagation to provide desirable medicinal orchid seedlings, but unless there are political and popular will, financial input by government, organizations or individuals, seedlings taken into the field and cared for and education and supervision provided for farmers, Earth will still be ravaged, and precious wild plants will disappear.

Perhaps it should be stressed that whereas many conservationists have focused on micropropagation involving explants (Chang 2007; Pant 2011; Behera et al. 2013), such efforts actually narrow the gene pool of the particular species. Conservationists should work with plants raised from seed from diverse habitats. Micropropagation should be reserved for those plants which are so rare or so difficult to cultivate that it would be extremely unlikely that one could obtain seeds from them. For orchid conservation, propagation employing seed germination is superior to meristematic clonal propagation because it ensures survival of the widest gene pool. When selecting a plant for clonal propagation, it is essential to ensure that the plant must be the toughest, fastest-growing, disease-resistant and best adapted to the environment for which it is destined, but it is doubtful that scientists working in isolation have the resources to identify such clones. There is a place for making clonal propagations on beautiful representatives of a species because when these are offered on the market, it will satisfy the demand by collectors to own plants of the species.

Finally, unless there are political and popular will, financial input by government, organizations or individuals, seedlings taken into the field and cared for and education and supervision provided for farmers, Earth will still be ravaged, and precious wild plants will disappear.

Fig. 12.40 *Habenaria latilabris.* From: *Annals of the Royal Botanic Gardens, Calcutta* vol. 8 (3): t. 423 (1891) [R Pantling]

HABENARIA LATILABRIS, Hook fil.

References

Abraham A, Vatsala P (1981) Introduction to orchids, with illustrations and descriptions of 150 South Indian orchids. TPGRI, Trivandrum

Akarsh (2004) Newsletter of ENVIS NODE on Indian medicinal plants 1(2): June 2004

Arditti J, Flick BH, Ehmann A, Fisch MH (1975) Orchid phytoalexins. Part 2. Isolation and characterization of possible sterol companions. Am J Bot 62:738–742

Balasubramaniam P, Prasad SN (1996) Ethnobotany and conservation of medicinal plants by Irulas of Nilgiri Biosphere Reserve. In: Jain SK (ed) Ethnobiology in human welfare. Deep Publications, New Delhi

Behera D, Rath CC, Mohapatra U (2013) Medicinal orchids in India and their conservation: a review. Floricult Ornament Biotechnol 7:53–59

Bose TK, Bhattacharjee SK (1980) Orchids of India. Naya Prokash, Calcutta

Caius JF (1936) The medicinal and poisonous orchids of India. J Bombay Nat Hist Soc 38(4):791–799

Chang DCN (2007) The screening of orchid fungi (OMF) and the applications. In: Chen WH, Chen HH (eds) Orchid biotechnology. World Scientific, New Jersey

Chase MW, Cameron KM, Freudenstein JV et al (2015) An updated classification of Orchidaceae. Bot J Linn Soc 177(2):151–174

Chen XQ, Wood JJ (2009) Pholidota Lindley ex Hooker, Exot. Fl. 12 ad t. 138. 1825. In: Chen XQ, Liu ZJ, Zhu GH et al (eds) Flora of China—Orchidaceae. Science Press, Beijing

Chopra RN (1933) The indigenous drugs of India. The Art Press, Calcutta. Republished as Chopra's Indigenous Plants of India, 2nd ed. Academic Publishers (1986), Kolkata

Chowlu K, Mahar KS, Das AK (2017) Ethnobotanical studies on orchids among Kamti community of Arunachal Pradesh, India. Indian J Nat Prod Resour 8 (1):89–93

Comber JB (2001) Orchids of Sumatra. Natural History Publications (Borneo), Kota Kinabalu

Dagar HS, Dagar JC (2003) Plants used in ethnomedicine by the Nicobarese of Islands in Bay of Bengal, India. In: Singh V, Jain AP (eds) Ethnoboany and medicinal plants of India and Nepal. Scientific Publishers, Jodhpur, pp 773–778

Dash PK, Sahoo S, Bal S (2008) Ethnobotanical studies on orchids of Niyamgiri Hill Ranges, Orissa, India. Ethnobot Leaflets 12:70–78

Deb CR, Imchen T (2008) Orchid diversity of Nagaland. Scichem Publishing House, Udaipur

Deorani SC, Sharma GD (2007) Medicinal plants of Nagaland. Bishan Singh Mahendra Pal Singh, New Delhi, p 291 (quoted by Nongdam 2014)

Duggal SC (1972) Orchids in human affairs (A review). Pharm Biol 11(2):1727–1734

Dutt UC (1900) The material medica of the Hindus. Rev ed. D Mukerjee, Calcutta

Dymock W, Warden CJH, Hooper D (1893) A history of the principal drugs of vegetable origin met with in British India. Education Soc. Press, Bombay

Ganesan S, Kesavan L (2003) Ethnomedicinal plants used by the ethnic group Valaiyans of Vellimalai Hills (Reserve Forest), Tamil Nadu, India. In: Singh V, Jain AP (eds) Ethnobotany and medicinal plants of India and Nepal. Scientific Publishers, Jodhpur, pp 754–760

Gupta AK, Tandon N (2004) Reviews on Indian medicinal plants, vol 3(Are-Azi). Indian Council of Medical Research, New Delhi

Hu XM, Zhang WK, Zhu QZ et al (2000) Zhonghua Bencao, vol 8. Shanghai Science and Technology Publication, Shanghai

Jalal JS, Kumar P, Tewari L, Pangtey YPS (2010) Orchids: uses in traditional medicine in India. In: National seminar on medicinal plants of Himalayas. Regional Res Institute Himalayan Flora, Tariket

Jayaweera DMA (1981) A revised handbook of the flora of Ceylon, vol II. A.A. Balkema, Rotterdam

Jin H, Xu ZX, Chen JH, Han SF, Ge S, Luo YB (2009) Interaction between tissue cultured seedlings of Dendrobium officinale and mycorrhizal fungus (Epulorrhiza sp.) during symbiotic culture. Chin J Plant Ecol 33(3):433–441

Jonathan KH, Raju AJS (2005) Terrestrial and epiphytic orchids of Eastern Ghats. EPTRI-ENVIS Newsl 11 (3):2–4

Khajuria AK, Kumar G, Bisht NS (2017) Diversity with ethnomedicinal notes on orchids: a case study of Nagdev range, Pauri Garhwal, Uttarakhand, India. J Med Plant Stud 5(1):171–174

Kumar V (2003) Wild edible plants of Surguja district of Chhattisgarh state, India. In: Singh V, Jain AP (eds) Ethnobotany and medicinal plants of India, vol 1. Scientific Publications, Jodhpur

Lawler LJ (1984) Ethnobotany of the Orchidaceae. In: Arditti J (ed) Orchid Biology Reviews & Perspectives, vol 3. Cornell University Press, Ithaca

Luo YB (2016) The development of traditional Chinese medicinal Dendrobium business. Abstracts, The Third Shanghai Chenshan International Conference, 22–23rd April, 2016. Shanghai: Shenshan Plant Science Research Centre, Chinese Academy of Sciences, Shanghai Chenshan Botanaical Gardens, 3888 Chenhua Road, Songjiang District, Shanghai

Majumder PL, Sarkar AK (1982) Imbricatin, a new modified 9,10-dihydrophenanthrene derivative of the orchid Pholidota imbricata. Indian J Chem 21B:829–831

Manandhar NP, Manandhar S (2002) Plants and people of Nepal. Timber Press, Portland

Manilal KS (2003) Hendrik van Rheede's Hortus Malabaricus with annotations and modern botanical nomenclature, vol 11 & 12. University of Kerala, Thiruvananthapuram

Maridass M, Ramesh U (2010) Investigation of phytochemical constituents from Eulophia enpidendraea. Int J Biol Tech 1(1):1–7

Matthew KM (1995) An excursion flora of Central Tamilnadu, India. A.A. Balkaema, Rotterdam

Medhi RP, Chakrabarti S (2009) Traditional knowledge of NE people on conservation of wild orchids. Indian J Tradit Knowl 8(1):11–16

Mohanty JP, Pal P, Barma AD (2015) An overview on orchids. UJPSR 1(1):45–50

Musharof Hossain M (2009) Traditional therapeutic uses of some indigenous orchids of Bangladesh. Med Aromat Plant Sci Biotechnol 3:100–106

Nadkarni AK (1954) Dr. K.M. Nadkarni's Indian Materia Medica, vol 2, 3rd edn. Popular Book Depot, Bombay

Nautiyal S, Saxena KG, Rao KS, Maikhuri RK (2003) Transhumant pastoralism in the Nanda Devi Biosphere Reserve, India: a case study in the buffer zone. Mt Res Dev 23(3):255–262

Nongdam P (2014) Ethno-medicinal uses of some orchids of Nagaland, Northeast India. Res J Med Plant 8:126–139

Panda AK, Debasis Mandal D (2013) The folklore medicinal orchids of Sikkim. Anc Sci Life 33(2):92–96

Pandey NK, Joshi GC, Mudaiya RK et al (2003) Management and conservation of medicinal orchids of Kumaon and Garhwal Himalaya. J Econ Taxon Bot 27(1):114–116

Pant B (2011) Medicinal orchids of Nepal and their conservation by in-vitro technique. In: Proceedings of the 20th World Orchid Congress, Singapore

Pant B, Raskoti BB (2013) Medicinal orchids of Nepal. Himalayan Map House, Kathmandu

Puri HS (1970) Vegetable aphrodisiacs of India. Q J Crude Drugs Res 11:1742–1752

Rajendran A, Rao NR, Kumar KR, Henry AN (1997) Some medicinal orchids of Southern India. Anc Sci Life 17(1):10–14

Rao AN (2004) Medicinal orchid wealth of Arunachal Pradesh. Newsl ENVIS Node Indian Med Plants 1 (2):1–5

Rao TA (2007) Ethnobotanical data on wild orchids of medicinal value as practiced by tribals at Kudremukh National Park in Karnataka. Orchid Newsl 2(2):1–7

Rao TA, Sridhar S (2007) Wild orchids in Karnataka. A pictorial compendium. Institute of Natural Resources Conservation, Education, Research and Training (INCERT), Bangalore

Reddy CH, Nagesh K, Reddy KN, Vatsavaya SR (2003) Plants used in ethnoveterinary practices by gonds of Karimnagar district, Andhra Pradesh, India. In: Singh V, Jain AP (eds) Ethnoboany and medicinal plants of India and Nepal. Scientific Publishers, Jodhpur, pp 679–685

Reddy KN, Subharaju GV, Reddy CS, Raju VS (2005) Ethnobotany of certain orchids of Eastern Ghats of Andhra Pradesh. EPTRI-ENVIS Newsl 11(3):5–9

Ridley HN (1906) Malay drugs. Agric Bull Straits Settlements FMS 5:245–254

Santapau H, Kapadia Z (1966) The orchids of Bombay. Govt. of India Press, Calcutta

Seidenfaden G (1999) 149. Orchidaceae. In: Matthew KM (ed) The flora of the Palni Hills, South India, Part 3. The Rapinat Herbarium. St. Joseph's College, Tiruchirapalli

Shanavaskhan AE, Sivadasan M, Alfarhan AH, Thomas J (2012) Ethnomedical aspects of angiospermic epiphytes and parasites of Kerala, India. Indian J Tradit Knowl 11(2):250–258

Shriram V, Kumar V, Kishor PBK et al (2010) Cytotoxic activity of 9.10-dihydromethoxyphenanthrene-1,7-diol from Eulophia nuda against human cancer cells. J Ethnopharmacol 128(1):251–253

Singh A, Duggal S (2009) Medicinal orchids: an overview. Ethnobot Leaf 13:351–363

Sivakumar A, Subramanian MS, Karunakaran M, Burkanudeen A (2003) Ethnobotany of Poliyars of Anaimalai Hills, Tamil Nadu. In: Singh V, Jain AP (eds) Ethnoboany and medicinal plants of India and Nepal. Scientific Publishers, Jodhpur, pp 679–685

Sivakumar A, Murugesan M (2005) Ethnobotanical studies of wild edible plants used by tribals of Anaimalai Hills, the Western Ghats. Anc Sci Life 25(2):68–73

Subramani SDP, Goraya GS (2003) Some folklore medicinal plants of Kolli Hills. Record of a Natti Vaidyas Sammelan Ibid: 665–669

Teoh ES (2011) Medicinal orchids: the issue of conservation. Malayan Orchid Rev 45:105–113

Teoh ES (2016) Medicinal orchids of Asia. Springer International, Switzerland

Trivedi VP, Dixit RS, Lal VK (1980) Orchids in the drug markets of Bareilly, Kanpur and nearby districts. Nagarjun (Calcutta) 23(8):157–163

Tsering J, Tam N, Tng H et al (2017) Medicinal orchids of Arunachal Pradesh: a review. Bull Arunachal Forest Res 32(1&2):1–16

Uphof JC Th (1968) Dictionary of economic plants. Verlag von J. Cramer, Lehre

Van Rheede HA (1703) Hortus Indicus Malabaricus, vol 11 & 12. Joannis van Someran, Haeredum Johannis van Dyck, Henrici & Viduae Theodori Bom, Amstelaedami

Van Steenis CGGJ (ed) (1948) Flora Malesiana, vol 4, part 1. Noordhoff-Kolff N.V.H, Batavia

Vanlairuati MT, Pradhan S, Das SK (2012) Habitat studies from conservation of medicinal orchids of Mizoram. In: Ghosh SN (ed) Proceedings of the international symposium on minor fruits and medicinal plants for health and ecological security (ISMF & MP), West Bengal, 19–22 Dec 2011. Bidhan Chandra Krishi Viswandyalaya, Mohanpur, pp 120–123

Wood JJ, Beaman TE, Lamb A et al (2011) The orchids of Mount Kinabalu, vol 1. Natural History Publications (Borneo), Kota Kinabalu

Yoganarasimhan SN, Chelladurai V (2000) Medicinal plants of India vol. 2 – Tamil Nadu. Regional Research Institute, Bangalore

Medicinal Orchids of Nepal and Bhutan 13

Political borders are porous to plants. Nepal being adjacent to India, much of its orchid vegetation is also present in India. The unique character of Nepal lies in its vertical ascent. Travelling across a short stretch of land which can be covered in a day, one can comfortably traverse tropical lowland, low montane, high montane (temperate), subalpine and alpine forests and study their vegetation.

Being so close to India, one would expect that Ayurvedic, Unani and other forms of Indian medicine would greatly influence the practice of traditional Nepali medicine, but that is not quite the case. Nepal has 18 different tribes, each with its own customs and unique knowledge of the terrain that it occupies, including the natural vegetation. Medicinal orchid usage varies from one area of Nepal to another. However, the species employed are generally, albeit not entirely, similar to those employed in India. Almost a quarter of the orchid species that occur in Nepal are employed medicinally (94 out of 388 species) (Acharya and Rokya 2010; Raskoti 2012; Subedi et al. 2013), whereas in Bhutan, which is Nepal's immediate neighbour on the east, only two orchids are employed medicinally at the National Institute of Traditional Medicine: *Dactylorhiza hatagirea* as a tonic and aphrodisiac and a species of *Otochilus* to treat rheumatism (Gyeltshen, personal comm. 2013) (Figs. 13.1 and 13.2).

Two thirds of Nepal's population rely on traditional medicine to treat their illnesses because this is traditional, readily available and free or relatively inexpensive. In the cities people also make use of traditional medicine, viewing it as being devoid of major side effects. There is nothing to lose by the addition, as it were.

Whereas in neighbouring Bhutan the percentage of land covered by forest has remained unchanged over the decades (around 70%), Nepal witnessed a rapid annual loss rate of 1.7% over the last two decades. By 2006, only 29% of the land was covered by forest against a desired figure of 33% (Baral and Kurmi 2006). This loss is aggravated by the fact that maximum distribution of medicinal orchids occurs between 500 and 2500 m, whereas most protected areas (e.g. national parks) lie between 3000 and 3500 m (Acharya and Rokya 2010). Cattle grazing and trampling in the area pose the principal threat to the survival of the ground orchids (Acharya and Rokya 2010; Bhattarai et al. 2014). Illegal trade, over-collection of medicinal species and widespread ignorance pose additional threats. The species most threatened is *Dactylorhiza hatagirea* because Ayurvedic practitioners value it for its alleged aphrodisiacal properties (Bhattarai et al. 2014).

For rural folk living from hand to mouth in remote areas, collection of orchids and medicinal plants for sale provides an important source of income to sustain their livelihood. In a recent survey, the investigators discovered that collectors earn an average of US$2 per kilogram

© Springer Nature Switzerland AG 2019
E. S. Teoh, *Orchids as Aphrodisiac, Medicine or Food*, https://doi.org/10.1007/978-3-030-18255-7_13

Fig. 13.1 *Dactylorhiza hatagirea* at 3450 m, Darchula, Nepal. Photo: SK Ghimire

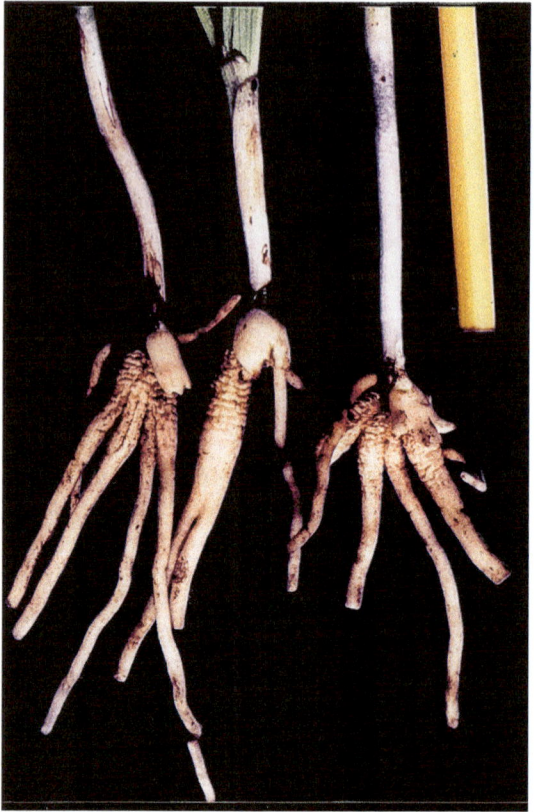

Fig. 13.2 Digitate tubers of *Dactylorhiza hatagirea*. Photo: SK Ghimire

of medicinal orchids collected, but this could increase if the species is rare and much sought after (Subedi et al. 2013). Apart from those species which are employed medicinally in Nepal, a hundred more feature in the traditional medicine of other countries. When their own resources become scarce, neighbours might look to Nepal for supply. In 1985 around 100 trucks of 8 ton capacity were loaded with wild orchids extracted from eastern Nepal for transshipment to India as materials for Ayurvedic preparations (Bailes 1985). Five tons of *Dactylorhiza hatagirea* (syn. *Orchis latifolia*) worth US$4500 were harvested annually for preparation of salep (Vaidya et al. 2002). *Dactylorhiza hatagirea* and *Gastrodia elata* were also exported to China (Bhattarai et al. 2002).

Using DNA barcoding to identify the medicinal orchid species, Subeidi, Kunwar, Choi and their team which involved committed conservationists from Uppsala University in Sweden identified 60 species that were employed to cope with 38 different ailments, with 80% being species included in the review by Acharya and Rokya (2010). The 12 species from this study brings the total number of Nepali medicinal species to 94. Species in five epiphytic genera, *Coelogyne*, *Dendrobium*, *Cymbidium*, *Bulbophyllum* and *Pholidota*, and two terrestrial genera, *Habenaria and Malaxis*, are most commonly employed in the traditional medicine of Nepal (Subedi et al. 2013).

Table 13.1 provides a list of plants classified by genera employed for some common conditions, and it separately tabulates the genera of orchids used for the same purposes. Table 13.2 lists the medicinal orchids by species. Many plants can be used for a particular complaint, and often several among these would be more readily available than an orchid. Local tribal

Table 13.1 A sampling of the usages of medicinal plants and orchids in Nepal

Usage	Total plant genera	Orchids
Food (vegetable)	238	8 (*Brachycorythis, Cypripedium, Dactylorhiza, Epipactis, Habenaria, Malaxis, Platanthera, Satyrium*)
Aphrodisiac	13	3 (*Calanthe, Coelogyne, Dendrobium*)
Tonic *Crepidium, Dendrobium*	54	22 (*Aerides, Brachycorythis, Calanthe, Cypripedium, Dactylorhiza, Dienia, Flickingeria, Habenaria, Herminium, Neottianthe, Otochilus, Phaius, Pholidota, Platanthera, Pleione, Ponerorchis, Satyrium, Smitinandia, Spiranthes, Zeuxine*)
Backache	21	2 (*Arundina, Coelogyne*)
Blood disorders		1 (*Eulophia*)
Boils	134	6 (*Coelogyne, Cymbidium, Dendrobium, Luisia, Pholidota, Vanda*)
Burns		4 (*Bulbophyllum, Coelogyne, Dactylorhiza, Dendrobium*)
Cough, cold, sore throat	21	6 (*Brachycorythis, Cymbidium, Cypripedium, Dendrobium, Eulophia, Habenaria*)
Cuts	138	5 (*Aerides, Gymnadenia, Pleione, Rhynchostylis, Vanda*)
Demulcents	10	2 (*Cymbidium, Dactylorhiza*)
Diarrhoea	109	4 (*Brachycorythis, Malaxis, Nervilia, Satyrium*)
Fractures/dislocations	34	9 (*Bulbophyllum, Calanthe, Cymbidium, Dactylorhiza, Dendrobium, Otochilus, Papilionanthe, Pholidota, Thunia, Vanda*)
Earache		1 (*Vanda*)
Expectorants	19	2 (*Brachycorythis, Dactylorhiza*)
Fever	174	5 (*Calanthe, Coelogyne, Crepidium, Dendrobium, Malaxis*)
Gynaecological conditions	71	1 (*Bulbophyllum*)
Headache	98	3 (*Coelogyne, Epipactis, Pinalia*)
Indigestion	100	1 (*Coelogyne*)
Liver ailments		3 (*Gymnadenia, Oberonia, Pleione*)
Malaria		1 (*Satyrium*)
Muscle pain		1 (*Luisia*)
Nerve disorder (hysteria, madness, epilepsy)		4 (*Cypripedium, Dendrobium, Epipactis, Nervilia*)
Nose bleed		1 (*Calanthe*)
Pimples	40	1 (*Dendrobium*)
Purgative	35	1 (*Cymbidium*)
Respiratory disorder		6 (*Conchidium, Crepidium, Cypripedium, Dendrobium, Eulophia, Nervilia*)
Rheumatism	54	5 (*Acampe, Otochilus, Pholidota, Rhynchostylis, Vanda*)
Skin lesions		1 (*Pholidota*)
Snake bites		2 (*Habenaria, Dendrobium*)
Stomachache, etc.	99	4 (*Coelogyne, Dactylorhiza, Eria, Gymnadenia, Liparis*)
Tuberculosis		4 (*Anoectochilus, Bulbophyllum, Crepidium, Eulophia*)
Urinary disorders		3 (*Gymnadenia, Herminium, Nervilia*)
Worms	109	1 (*Luisia*)
Wounds	206	9 (*Aerides, Coelogyne, Dactylorhiza, Gymnadenia, Habenaria, Luisia, Pleione, Rhynchostylis, Vanda*)
Veterinary usage	62	2 (*Coelogyne, Dendrobium*)

Table 13.2 Medicinal orchid species from Nepal

Acampe praemorsa (Roxb) Blatt & McCann (syn.
Acampe papillosa Lindl.) Roxb.

Aerides odorata Lour.

Anoectochilus roxburghii (Wall.) Lindl.

Arundina graminifolia (D. Don) Hochr.

Brachycorythis obcordata (Lindl.) Summerh.

Bulbophyllum careyanum (Hook.f.) Spreng

Bulbophyllum leopardinum (Wall.) Lindl. ex Wall

Bulbophyllum odoratissimum (Sm.) Lindl.

Bulbophyllum umbellatum Lindl.

Calanthe plantaginea Lindl.

Calanthe puberula Lindl.

Calanthe sylvatica (Thou.) Lindl.

Coelogyne corymbosa Lindl.

Coelogyne cristata Lindl.

Coelogyne elata Lindl. [syn. *Coelogyne stricta* (D. Don)
Schltr.]

Coelogyne fimbriata Lindl.

Coelogyne flaccida Lindl.

Coelogyne fuscescens Lindl.

Coelogyne nitida Lindl. (syn. *Coelogyne ochracea* Lindl.)

Coelogyne ovalis Lindl.

Coelogyne prolifera Lindl.

Crepidium acuminatum (D. Don.) Szlach. (syn. *Malaxis
acuminata* D. Don)

Cymbidium aloifolium (L.) Sw.

Cymbidium elegans Lindl.

Cymbidium devonianum Paxt.

Cymbidium iridioides D. Don.

Cymbidium cordigerum D. Don

Cymbidium elegans Rchb. f.

Cypripedium himalaicum Rolfe

Dactylorhiza hatagirea (D. Don) Soo

Dendrobium amoenum Wall. ex Lindl.

Dendrobium densiflorum Lindl.

Dendrobium eriiflorum Griff.

Dendrobium fimbriatum Hook, f. var. *oculatum* Hook.f.

Dendrobium heterocarpum Wall. ex Lindl.

Dendrobium longicornu Wall ex Lindl.

Dendrobium monticola P.F. Hunt & Summerh.

Dendrobium nobile Lindl.

Dendrobium transparens Wall ex Lindl.

Dienia cylindrostachya (Lindl.) Kuntze.

Epipactis helleborine (L.) Crantz.

Eria muscicola Lindl.

Eria spicata (D. Don)

Eulophia dabia [D. Don (Hoch.)]

Eulophia nuda Lindl. [syn. *Eulophia spectabilis* (Dennst.)
Suresh]

(continued)

Table 13.2 (continued)

Flickingeria fimbriata (Bl.) Hawkes

Gastrodia elata Blume

Gymnadenia orchidis Lindl.

Habenaria arietina Hook.f. [syn. *Habenaria intermedia*
D. Don. var. *arietina* Hook. f.]

Habenaria commelinifolia (Roxb.) Wall ex Lindl.

Habenaria furcifera Lindl.

Habenaria intermedia D. Don

Habenaria pectinata D. Don

Liparis nervosa (Thunb.) Lindl.

Luisia trichorrhiza (Hook. f.) Blume

Luisia tristis (G. Forst.) Hook f. (syn.*Luisia teretifolia*
Gaud.)

Malaxis cylindrostachya (Lindl.) Kuntze

Malaxis monophyllos Sw.

Nervilia concolor (Blume) Schltr. (syn. *Nervilia
aragoana* Gaud.)

Oberonia caulescens Lindl. ex Wall.

Otochilus porrectus Lindl.

Papilionanthe teres (Roxb.) Lindl. (syn. *Vanda teres*
Roxb.)

Pholidota articulata Lindl.

Pholidota griffithii Hook, f.

Pholidota imbricata Hook.

Pinalia spicata (D. Don) S.C. Chen & J.J. Wood [syn.
Eria spicata (D. Don) Hand.-Mazz.]

Platanthera sikkimensis (Hook f.) Kraenzl.

Pleione humilis (Sm.) D. Don

Pleione maculata (Lindl.) Lindl. & Paxton

Pleione praecox (Sm.) D. Don

Ponerorchis chusua (WW Sm). Soo.

Rhynchostylis retusa (L) Blume)

Satyrium nepalense D. Don var. *nepalense*

Smitinandia micrantha (Lindl.) Holtt.

Spiranthes sinensis (Pers.) Ames

Thunia alba (Lindl.) Rchb.f.

Vanda cristata Lindl. [syn. *Trudelia cristata* (Wall ex
Lindl.) Senghas]

Vanda tessellata (Roxb.) Hk. ex D. Don

Vanda testacea (Lindl.) Rchb.f.

Zeuxine strateumatica (L.) Schltr.

usage of orchids, therefore, does not pose a threat to orchid survival, whereas trade in orchids, especially those marketed as aphrodisiacs, does.

There appears to be a great interest in tonics and orchids in 19 genera being employed for its preparation. This is probably an Indian cultural influence because a similar interest exists in India

Fig. 13.4 *Gymnadenia orchidis*. Photo: SK Ghimire

Fig. 13.5 *Cypripedium himalaicum* at 3900 m, Upper Chamelia Valley, Darchula. Nepal. Photo: K Ghimire

(Teoh 2016) and in Thailand (Chuakul 2002) (Figs. 13.3, 13.4, 13.5 and 13.6).

Nepali tribes are not conversant with the bacterial theory of disease. Whereas they may find a plant useful for treating 'boils and pimples' for which they apparently note a similarity, they do not use the same for cuts and wounds. However, phytoalexins being only bacteriostatic and not bactericidal, orchids would be unsuitable for treating carbuncles.

Headache, stomachache, diarrhoea, worm infestation and gynaecological problems are common, and a vast array of herbs are available to

Fig. 13.6 *Habenaria intermedia* at 2300 m, Darchula, Nepal. Photo: SK Ghimire

treat these conditions which possibly renders it unnecessary to resort to orchids, and relatively few orchids are used to treat these conditions: no orchid is employed for gynaecological disorders. On the other hand, several orchid species appear to have a unique place in the treatment of fractures.

Another point to note is that although *Artemisia* (not an orchid) is widely employed in Nepali tribal medicine, all tribes have not learnt to use it to treat malaria. Instead the illness is treated with tubers of *Satyrium nepalense* (Pant and Raskoti 2013). This shows a lack of familiarity with Traditional Chinese Medicine (TCM). Perhaps there was little or no contact with China in the past. Thus with respect to *shihu* (medicinal *Dendrobium* employed in Traditional Chinese Medicine) although two classic species (*Dendrobium moniliforme*, *D. nobile*) occur in Nepal, they are not employed medicinally. Instead, pseudobulb of *Dendrobium densiflorum* (local name *sungabha*) is used, but only to treat boils and pimples. Children suffering from fever are made to take a bath with lukewarm water to which plant juice of *Dendrobium longicornu*

(Nepali *bawar*) has been added. Boiled root of *Dendrobium longicornu* is fed to livestock suffering from cough. Readers are referred to Manandhar and Manandhar's *Plants and People of Nepal* (2002), Subedi et al.'s *Collection and trade of wild harvested orchids in Nepal* (2013) and Pant and Raskoti's *Medicinal Orchids of Nepal* (2014) for more details of medicinal orchid usage in the country. Table 13.1 provides a summary of their findings (Figs. 13.7, 13.8 and 13.9).

Bhutan

Very little has been published about the usage of orchids in Bhutan, apart from the one high altitude species, *Dactylorhiza hatagirea* (local name *dhang-lag*). As in India and Nepal, it is employed as aphrodisiac and tonic. It is said to impart a glow to the skin (Wangchuk et al. 2008) and promote the generation of sperms (Yeshi et al. 2017). None of the other 124 high altitude medicinal plants were employed as aphrodisiac (Wangchuk et al. 2016).

Fig. 13.7 Amchi Tshampa Nawang Gurung in his clinic at Jomson, Mustang, Nepal. Buddha in the picture is the Medicine Buddha Bhaisajyaguru. He is depicted holding a sprig of myrobalan (*Citrus medica*), in a style that is also adopted in the depiction of Shennong (China's legendary herbalist) and numerous Renaissance physicians and apothecaries. Photo: SK Ghimire

Fig. 13.8 Amchi Tangyal Sangbo at his clinic in Pungmo, Dolpa, Nepal. Photo: SK Ghimire

Fig. 13.9 Unsupervised livestock grazing in the Himalayas significantly alters the ecology of alpine meadows which are the prime habitat of *Dactylorhiza hatagirea*. Photo of Pilkanda, Darchula, Nepal by SK Ghimire

The Food and Agriculture Organization (FAO) of the United Nations listed three more orchids but did not explain their usage: *Gymnadenia crassinervis* (*wangla*, *wanpoilakpa*), *Dendrobium crumenatum* (*pushelts-chhog*) and *Coelogyne occulata* (*pusheltsey meonpa*) (Nawang 1996). More details are provided on *Dendrobium crumenatum* in *The Quintessence Tantras of Tibetan Medicine* which states that it grows near river banks and its roots emit the scent of camphor. The plant is administered to stop vomiting, relieve urine retention or haemorrhoids and treat fever (Clark 1995). When I visited Bhutan a few years ago, I learnt that an *Otochilus* species was employed medicinally, but I was unable to discover how it was employed (Figs. 13.10 and 13.11).

Fig. 13.10 *Dendrobium hookerianum* From: *Annals of the Royal Botanic garden, Calcutta* vol. 8(2): t. 83 (1891) [R Pantling]

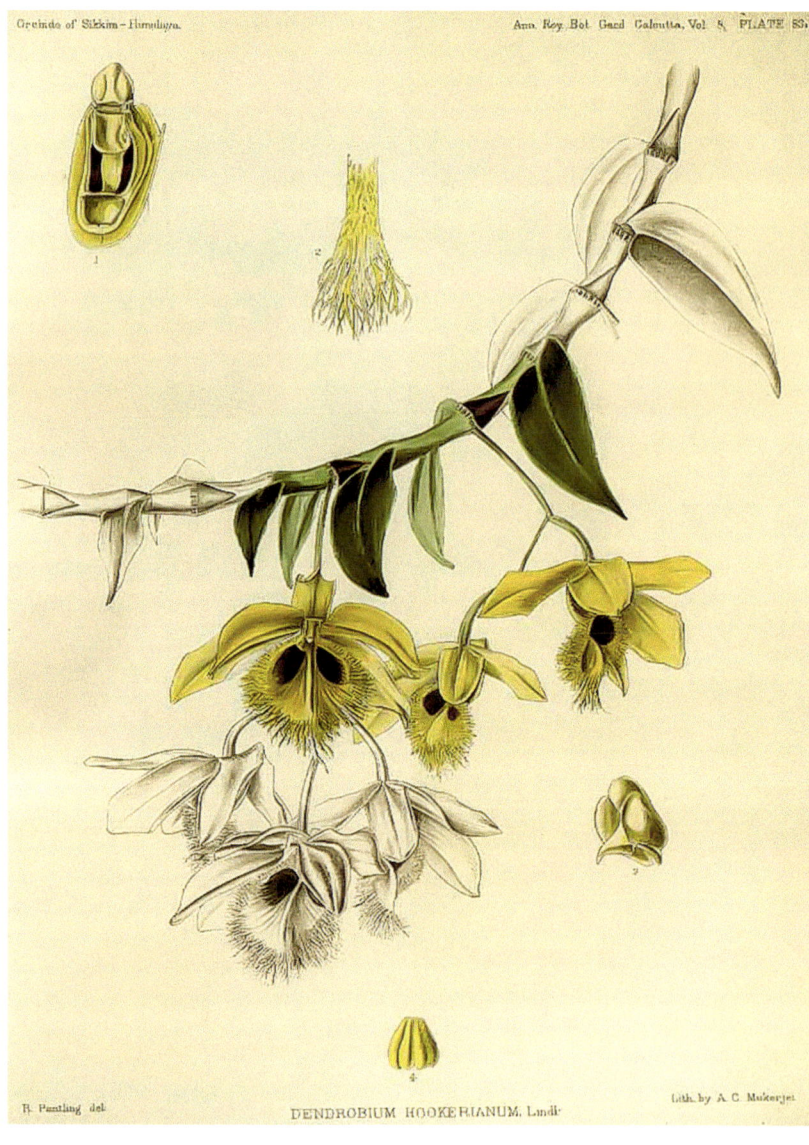

Orchids of Sikkim-Himalaya.   Ann. Roy. Bot. Gard. Calcutta, Vol. 8,   PLATE 83.

R. Pantling del.   DENDROBIUM HOOKERIANUM, Lindl.   Lith. by A. C. Mukerjei

In the markets of Bhutan, the inflorescence of several orchids is sold as vegetable, and presumably they are all consumed as food—*Calanthe plantaginea, C. veratrifolia, Coelogyne cristata, Cymbidium erythraeum, C. hookerianum, C. iridioides, Dendrobium hookerianum* and *Arachnis cathcartii* (syn. *Esmeralda cathcartii*). *Cymbidium* floral shoots are called *ola-choto* or *ola-tshae*. They impart a slightly bitter taste to pork stew. Sometimes they are cooked with hot capsicum and yak cheese (Masters 2015) (Figs. 13.12, 13.13, 13.14 and 13.15).

Fig. 13.12 *Arachnis cathcartii [*as *Esmeralda cathcartii]* From Hooker JD, Fitch WH, *Illustrations of Himalayan Plants*, t. 23 (1855) [WH Fitch]

Fig. 13.13 *Coelogyne cristata*. From *Lindenia, Iconographie des orchidees* [E von Lindemann] Plates 145–192, vol. 4: t. 173 (1888) [P de Pannemeker]

Fig. 13.14 *Cymbidium elegans* var. *elegans*. From: Lindley J, *Sertum Orchidaceum* t. 14 (1818) [Miss Drake]

Fig. 13.15 *Calanthe plantaginea*. From Lindley J, *Sertum Orchidaceum*, t. 24 (1883) [Miss Drake]

References

Acharya KP, Rokya MB (2010) Medicinal orchids of Nepal: are they well protected? Our Nat 8:82–91

Bailes CP (1985) Orchids in Nepal, the conservation and development of a natural resource. Advisory report and recommendations. Royal Botanic Gardens, Kew, Richmond

Baral SR, Kurmi PP (2006) A compendium of medicinal plants of Nepal. IUCN (Published by Mrs. Rachana Sharma), Kathmandu

Bhattarai S, Chaudhary RP, Quave CL, Taylor RS (2002) Prioritization and trade of ethnomedicinal plants by the people of Manang district, Central Nepal. In: Chaudhary RP, Subedi BP, Vetaas O (eds) Vegetation and society, their interaction in the Himalayas. Tribhuvan University/University of Bergen, Nepal/Bergen, pp 151–169

Bhattarai P, Pandey B, Gautam RK, Chhetri R (2014) Ecology and conservation status of threatened orchid *Dactylorhiza hatagirea* (D. Don) Soo in Manaslu conservation area, Central Nepal. Am J Plant Sci 5:3483–3491

Chuakul W (2002) Ethnomedical uses of Thai orchidaceous plants. Mahidol Univ J Pharm Sci 29 (3–4):41–45

Clark B (translator) (1995) The quintessence tantras of Tibetan medicine. Snow Lion Publications, Ithaca

Manandhar NP, Manandhar (2002) Plants and people of Nepal. Timber Press, Portland

Masters S (2015) Orchids of Bhutan. Orchid Rev 49:96–105

Nawang R (1996) Medicinal plants. In: Non-wood products of Bhutan. FAO, Bangkok

Pant B, Raskoti BB (2013) Medicinal orchids of Nepal. Himalayan Map House Pvt. Ltd., Kathmandu

Pant B, Raskoti BB (2014) Medicinal orchids of Nepal. Himalayan Map House (Pte), Kathmandu

Raskoti BB (2012) The orchids of Nepal. Bhaktar Bahadur Raskoti & Rita Ale, Kathmandu

Subedi A, Kunwar B, Choi Y et al (2013) Collection and trade of wild harvested orchids in Nepal. J Ethnobiol Ethnomed 9:64

Teoh ES (2016) Medicinal orchids of Asia. Springer, Cham

Vaidya BN, Shrestha M, Joshee N (2002) Report on Nepalese orchid species with medicinal properties. The Himalayan plants. Can they save us? In: Watanabe T, Bista MS, Saiju HK (eds) Proc. Nepal-Japan Symposium on conservation and utilization of Himalayan medicinal resources. Japan Society for the Conservation and Development of Himalayan Medicinal Resources (SCDHMR), pp 146–152

Wangchuk P, Ugyen S, Thinley J, Afaq SH (2008) High altitude plants used in Bhutanese traditional medicine. Ethnobotany 20:54–64

Wangchuk P, Namgay K, Gayleg K, Dorji Y (2016) Medicinal plants of Dagala region in Bhutan: their diversity, distribution, uses and economic potential. J Ethnobiol Ethnomed 12:28

Yeshi K, Kashyap S, Yangdon P, Wangchuk P (2017) Taxonomical identification of Himalayan edible medicinal plants in Bhutan and the phenolic contents and antioxidant activity of selected plants. J Biol Active Prod Nat 7(2):89–106

Medicinal Orchids of Thailand and Myanmar

14

Thailand lies between China and the Malay Archipelago, so it is interesting to discover whether its medicinal orchid usage has been influenced by the practices in either of these two large regions. That question was answered by a survey conducted by Wongsatit Chuakul of Mahidol University in Bangkok who interviewed herbalists and collected plant specimens from nine provinces in Thailand, seven stretching right across central Thailand and two from the southern peninsula. Forty-two medicinal orchid species belonging to twenty-five genera were obtained during the study period. *Dendrobium*, *Bulbophyllum*, *Cymbidium* and *Luisia* species were more common than species in other genera (Chuakul 2002).

Medicinal orchid usage in Thailand is distinctive. It is not influenced by Traditional Chinese Medicine (TCM), nor by Ayurveda, or by practices in the neighbouring Malay Archipelago. Of the 42 species employed medicinally, only 13 are employed medicinally outside the country. *Habenaria dentata* is used to treat abscesses in Thailand and China, but the Chinese have other medicinal uses for this orchid, whereas Thais limit it to this single usage. Thai usage for the remaining 12 commonly used species are totally different from their medicinal usage in other countries. Species that are endemic and those with distribution confined to either Thailand and Indochina (e.g. *Trias*) or Thailand and Malaysia (e.g. *Phalaenopsis pulcherrima*) are not employed medicinally outside Thailand.

In Thailand, ten orchid species are employed as tonics, the magic potion recommended by tribal practitioners in lieu of health supplements offered by the western pharmaceutical industry (Table 14.1). Six species are used to treat liver dysfunction, four for asthma and fever. Of the four species employed to treat fever, three (*Dendrobium draconis*, *D. trigonopus* and *Grammatophyllum speciosum*) are also used to treat worm infestation. Three species are employed for the treatment of ear infection and abscesses, and another three are used to relieve body ache. *Habenaria dentata* is used to treat either body ache or abscesses. Seven species have single usage (Fig. 14.1).

Coelogyne fuscescens var. *brunnea* is the only aphrodisiac orchid, and apart from such usage, it is also employed to treat burns and ear infection (Chuakul 2002). To handle venomous bites by snakes, scorpions or centipedes, a Thai herbalist remedy recommends crushing the stem of *Grammatophyllum speciosum* in distilled rice wine to make a potion to be drunk as well as for application on the bite. However, as snake bite may be extremely serious and proof of this remedy's efficacy was lacking, the authors of the ethnobotanical publication recommended every victim to seek immediate help from a hospital (Chuakul et al. 1997) (Figs. 14.2 and 14.3).

© Springer Nature Switzerland AG 2019
E. S. Teoh, *Orchids as Aphrodisiac, Medicine or Food*, https://doi.org/10.1007/978-3-030-18255-7_14

Table 14.1 Medicinal Thai orchids listed according to their usage (after Chuakul 2002)

Illness/usage	Orchid species employed
Tonic	*Acampe ochracea* (Lindl.) Hochr.; *Acampe rigida*(Buh-Ham ex Sm.) P.F.Hunt; *Apostasia wallichii* R.Br.; *Bulbophyllum retusiusculum* Rchb.f.; *Bulbophyllum rufinum* Rchb.f.; *Calanthe cardioglossa* Schltr.; *Geodorum recurvum* (Roxb) Alston (syn. *Geodorum attenuatum* Griff.); *Robiquetia succisa* (Lindl.) Seidenf.; *Bulbophyllum lopalanthum* Verm. (syn. *Sunipia grandiflora* (Rolfe) P.F.Hunt); *Bulbophyllum densiflorum* [syn. *Trias disciflora* (Rolfe) Rolfe]
Liver dysfunction	*Bulbophyllum flabellum-veneris* (J.Koenig.) Aver. *Cymbidium ensifolium* (L.) Sw.; *Luisia thailandica* Seidenf.; *Vanilla aphylla* Bl.
Kidney problems	*Cymbidium aloifolium* (L.)Sw.
Heart problems	*Nervilia aragoana* Gaudich.
Faintness	*Nervilia crociformis* (Zoll. & Mor.) Seidenf.
Asthma	*Dendrobium cumulatum* Lindl.; *Thrixspermum centipeda* Lour. *Bulbophyllum nasutum* Rchb.f. [syn. *Trias nasuta* (Rchb.f.) Stapf.]
Fever	*Dendrobium draconis* Rchb.f.; *D. trigonopus* Rchb.f.; *Grammatophyllum speciosum* Bl.; *Spathoglottis eburnea* Gagnep.
Worms	*Dendrobium draconis* Rchb.f.; *Dendrobium trigonopus* Rchb.f.; *Grammatophyllum speciosum* Bl.
Snakebites, scorpion and centipede bites	*Grammatophyllum speciosum* Bl.
Insect bites	*Eulophia herbacea* Lindl.; *Habenaria hemistrata* Rolfe
Ear infection	*Coelogyne fuscescens* var. *brunnea* Lindl. *Cymbidium aloifolium* (l) Sw., *Cymbidium finlaysonianum* (l.) Sw.; *Phalaenopsis pulcherrima* (Lindl.) (syn. *Doritis pulcherrima* Lind
Abscess	*Habenaria dentata* (Sw.) Schltr.; *Spathoglottis affinis* de Vriese
Fractures	*Coelogyne trinervis* Lindl.
Headache	*Dendrobium indivisum* Miq. var. *pallidum* Seidenf.
Body ache	*Habenaria dentata* (Sw.) Schltr.; *Gastrochilus obliquus* (Lindl.) Kuntze., *Mycaranthes pannea* (Lindl.) S.C.Chen & J.J.Wood (syn. *Eria pannea* LIndl.)
Burns	*Bulbophyllum lobbii* var. *siamensis*; *Coelogyne fuscescens* var. *brunnea* Lindl
Diabetes	*Cleisostoma fuerstenbergianum* F. Kraenzl.; *Luisia thailandica* Seidenf., *Luisia trichoriza* (Hook.f.) Bl.
Aphrodisiac	*Coelogyne fuscescens* var. *brunnea* Lindl.

Fig. 14.1 *Grammatophyllum speciosum* (tiger orchid) (©Teoh Eng Soon 2019. All Rights Reserved)

Fig. 14.2 *Coelogyne fuscescens* var. *brunnea*. From: *Curtis Botanical Magazine* vol. 91 [ser. 3 vol. 21] t. 5494 (1865) [WH Fitch]

Today, many *Dendrobium* species are plundered from the forests of Thailand and Indochina for sale to Chinese herb dealers, but their designated usage is as *shihu* which is a Chinese medicinal preparation, not Thai. These orchids have no role in Thai herbal medicine. Many Thai orchids that play a role in Chinese herbal medicine were not mentioned by Thai herbalists (Figs. 14.4, 14.5, 14.6, 14.7, 14.8 and 14.9).

Medicinal Orchids of Myanmar

Likewise, in Myanmar, many orchid species are employed to make a tonic. In the course of their study on medicinal orchid usage in various parts of Myanmar, Myint Myint San and his colleagues from the Forest Department of Myanmar discovered that in Kutkai Township, Northern Shan State, nine species of *Dendrobium* were valued as a tonic. Pseudobulbs of *Dendrobium cariniferum* (local name *Mahar deiwi*), *Dendrobium chrysotoxum* (*Shwe tu, Mouk Khan*

Fig. 14.3 *Coelogyne cristata.* From *Edward's Botanical Register* vol. 27: t. 57 (1841) [SA Drake]

War), *Dendrobium crepidatum* (Ganaing *Na Bay Pauk*), *Dendrobium dixanthum* (*Shwe War Ga Lay*), *Dendrobium falconeri* (*Myet Thit Khwa*), *Dendrobium fimbriatum* (*Arme Let Tan To*) and *Dendrobium nobile* (*Dawn Mee Thitkhwa*) were employed for the preparation of the tonic. The fruit of *Dendrobium cariniferum* and the flowers of *Dendrobium chrysotoxum* were also employed in the preparation of a tonic. Leaves of *Dendrobium moschatum* and *Dendrobium pendulum* (*Mya-sit-kyoe*) were a specific tonic for the liver (Table 14.2) (San and Myint 2009). At Matu Pe Township, *Malaxis biaurita* is employed to prepare a tonic (San et al. 2015) (Figs. 14.10 and 14.11).

Usage of orchids as tonic appears to be a common practice among tribals. In Arunachal Pradesh Jambey Tsering and his colleagues discovered that 32 species of orchids were employed as tonic (Tsering et al. 2017).

Several *Dendrobium* species have additional uses. *Dendrobium falconeri* is consumed by patients suffering from lung cancer. *Dendrobium crepidatum* is used to treat arthritis and rheumatism. Leaves of *Den. moschatum* are employed to make a paste to apply on fractures. Usage of *Dendrobium nobile* is the most varied: it is employed as aphrodisiac or to boost a person's immunity, relief pain and treat stomachache,

Fig. 14.4 *Bulbophyllum lobbii* (syn. *Bulbophyllum siamensis*) (©Teoh Eng Soon 2019. All Rights Reserved)

Fig. 14.6 *Mycaranthes pannea* (syn. *Eria pannea*) (©Teoh Eng Soon 2019. All Rights Reserved)

Fig. 14.5 *Phalaenopsis pulcherrima* forma *alb*a (syn. *Doritis pulcherrima* var. *alba*) (©Teoh Eng Soon 2019. All Rights Reserved)

pulmonary tuberculosis, fevers and lumbago (San and Myint 2009.

At Matu Pe Township in the southwestern Chin State, women prepare a toothpaste or shampoo with pseudobulbs of *Dendrobium gregulus* and *Dendrobium laterale* (San et al. 2015). In the Ayeyarwady Delta and in Tanintharyi Region, the

Fig. 14.7 *Luisia thailandica* (©Teoh Eng Soon 2019. All Rights Reserved)

Fig. 14.8 *Geodorum recurvum* (©Teoh Eng Soon 2019. All Rights Reserved)

Fig. 14.9 *Calanthe cardioglossa* (©Teoh Eng Soon 2019. All Rights Reserved)

Table 14.2 Medicinal orchids of Myanmar

Orchid species	Medicinal usage
Bulbophyllum sp.	Dandruff
Coelogyne cristata	Aphrodisiac wounds and sores
Coelogyne nitida	Headache, fever; burns
Coelogyne prolifera	Headache, fever
Coelogyne stricta	Headache, fever
Cymbidium aloifolium	Gonorrhoea, conjunctivitis; stomachache, dysentery, earache, fractures
Dendrobium aphyllum	Tonic
D. cariniferum	Tonic
D. chrysotoxum	Tonic
D. crepidatum	Tonic
D, crystallinum	Tonic
D, dixanthum	Tonic; arthritis and rheumatism
D. falconeri	Tonic; lung cancer
D. fimbriatum	Tonic
D. formosum	Tonic
D. moschatum	Fractures
D. nobile	Tonic; aphrodisiac; enhance immunity; pain relief, stomachache, lumbago; fever, tuberculosis
D. ochreatum	Tonic
D. parishii	Tonic
D. pendulum	Tonic for liver
D. pulchellum	Tonic
Eulophia spp.	Unspecified
Geodorum spp.	Promotes longevity
Habenaria spp.	Unspecified
Malaxis biaurita	Tonic
Mycaranthes pannea	Malaria; bone aches
Pecteilis hawksiana	Unspecified
Pholidota articulata	Tonic
Pholidota imbricata	Tonic; arthritis
Pleione praecox	Tonic
Rhynchostylis retusa	Rheumatism; cuts and sores; emetic; insect repellent
Vanda cristata	Tonic

women use *Bulbophyllum* pseudobulbs to make the shampoo for they believe that the natural shampoo enhances hair growth and colour and it also removes dandruff (Kurzweil and Lwin 2014). Here, paste made with pseudobulbs of *Coelogyne prolifera* is consumed for headaches or fever; also it is applied externally on burns. A poultice made with pseudobulb of *Pholidota imbricata* is applied over swollen joints. Fibres of *Thunia alba* are used to treat cracks on the heel (San et al. 2015).

Current database lists 800 species in 150 genera for orchids of Myanmar (Grant 1895; Kurzweil and Lwin 2014), but there are probably many more species which remain undiscovered. The orchid flora of the country is very diverse, and there are affinities with neighbouring countries such as Thailand, Yunnan (China), northeast India, Bangladesh and Bhutan. Thirty one are stated to be medicinal but the primary sources are not indicated. Tubers of *Eulophia*, *Habenaria*, *Geodorum* and *Pecteilis hawksiana* (syn. *Pecteilis sagarikii*) are employed as medicinal herbs by native practitioners of Mount Popa in Central Myanmar. In the Shan State in northern Myanmar, it is believed that eating the tubers of *Geodorum* will ensure longevity (San et al. 2015).

In the southern Chin State, the following are employed medicinally with applications similar to those of the northern Shan; only the species are different.

For making tonic, pseudobulbs of *Dendrobium aphyllum*, *D. ochreatum*, *D.parishii*, *D. pulchellum*, *D. crystallinum*, *D. formosum*, *Pholidota articulata*, *Pholidota imbricata* and *Pleione praecox* and leaves of *Vanda cristata* are employed. For toothpaste and shampoo, pseudobulb of *Dendrobium incurvum* is an alternative. For headache or fever, use *Coelogyne stricta* and *Coelogyne nitida*, the latter also for burns. *Coelogyne cristata* has more uses: besides being an aphrodisiac, juice extracted from pseudobulbs is applied on wounds and boils; a gum made with mashed pseudobulb is applied on sores. Powdered leaves of *Rhynchostylis retusa* are employed to treat rheumatism, juice extracted from the roots is applied to cuts and wounds and dried flowers to induce vomiting. Flowers of *Rhynchostylis retusa* also acts as an insect repellent. The entire plant of *Eria pannea* is used to treat ague and bone aches. Rhizome and flowers of *Cymbidium aloifolium* are employed to treat gonorrhoea and conjunctivitis (San et al. 2015). In the Kachin State where it is a common epiphyte, decoction of pseudobulbs of *Cymbidium aloifolium* is used

Fig. 14.12 *Dendrobium parishii* (©Teoh Eng Soon 2019. All Rights Reserved)

Fig. 14.14 *Dendrobium dixanthum* (©Teoh Eng Soon 2019. All Rights Reserved)

Fig. 14.13 *Dendrobium formosum* (©Teoh Eng Soon 2019. All Rights Reserved)

Fig. 14.15 *Dendrobium falconeri* (©Teoh Eng Soon 2019. All Rights Reserved)

Fig. 14.16 *Pleione praecox* (©Teoh Eng Soon 2019. All Rights Reserved)

Fig. 14.17 *Coelogyne nitida* (©Teoh Eng Soon 2019. All Rights Reserved)

Fig. 14.18 *Pecteilis hawksiana* (syn. *Pecteilis sagarikii*) (©Teoh Eng Soon 2019. All Rights Reserved)

However, it is interesting to note that in the WHO's *Medicinal Plants of Myanmar* 59 medicinal plants are described but no mention is made of any orchid.

References

Chuakul W (2002) Ethnomedical uses of Thai orchidaceous plants. Mahidol Univ J Pharm Sci 29(3–4):41–34

Chuakul W, Saralamp P, Paonil W et al (1997) Medicinal plants in Thailand, vol 2. Dept of Pharmaceutical Botany, Faculty of Pharmacy, Mahidol University, Bangkok

Grant B (1895) The orchids of Burma. Hanthawaddy Press, Rangoon

Kurzweil H, Lwin S (2014) A guide to orchids of Myanmar. Natural History Publications, Kota Kinabalu

San MM, Myint KW (2009) Preliminary survey on medicinal values of some *Dendrobium* species occurring in Kutkai Township, Northern Shan State. The Republic of Myanmar Ministry of Environmental Conservation and Forestry, Forest Department Leaflet No. 14/2009

to treat stomachache and dysentery. The pseudobulb is also employed to treat earache, whereas a poultice made from the leaves is employed for fractures (Kurzweil and Lwin 2014) (Figs. 14.12, 14.13, 14.14, 14.15, 14.16, 14.17, and 14.18).

San MM, Aung NMM, Soe HS, Kyaw YMM (2015) Study on distribution and medicinal values of wild orchids in Matu Pe Township, Southern Chin State. The Republic of Myanmar Ministry of Environmental Conservation and Forestry, Forest Department Leaflet No. 30/2015

Tsering J, Gogoi BJ, Hui PK et al (2017) Ethnobotanical appraisal of wild edible plants used by the Monpa community of Arundacal Pradesh. Indian J Tradit Knowl 16(4):621–637

Medicinal Orchids in the Malay Archipelago

15

Interest in securing the sources of valuable spices had propelled European maritime traders to fund seafaring expeditions to India and the Far East. When the Verenigde Oost-Indische Compagnie (VOC, United East India Company or better known as the Dutch East India Company) was founded in 1602 in the Netherlands, instructions drafted by Carolus Clusius (Charles de l'Ecluse) (1520–1609) were handed to apothecaries and surgeons on board '*that they bring along branchlets with their leaves, laid between paper. ... Especially of the searched after spices: pepper, nutmeg, mace, cloves and cinnamon, but also of any other interesting plant. To make illustrations, and to record local names and uses, and how and where they grow*' (Baas and Veldkamp 2013).

Clusius was first prefect of the Leiden Botanic Gardens, and he would later be made an honorary professor of botany at Leiden University. He translated into Latin the botanical works of the Jewish physician and naturalist, Garcia da Orta (c. 1490–1570) who worked in Goa, India. da Orta's *Coloquios dos simples e drogas da India* became an important medical text which was widely read in Europe. By invitation, Clusius was also able to study and describe the botanical specimens collected during Francis Drake's circumnavigation of the world. However, da Orta and Clusius did not see the living plants in Malesia nor had they spoken to the local people.

Thus, when da Orta made mistakes in his *Coloquios dos simples e drogas da India*, Clusius did not correct them. These mistakes were subsequently pointed out by Jacobus Bontius (1592–1631) who served in Batavia as personal physician to its first governor-general, Jan Pieterszoon Coen (1587–1629). Nevertheless, Clusius's influence on the VOC played a prominent positive role in the development of Asian tropical botany. During the seventeenth century, tropical medicinal plants and spices were cultivated in the Botanic Gardens in Leiden and Amsterdam, in the private gardens of directors of the VOC and in the garden of King-Stadhouder William III (1650–1702). Such interest paved the way for the monumental studies undertaken by van Rheede in Southern India, Bontius in Java and Rumphius in Ambon. It partially explained the inordinately generous support bestowed on Rumphius.

Rumphius (1621–1702), the Indomitable, Blind Naturalist

One could hardly find a scientist more persevering than Rumphius (1621–1702). He was baptized as Georg Eberhard Rumpf but is better known by the anglicized name, Rumphius. The last name was actually derived from the Latin version Georgius Everhardus Rumphius adopted when his magnum opus, *Herbarium Amboinense*, was translated from Dutch into Latin by Johannes Burman (1707–1779) who had it printed as an expensive limited edition of 500 copies between 1741 and 1755. The set cost 1000 guilders (2000 Dutch shillings, now Euro 20,000), 'about a third of the income of a physician in Amsterdam' (Veldkamp 2011). When Rumphius was elected a member of the Academia Naturae Curiosorum of the German Roman Empire, he was given the nickname Plinius [after the famous Roman encyclopaedist, Gaius Plinius Secundus (23–79 CE) who praised the aphrodisiac properties of orchids], but he never used that name. Later, his own name would itself denote a mind overflowing with knowledge: Blume, the second director of Bogor Botanic Gardens, a medical doctor, was nicknamed 'Rumphius' (Veldkamp 2011).

Rumphius was born in Wolfersheim, a small town some 25 miles north of Frankfurt in 1627 in the midst of the catastrophic Thirty Years' War (1618–1648). The war started as a religious war between Catholics and Protestants in the Holy Roman Empire, but it soon involved large areas of Europe and resulted in 8 million casualties. Rumphius' father, August Rumpf (d. 1666), was a *baumeister*, a building contractor who also designed buildings and effected their repair. The reconstruction August Rumpf undertook during the war brought him into contact with local aristocrats. This gave young Rumphius the opportunity to enjoy the privilege of instruction by a private tutor in Latin, Greek and Hebrew. August taught his son mathematics, the principles of construction, drafting and the construction of defensive structures like fortifications and redoubts. Admitted to the Gymnasium in Hanau when he was 10 or 11, Rumphius acquired knowledge of grammar, rhetoric, logic, arithmetic, geometry, music and astronomy. Rumphius was fluent in Dutch, his mother's native tongue, and wrote entirely in this language (Beekman 2003).

When he was 18, Rumphius wanted to visit Italy. He signed up as a mercenary not knowing that the recruitment was a trick to provide a relief force for Brazil. A storm changed his fortune. Shipwreck brought him to Portugal where he spent the next 4 years as a mercenary soldier. When the war ended, Rumphius returned home, and for a period of 18 months, he worked as a *bauschreiber* for the local count, drawing plans for construction, seeing to the job and instructing his employer's children in arithmetic, geometry, architecture and related arts. He quit his job because he did not share the religious views of the count. He signed on again as an *adelborst* (gentleman soldier, perhaps 'officer'), this time with the multinational Dutch East India Company.

He arrived in Batavia (now Jakarta) in the Dutch East Indies in 1653 as a midshipman. That December he was sent to Moluccas as part of a military force to deal with the Fifth Ambonese War (1651–1656). Moving up to ensign and engineer, Rumphius, now 30 years old, managed to transfer to a civilian post in 1657, and shortly afterwards he obtained dispensation from the governor to undertake full-time study of the fauna and flora of the Spice Islands. His progress in the Dutch East Indies showed that he must have impressed the people who mattered. At one stage, he even drew a higher salary than the governor of the island. Rumphius lived well for many years (Fig. 15.1).

In his *Herbarium Amboinense* published posthumously between 1741 and 1755, Rumphius described 1200 species of plants. It was an amazing achievement. Any single one of the catastrophes that he met would have destroyed a lesser man. Perhaps his stint in the army had taught him to keep fighting after losing friends and comrades, and he had become resourceful, learning to adapt when he worked as a *bauschreiber*.

In 1670, 13 years after he started working as a self-taught naturalist, Rumphius became blind as a result of glaucoma. It was only through the help of his wife, son and associates that he managed to continue with his work. His wife, Susanna, two daughters and a maid died when a wall collapsed on them during an earthquake that was followed by a *tsunami* in 1674 (Veldkamp 2011). Then, when the six-volume work was nearing completion, a fire which razed the town destroyed his library. He lost a large portion of the text and half

Fig. 15.1 Rumphius
(Georg Eberhard Rumpf)
(1621–1702)

the illustrations. That happened in 1687. Rumphius managed to get the drawings redrawn with the help of his son, Paulus. Everything that he had compiled in Latin in the original work had to be rendered in Dutch for his associates did not know Latin. For portions of the text that were lost, he dictated from memory (Veldkamp 2011)!

In 1690 the work was finally completed. Six volumes were dispatched by sea to the Netherlands, but the boat carrying it was sunk by the French, forcing Rumphius and his colleagues to start all over again. Fortunately, he had retained a working copy in Ambon, and thus he managed to complete a new set, expanded to nine volumes. Six volumes reached the

Netherlands in 1896, and the last three books were dispatched a year later. At this juncture, the Dutch East India Company (VOC) decided that the work carried too many sensitive items that might adversely affect their monopoly of the spice trade, in particular the data on nutmeg and cloves. The publication was put on hold. In 1702 the company softened its stance, but they still wanted to remove passages that would affect their trade, and secondly, the VOC would not contribute any money to its publication (Beekman 2011). Rumphius never got to see the book in print because it was only published 40 years after his death. An English translation of all six volumes of *Herbarium Amboinense* was carefully translated with annotations and an introduction by C Monty Beekman and published by Yale University Press in 2011.

Medicines shipped from the Netherlands were often spoilt and ineffective when they arrived in the East Indies. This forced Rumphius to rely on local remedies even though he realized that most of the treatments were only based on anecdotes or 'old wives' tales'. Nevertheless, he believed that some might still be efficacious, so he set out to document all that he heard, saw or tested. In his preface to the *Ambonese Herbal*, Rumphius stated that it was addressed 'particularly to those who lived in the Indies' (Beekman 2003). He established a medicinal garden and a forest garden, the latter now known as *Dusun Rumphius*.

Over 35 species of orchids are described, and many are illustrated in 55 pages in Volume 5 of *The Ambonese Herbal*. As in the case of orchids described in the *Hortus Indicus Malabaricus*, the names Rumphius gave to most of these Indonesian orchids are very different from the names applied to them today. One is required to rely on the detailed descriptions of Rumphius and on the line drawings to identify the species. Vegetal form is well presented in the illustrations, but single flowers are often not featured with the necessary details for the modern taxonomist. Although some identities are uncertain, whatever could be identified with some degree of certainly were named by J. J. Smith who worked at Buitenzorg (Bogor Botanic Gardens) from 1891 to 1924. Smith was an expert on Indonesian

orchid species and familiar with the work of Rumphius (de Wit 1977; Arditti 1989). Nevertheless, Smith's is not the final word; for instance, recently, Jim Comber suggested that the correct identity of the principal 'yellow *Angrec*' is *Dendrobium bicaudatum*, whereas J.J. Smith specified *Dendrobium mirbelianum*, and de Wit thought it was *Dendrobium strebloceras* (Fig. 15.2); however, the last species is endemic in Halmahera and thus unknown in Ambon (Beekman 2003). All the three suggested species belong to Section Spathulata of *Dendrobium*, i.e. their flowers have narrow twisted petals, their sepals forming a triangle and lip is white with purple keels. *Dendrobium bicaudatum* and *Dendrobium strebloceras* are fragrant, which fits the description of Rumphius. They differ in that petals of *Dendrobium bicaudatum* are only slightly twisted, whereas there are three twists on the petals of *Dendrobium strebloceras*. Therefore, Comber's *Dendrobium bicaudatum* fits Rumphius's description better than the other species (Fig. 15.2).

Rumphius commented that leaves of his yellow *Angrec* 'had a sour taste, mixed with saltiness, and it blunts the teeth' (Beekman 2003).

Among the yellow *Anggreks*, Rumphius described a flower that resembles a flying horsefly, ... 'somewhat thin, many on top of each other, on longish stems'. The flower is 'fashioned from five leaflets, of which the top is slightly curled forward, resembling [the horsefly's] back; the other two would be the wings; two others on the side are shorter, and cover a central leaflet which has two flaps on the side, which encircle the little central pillar; the outer leaves are somewhat striped like the Kananga flower, yellow on the outside, a bright yellow inside, like the *Tsjampacca* flower (*Magnolia chempaca*); the little helmet has a rim like a tongue, bent upwards, and striped with purple lines. The lower part of the flower resembles a horn, like Larkspur, representing the horsefly's head, almost odorless' (Beekman 2003). Among the *Spatulata* section of *Dendrobium*, the species *Dendrobium sutiknoi* from Maluku is the closest fit for the description, but it was only described by Peter

O'Byrne in 2005 and therefore unknown to Smith, de Wit, Comber and Beekman (Fig. 15.3).

A *Vanda* plant with flowers is illustrated in Plate 46 and referred to by Rumphius as *Angraecum furvum* or, in Malay, *Angrek kitsjil glap*. He described the flowers as yellow on the outside and dark inside, of a russet or smoky colour, with yellow at the edges. They emitted a faint scent. This is probably *Vanda lindenii* which is native to the Spice Islands and Mindanao (O'Byrne 2001) (Fig. 15.4).

Plate 49 features another *Vanda*, a wiry plant with open, well-spread flowers. It scrambles over rocks, and Rumphius referred to it as *Angraecum saxatile*. The illustration and description leave no doubt that this is *Vanda saxatilis* which is endemic in Maluku. The current name was assigned by J.J. Smith for the saxicolous orchid, a lithophyte (Fig. 15.5).

Whereas for many orchids identification is uncertain, from the excellent drawing in the *Ambonese Herbal*, there can be no doubt that the Purple *Angrek* is *Dendrobium purpureum* (Fig. 11.4). It had two Malay names, *Angrek jambu* and *Angrek cassamba*. Growing on its common host plants, *Waringin* (*Ficus benjamina*), *Sumaria* (*Casuarina equisetifolia*) and clove (*Syzygium aromaticum*), its canes could be 7 to 8 feet long. It was used to treat *matta icaro* (whitlow) without the addition of other ingredients. Stems were crushed, heated and then smeared on the affected parts of the fingers. The heat caused whitlow abscesses to rupture, releasing the pus and accelerating healing. *Phaius amboinensis* (*Angraecum*

terrestre alterum; Malay *Angrec tana*) is included in the illustration for *Dendrobium purpureum*. The *Phaius* was fully described, but Rumphius was unaware of any usage (Fig. 15.6).

The section on yellow orchids concludes with a detailed description of *Luisia confusa* (syn. *Luisia amboinensis*) labelled by Rumphius as *Angraecum decimum* or *angustifolium* and referred to by Jacobus Bontius as *Sedum arborescens*. The species is widespread in Indonesia, including Ambon. Rumphius reported that the renowned physician Jacobus Bontius ascribed many excellent virtues to the plant. Javanese used conserved leaves to treat disorders of the brain and nerves, bloody diarrhoea and poisoned wounds inflicted by krisses and pikes. It was believed that the preparation fortified the heart.

Also well described and illustrated is *Dendrobium anosmum* (Figs.15.7 and 15.8) whose strong fragrance did not delight Rumphius.

He called it the Dog Orchid, *Angraecum caninum*, in Malay *Angrek andjing*, on account of the canine odour of its flowers. It grew on trees with short, thick, mossy trunks like *kinar* (*Kleinhovia hospita*) near beaches and in the lowland (Beekman 2003). Based on the description in the *Ambonese Herbal*, J.J. Smith identified *Angraecum album minus* (plant on left of Fig. 11.5) as *Dendrobium ephemerum* which is found in the lowlands of Sulawesi and the Moluccas. However, plant and flower with its pointed petals included in the illustration could fit the fragrant pigeon orchid, *Dendrobium crumenatum*, a close relative, also found in the lowlands and widespread throughout Southeast Asia.

Among the best known illustrations in *Ambonese Herbal* is the one of *Grammatophyllum scriptum* which Rumphius referred to as *Angraecum scriptum* (inscribed orchid) (Figs. 15.9 and 15.10). It was 'an aristocrat of wild plants that lived high on the trees like aristocrats flaunting their finery, also just like majestic castles or fortresses situated on high'. In those days, Moluccan princesses would not permit common women to deck their hair with its flowers unless they were royal wives, sisters or daughters. Whereas it could be found on several species of trees, it was on the *Kalappa* tree (*Cocos nucifera*, coconut) that it grew best and readily fruited. Therefore, Rumphius also called it *Angrec calappa*. *Angrek lidah* was an alternative name which alluded to the stiff leaves that resembled huge tongues. Another name used by Rumphius was *Helleborine molucca*. An outstanding feature of this *Angrek* was the stiff, upward-pointing, white roots surrounding the plant and forming a nest to capture nutrients for the orchid. Rumphius recorded that the Javanese called its flowers *Rangrec*, whereas the Portuguese gave it the names, *Fulha a lacra* or *Fulha lacre*.

The inscribed orchid was used to treat whitlow in the following manner:

> Take marrow of the stem; pound it with some *Curcuma* (turmeric), adding some salt water; then wrap the affected finger with the poultice. The

Fig. 15.4 (right) *Vanda lindenii* and (left) a *Dendrobium*. From: Rumphius, *Herbarium Amboinense*, vol. 6, t. 46 (1750)

abscess will quickly ripen. Early lesions may simply dry up (Beekman 2003).

It was also used to treat worm infestations:

'Peel the stems, mesh the marrow with a little ginger and smear the poultice on the stomach. Some itching may occur but this is transient. Worms will be killed and bad intestinal humour will be immediately expelled'. It will even shrink a *Tehatu* (enlarged spleen). The poultice is wrapped around swollen legs to relieve oedema (Beekman 2003).

Rumphius stated that if one chewed the marrow of the pseudobulbs and rinsed one's mouth with the sap help to cure thrush. The taste of the

Angrec was bland but distinctly 'cooling'. *Grammatophyllum scriptum* was eaten with food to stop 'bloody flux' (bloody diarrhoea. In the seventeenth century it was most likely due to amoebic or bacterial dysentery).

Finally, he revealed a surreptitious Ambonese practice: when a man desired a woman he would add seeds of *Grammatophyllum scriptum* to her food or drink. If she consumed the adulterated item, she was bound to follow him (Beekman 2011). Caring lovers employed a different orchid which Rumphius called *Herba supplex minor*, the small supplicant herb, identified by J.J. Smith as

Fig. 15.5 (left) *Vanda saxatile* and (right) *Arachnis* sp. (*Arachnis flos-aeris*). From: Rumphius, *Herbarium Amboinense*, vol. 6, t. 49 (1750)

Dendrobium moluccense, now renamed *Oxystophyllum moluccense*. This orchid bore stiff, pointed leaves arranged in alternating rows and sheathing the short stem at their base. It resembled a pair of hands held in supplication (herb at bottom of Fig. 11.6). Men in Hitu and Ternate sent these orchid leaves to their lovers whenever they needed to beg forgiveness. If a woman forgave the man, she would send back the succulent, though less stiff, leaves of *Herba simplex major secunda* (*Dendrobium acinaciforme*) (Fig. 15.11, left). Rumphius explained that in the Indies, it was essential to understand the hieroglyphic grammar (Beekman 2003).

Elsewhere, *Oxystophyllum moluccense* had a sinister usage. Alfurs on Ceram wore leafy stems of the orchid as armbands when they went head-hunting, believing that the stiff orchid leaves conferred courage.

In Ternate, natives employed the pseudobulbs of *Orchis amboinica major*, *radice digitata* or *Daun cora cora* (tentatively identified as *Spathoglottis plicata*) to treat 'large, bluish swellings that contained little pus'. (Rumphius was probably referring to bruises.) The crushed tuber was wrapped with a leaf of *Buro malacco* (*Cissus aristata*), heated over a fire and placed over the swelling.

Fig. 15.6
(top) *Dendrobium purpureum*; (centre) *Spathoglottis plicata*; (bottom) *Oxystophyllum moluccense*. From: Rumphius, *Herbarium Amboinense*, vol. 6, t. 1 (1750)

Liparis condylobulbon (syn. *Liparis treubii*) (*Angraecum gajang*) flowers during the rainy season. Rumphius reported that to treat a swollen or hardened abdomen, the local people withered the leaves over a fire until they became limp before rubbing them over abdomen. The patient also had to chew and swallow the marrow of sliced pseudobulbs (de Wit 1977).

The entire plant of *Calanthe veratrifolia* (syn. *Calanthe triplicata*) is quite sharp, and Rumphius warned that one should handle it carefully. Nevertheless, it had medicinal usage. To treat swollen hands, one took the roots of the orchid, some nutmeg, bangle (*Zingiber purpureum*) and ginger,

rubbed them together, and applied the mix to the affected parts. Rumphius commented that natives living on this island possessed such tough mouths, they would chew roots of *Calanthe triplicata* with *pinang*, nutmeg and ginger whenever they suffered from persistent diarrhoea 'caused by cold or raw dampness' (Fig. 15.12).

During the seventeenth century, salep, a powder derived from the paired tubers of Mediterranean terrestrial orchids (*Satyrium*, *Orchis*, *Anacamptis*, *Dactylorhiza*, *Himantoglossum*, *Ophrys*, *Serapia* species) which looked like testicles was much touted as an aphrodisiac. Salep bars flourished in major European cities.

Fig. 15.7   (right)
Dendrobium anosmum and
(left) a *Dendrobium* from
Section *Crumenata*,
e.g. *Dendrobium
crumenatum*. From:
Rumphius, *Herbarium
Amboinense*, vol.
6, t. (1750)

Fig. 15.7   (right) *Dendrobium anosmum* and (left) a *Dendrobium* from Section *Crumenata*, e.g. *Dendrobium crumenatum*. From: Rumphius, *Herbarium Amboinense*, vol. 6, t. (1750)

Thus it is not surprising that Rumphius searched for *Satyrium* among the terrestrial *Angreks* hoping that they would have the same properties. He discovered that plants of *Eulophia* species, *Spathoglottis plicata* (*Orchis amboinica major*, *radice digitata* or *Daun cora cora*) and *Peristylus goodyeroides* had tubers (de Wit 1977; Beekman 2011). However, Rumphius found them unpalatable even after they were candied. He rather preferred the *small amboinensis orchid* (*Habenaria rumphii*; correct name, *Pecteilis susannae*) (Fig. 13.10). It was crumbly, crunchy, sweet and devoid of a nasty smell. Local people did not employ orchids as aphrodisiacs (Beekman

2011), and presumably the species were then more prevalent (Fig. 15.13).

The *Ambonese Herbal* also carried a description of *Phalaenopsis amabilis*, the moon orchid which is now the national flower of Indonesia (Fig. 15.14). Rumphius reported that it was called *Anggrek putih besar* in Malay or *Bombo terbang* (*Pombo terbang*, dove in flight), the latter a name adopted by the Dutch and rendered as *Vliegende Duive*. It was called *Angrec colan* in Bali and *Wanlecu* on Luhu. There was no mention of medicinal usage (Beekman 2011), but leaves of *Phalaenopsis amabilis* were eaten as food (Lawler 1984). Rumphius also mentioned two

Fig. 15.9 *Grammatophyllum scriptum* (©Teoh Eng Soon 2019. All Rights Reserved.)

Fig. 15.8 *Dendrobium anosmum* (©Teoh Eng Soon 2019. All Rights Reserved.)

small white orchids (*Angraecum album minus*), of which one has stems '4 or 5 feet long' and the other '5 or 6 inches long, coming at an angle from the tree and partly hang(ing) down' (Beekman 2011). Two common *Dendrobium* species that are widespread in Southeast Asia and whose stems fit the descriptions would be *Dendrobium ephemerum* and *Dendrobium crumenatum*, respectively. However, again Rumphius makes no mention of any medicinal applications although *Dendrobium crumenatum* was used to treat earache from India to Peninsular Malaysia and Indonesia.

A red *Angrec* is described and illustrated. From the drawing one may surmise that this is *Renanthera matutina* which is distributed from Sumatra to Malaysia, Kalimantan and Java

(Handoyo 2010), and probably it occurred in Ambon. 'It grew by the beach and along riverbanks in the valley. Flowers are deep yellow and densely overlaid with red lines and dots'. Its thick, young leaves pickled in salt and vinegar tasted rather like capers. Sometimes they were served in the company of other pickles (*Atsjaar*). The leaves are fibrous. Rumphius advised that the way to appreciate the pickled orchid was to chew the leaves and suck the juice. The thickest and fattest leaves were best for making pickles, and these should be obtained from plants growing in the forests (Figs. 15.15 and 15.16).

Rumphius went on to describe a variation of the red *Angrec*, this species possessing petals that were not pointed. There were three colour forms—red, orange and yellow, and they grew on the beach in thickets, but not on trees. He was describing *Renanthera elongata*. Rumphius also described *Vandopsis lissochiloides* (*Angraecum scriptum minus*) which he compared with *Grammatophyllum scriptum* although in

Fig. 15.10 *Grammato-phyllum scriptum.* From: Rumphius, *Herbarium Amboinense*, vol. 6, t. 42 (1750)

vegetative form the two are entirely different, *Macodes petola*, various species of *Dendrobium*, *Vanda* (perhaps *Vanda celebica*), *Coelogyne rumphii* and a few other species whose identities are still being discussed (Beekman 2011).

Jacobus Bontius (1592–1631)

Before Rumphius there was Jacobus Bontius (1592–1631), who was a personal physician to the governor of Batavia (Fig. 15.17). A brilliant doctor and a pioneer of tropical medicine, he was the first physician to describe beriberi, cholera

and several forms of dysentery. His was also a family reputation: his father was appointed to the first chair in medicine in Leiden University.

In 1626 the VOC appointed Bontius as apothecary, physician and surgical inspector to oversee all VOC territories in Asia. He stayed in Batavia (Jakarta) for 4 years during which time there were two sieges. He witnessed two dysentery epidemics, himself suffering an attack besides coming down with other serious illnesses including beriberi. His misfortunes during his posting were great and numerous. His wife died during the voyage out; the second wife whom he married in Batavia died from cholera less than 3 years

Fig. 15.11 *Dendrobium acinaciforme*. From: Rumphius, *Herbarium Amboinense*, vol. 6, t. 51. Fig. 1. (1750)

after their marriage; and his elder son succumbed to 'kinderpoxkens'. He also lost many friends. He was appointed to the Court of Justice in 1628 and became its Chief Officer in 1630 but was glad to give up the post after serving a year.

Bontius's main interests were medicine and medicinal botany, and despite his heavy duties, he relentlessly pursued his interests in depth. His manuscripts were written at night, after his official duties, and they were based on his personal observations. He had the opportunity to study actual samples of the herbs he commented on. He was a humble man who listened to the natives. To better understand the local illnesses, he performed autopsies with his colleague, the

Scottish surgeon, Andrew Durie. Bontius described 19 tropical diseases and the manner by which local physicians used herbal remedies to manage the diseases. For reference he had 2000 'important' books which he brought with him to the East, but he was seeing new maladies that he had not encountered in Europe. He published the first descriptions of beriberi and cholera, the former caused by vitamin B1 (thiamin) deficiency which was prevalent among people whose diet was composed principally of rice or tapioca.

Bontius appreciated that if the Dutch were to be successful in maintaining their colony in the East Indies, they must learn to employ local remedies to cope with illnesses. European simples

were ineffective against the tropical diseases, and they rapidly became mouldy and spoilt in the humid tropics. He mentioned tea as medicine in addition to it being a beverage: (in the presence of widespread water-borne diseases, hot tea was the safe drink). He recognized the importance of a healthy lifestyle. He emphasized the importance of fruit and vegetable as essential components of a healthy diet that played a role in the prevention of disease. Unfortunately, diet alone was insufficient to combat all the tropical diseases, and Bontius died at age 39 from dysentery (without antibiotics an incurable disease) before he could fulfil his ambition to produce a comprehensive *Natural History of the East Indies*. His four

books were published by his brother, and excerpts were included as an appendix to W. Piso's *De Indiae utriusque re naturali et medica.* (Amsterdam: Elzevier, 1658). In *Historiae naturalis et medicae Indiae orientalis* VI. *Historia plantarum*, pp. 87–160, Bontius described 70 plants, including the orchids *Luisia confusa, Dendrobium crumenatum* and possibly a third orchid (perhaps a *Vanilla* species) which he called *Sedum arborescens*. Bontius wrote that the plant had a fruit which was the length of one's middle finger, slimy inside and insipid. Conserved leaves were employed to treat nerve disorders, bloody diarrhoea and wounds inflicted

Fig. 15.13 *Pecteilis susannae* [as *Flos susannae*]. From: Rumphius, *Herbarium Amboinense*, vol. 6, t. 99 (1750)

by krisses (native daggers) or pikes, whereas its flowers fortified the heart (Beekman 2003).

Bontius did not restrict his interest to medicine. He gave the first account of the orangutan [man of the forest; Malay *orang hutan*] and was responsible for naming the animal. After tasting the mangosteen (*Garcinia mangostana*), Bontius compared it to nectar and ambrosia: he wrote that it surpassed the golden apples of the Hesperides and was 'of all the fruits of the Indies by far the most delicious, deservedly the Queen of Fruits, the finest fruit in the world' (Popenoe 1928). Rumphius commented on Bontius in his original manuscript which was unfortunately destroyed when his house and library burnt down in 1687.

Karel Heyne (1877–1947)

New information on medicinal orchids in the Dutch East Indies did not appear in the literature until the early part of the twentieth century. When K. Heyne was appointed head of the Museum for Economic Botany at Buitenzorg (1906–1927), he sent collectors throughout the Indonesian archipelago to gather plant specimens for him for many years. In his *De Nuttige Planten van Nederlandsch Indie* (the useful plants in Dutch East Indies) first published in 1913, Heyne quoted extensively from Rumphius, adding only meagre information that was obtained by his collectors

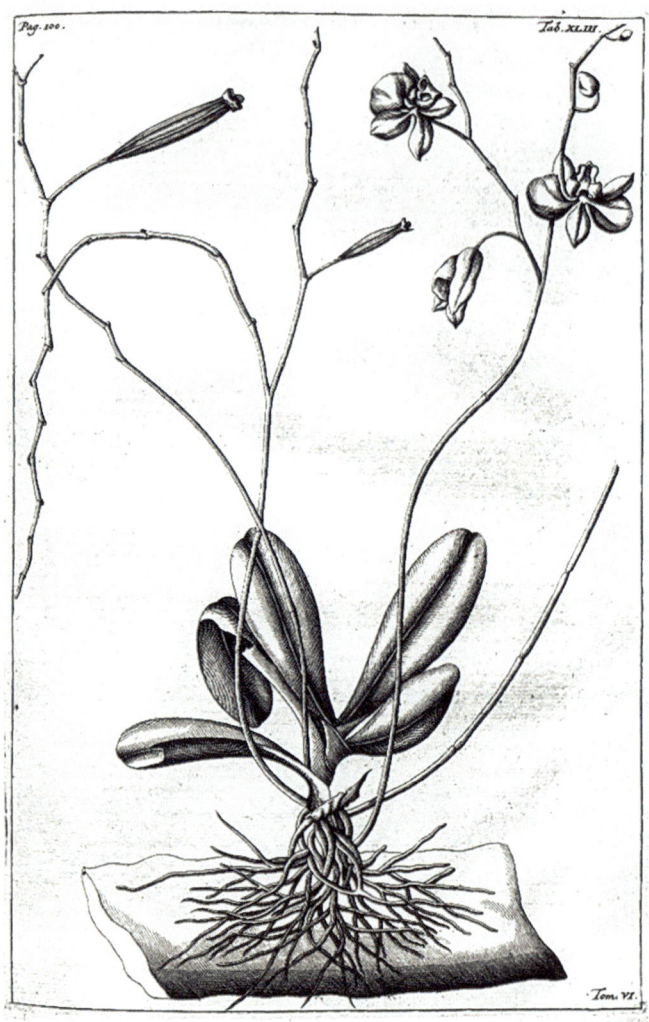

from other parts of Indonesia. Several orchids were widely used in Indonesia but most had no commercial value. The additional medicinal orchids that he described included the following:

1. *Vanilla planifolia* Andrews

   Heyne discussed this orchid in more detail than any other because it was an important cash crop. Although the plant was introduced by Marchal to Buitenzorg (Bogor Botanic Gardens) in 1819, commercial cultivation only commenced in 1846 in Java, and subsequent progress was slow. Vanilla growing gained importance during the period of recovery following World War I, and today, Indonesia is the second largest exporter of vanilla in the world.

2. *Vanilla abundiflora* J.J. Sm.

   One could obtain vanilla-like fragrance from ripe pods of *V. abundiflora*, but its aroma is much weaker than that of *Vanilla planifolia*.

3. *Liparis treubii* J. J. Sm.

   On the east coast of the Celebes, when people had constipation, they would chew the plant and swallow the juice while simultaneously rubbing their abdomen with its heated leaves.

Fig. 15.15 *Renanthera moluccana*. From: Rumphius, *Herbarium Amboinense*, vol. 6, t. 44 (1750)

The plant was known as *Angraecum gajang* var. Rumphius because it grew on the Gajang tree (*Inocarpus edulis*).

4. *Calanthe rubens* Ridl.

Heyne quoted De Clercq (1842–1906) who mentioned that *Calanthe rubens* was employed in native medicine. However, there was no specific information on its usage (Fig. 15.18).

5. *Calanthe veratrifolia* R. Br.

On the east coast of Sumatra, crushed flowers were employed to relieve pain from dental caries.

6. *Eulophia* sp.

At Ternate, tubers of a native, unidentified *Eulophia* species were crushed, wrapped in a leaf of *boero maloko* (*Vitus* sp.) and made into a 'hot porridge' to treat non-painful ulcers.

7. *Dendrobium crumenatum* Sw.

Heyne reported witnessing heated juice of *Dendrobium crumenatum* being instilled into the ear to treat earache in Batavia. A similar usage had been described by Ridley in the Malay Peninsula. Boorsma discovered traces of alkaloid in the pseudobulbs and leaves.

Ridley was quoted by Heyne on the medicinal usage of *Apostasia nuda*, *Acriopsis liliifolia* (syn.

Fig. 15.16 *Renanthera matutina*. From: Linden JJ, *Prescatorea*, t. 12 (1890)

Acriopsis javanica) and *Dendrobium pumilum*. Heyne described other non-medicinal uses for several Indonesian orchids—stems of *Dendrobium faciferum* for wickerwork; yellow green stems of *Dendrobium utile* to make baskets and mats for royalty (the rare species was simply too expensive for the common people); and leaves of *Dendrobium salaccense* to impart liquorice-like fragrance to cooked rice and to perfume women's hair: then quoting Rumphius, using the leaves of *Renanthera moluccana* to make pickles which tasted like capers (Heyne 1927) (Fig. 15.19).

Fig. 15.17 Jacobus Bontius (1592–1631)

J.J. Smith (1867–1947) and Reinier Cornelis Bakhuizen van den Brink

J.J. Smith, head of Buitenzorg Herbarium between 1913 and 1924, is well known as a leading orchid taxonomist and has his name attached to hundreds of orchid species. He described the orchids of Sumatra, Java, Kalimantan, Celebes, Amboina, Ceram, Moluccas, Talaud, Anambas, Natoena Islands and New Guinea and may be placed with R. Schlechter (who described the orchids of New Guinea) and Oak Ames (the authority on Philippine orchids) as the leading orchidologists of Southeast Asia in the first half of the twentieth century. However, he was principally interested in describing orchids and did not show any interest in their medicinal usage (van Steenis 1948, 1948–1954).

In 1937, Bakhuizen van den Brink observed that although the Dutch East Indies (Indonesia)

was home to thousands of orchid species, as a rule the people largely ignored these plants and seldom endowed them with distinctive names. Orchids were referred to simply as *Anggrek* or *Angkrek*, and it was only with the arrival of the Dutch that some effort was made to name some of the commoner orchids by either adopting a Dutch name (e.g. *Angkrek lotjis*, *Spatulotjis* or *Spatuklotis* for *Spathoglottis*, *Angkrek paneli* for *Vanilla* and *Angrek bulan* or 'moon orchid' for *Phalaenopsis amabilis* which was *Maan-orchidee* in Dutch) or by adding prefixes to describe the colour of its flowers, *merah* (red), *puteh* (white) and *kuning* (yellow), or its size, *besar* (big) and *kechil* (small). However, there were exceptions. At the foot of Mount Halimun, the juice of pseudobulb of *Acriopsis javanica* (now renamed *Acriopsis liliifolia*) was dropped into the ear to treat tinnitus and earache. The orchid was known as *Kiplengpeng* in Sundanese. Van den Brink concluded that the name was derived from the usage of the orchid: *ki* meaning 'herb', *pling* the ringing sound and *peng* rise and fall (of the sound). In Java, leaves of *Macodes petola* which carried a pattern that resembled ancient script were made into eye drops that were administered to children to improve their reading ability (Teoh 2016) (Fig. 15.20).

Papua New Guinea is home to a huge number of orchid species, but very few have been identified as medicinal plants. Ialibu women claimed that by eating the pale blue flowers of *Phaius tankervilleae* (native name: *kongimongo*) which have been heated over a wood fire, it would help them to conceive. Leaves of *Dendrobium* species were chewed to relieve cough in the Central Province (Holdsworth 1974). A *Diplocaulobium* species was employed to treat infected wounds in eastern New Guinea (Andre Millar quoted by Lawler 1984). The Chimbu believed *Dendrobium* had aphrodisiac properties. They also employed fresh juice of an orchid to treat respiratory disorders and *Dendrochilum* species for medicine (Sterly 1973 quoted by Lawler 1984). New Guinea natives used flowers of some orchids to treat contagious diseases (Cochran and Lucas 1958/1959, quoted by Lawler 1984).

Fig. 15.20 *Macodes petola* (©Teoh Eng Soon 2019. All Rights Reserved.)

The Malay Peninsula

With the arrival of the British and the establishment of the Straits Settlements of Penang, Malacca and Singapore from 1786 to 1819, many famous naturalists came to this region, and several hundred individuals contributed plants that they collected in the peninsula to various herbaria. Some of their names are immortalized in numerous orchid species: Anderson, Boxall, Burkill, Cuming, Curtis, Finlayson, Griffith, Henderson, Holttum, King, Kunstler, Lobb, Low, Maingay, Micholitz, Ridley, Roxburgh, Schlechter, Schomburg, Scortechini and Wallich, to name some. Alfred Russel Wallace (1823–1913) visited Singapore and Malacca and climbed Mount Ophir in 1854 making zoological and botanical studies.

In the old days, few families possessed a medicine kit for first-line treatment of common illnesses. Instead they relied on their own kitchen or garden to provide the herbs to relieve fever, headache, stomachache, diarrhoea, cuts and wounds, poisonous bites, etc. If the initial remedy did not work, they might try an alternative, still based on what was available at home or from bushes nearby. The next recourse would be the village healer, the *bomoh* or midwife (*bidan*) who possessed a wider range of remedies that included herbs from the village and nearby forests. *Bomohs* also possessed rare herbs which take the form of twigs, dried fruits, seeds and flowers or powders. Orchids being difficult to find, they were rarely employed, the exception being those ubiquitous orchids that flourished in sunlit surroundings along the roadside, on cattle-grazed ground or at the edge of forests. Orchids were more likely to be used by aboriginal tribes who depended on the forest for their food supply and medications. Tribals were familiar with forest plants and knew where to find specific plants. Surveys on the use of folk medicine must therefore include surveys on village practitioners of folk medicine as well as surveys on aboriginal tribes.

The Medical Book of Malayan Medicine

An early manuscript on Malay medicine was translated by one Inche Ismael, Moonshee in Penang around 1886, but it lay hidden in the library of the Pharmaceutical Society of Great Britain for over 40 years until it was noticed by J.D. Gimlette and I.H. Burkill in 1928. Written by hand on blue-lined foolscap, it contained texts in three different handwritings, the main text in one, margin notes in another and comments in pencil in a third. In 1930, it was edited with medical notes by Gimlette and determination of drugs by Burkill and published in the *Gardens Bulletin of the Straits Settlements*.

John Desmond Gimlette (1867–1934) worked as a doctor in the Federated Malay States before becoming residency surgeon in the northeastern border state of Kelantan. His interest and knowledge of Malay medicine was probably acquired in Kelantan because it was from here that he sent medicinal orchid specimens and notes to Singapore. Isaac Henry Burkill was a botanist. Their excellent collaborations are demonstrated

in the interesting details that characterize the book.

This *Medical Book of Malayan Medicine* described the uses of common plants, e.g. *limau purut* (*Citrus hystrix*) leaves, *chekur* (*Kaempferia galangal*) rhizome, *kayu manis* (bark of *Cinnamomum zeylandica*), *cabai jawa* (*Piper retrofractum*). The orchid was mentioned only once but in interesting detail:

> 521. Sections to explain the medicines for a swelling inside the ear which will not disperse (*sakit telinga bunting di-dalam sahaja tiada sampai keluar*).
> Take either the root or the fruit of a big epiphytic orchid (*Anggrek jantan*); *bawang merah* (the onion of *Allium cepa*); and *jintan manis si-dikit* (a small quantity of *Nigella salvia* seeds). Split the orchid root or fruit. Put a piece of onion and a few seeds inside it. Bury it in the hot ashes of a fire. When it is as hot as the patient can bear, squeeze the juice into the ear.
> 522. Moreover,
> Rub the remainder down on a stone. Apply the pulp outside the ear, and let him be relieved.

The book contained 543 prescriptions. Polypharmacy was the norm, the maximum being 24 items in a prescription for smallpox. In a lecture that he gave to the Straits Medical Society in 1894, Henry Ridley mentioned that one celebrated Malay medicine contained 100 different ingredients (Ridley 1906). For most conditions, alternative prescriptions were described which meant that if some ingredients for the remedy of a particular illness were not readily available, an alternative prescription could be employed. Furthermore, if the treatment failed to work, a second remedy could be tried. Orchids being uncommon items in a household, they were omitted with the exception of 'the big epiphytic orchid', which the translators believe to be probably the ubiquitous pigeon orchid, *Dendrobium crumenatum*. However, neither the fruit nor the root of this orchid is big. It is interesting to note that there was no mention of an alternative treatment for earache in the Moonshee book, which meant that this big orchid was really quite common, else any big orchid (*anggrek*) sufficed.

Henry Nicholas Ridley (1855–1956)

Henry Ridley is best known for recognizing the potential of rubber, promoting the planting of *Hevea brasiliensis* in Malaya and devising a convenient method for extracting and collecting the sap. However, he had wide-ranging interests. He undertook many collecting trips in the Malaya, compiled *The Flora of the Malay Peninsula*, was interested in Malay drugs and lectured on the Orchidaceae and Apostasiaceae to the Linnean Society in London and the Medical Society of the Straits Settlements (Ridley 1894). He was appointed director of the Singapore Botanic Gardens in 1888 (Fig. 15.21).

In the Malay community of that era, treatment from doctors or herbalists was sought mainly for illnesses which presented with dramatic symptoms like diarrhoea, fever, pain or trauma. Insidious conditions like heart disease or tuberculosis and madness were attributed to assault by evil spirits and approached as such. Charms, magic potions and rituals were very much a part of disease prevention and treatment. However, already in 1894, Ridley observed that Malayans were consulting Western-trained doctors rather than herbalists. The role of the latter was declining so rapidly that there was concern that their knowledge would be lost to posterity. Aborigines, the *Orang Asali* (Jakuns, Semangs and Sakais), possessed the best knowledge of native drugs because they depended on the forest for their food and livelihood. Medicinal items reached village herbalists in the form of twigs or roots, and most were quite unidentifiable. Native names provided a clue, but such names were not necessarily specific. *Angrek* could refer to a variety of orchids, *sakat* almost any epiphyte (Ridley 1894) (Fig. 15.22).

According to Ridley, *Corymborkis veratrifolia* was cultivated in Kelantan for use as a febrifuge to treat fever in children. Flowers of *Vanilla griffithii* were also used to treat fever, but since they last barely a day, it was a matter of chance whether they could be had. Sap from the leaves and stem was rubbed into hair to stimulate hair growth (Ridley 1906). Ridley was an authority on *Apostasia* (Ridley 1894), but it was left to de Vogel to discover that *Apostasia wallichii* was employed as an antidiabetic herb in Peninsular Malaysia (de Vogel 1969). *Materials for a Flora of the Malay Peninsula* published in 1907 when Ridley was still director of Singapore

Fig. 15.21 Henry Nicholas Ridley (1855–1956). Photo: courtesy of Singapore Botanic Gardens

Fig. 15.22 *Corymborkis veratrifolia* (©Teoh Eng Soon 2019. All Rights Reserved.)

Botanic Gardens morphed into his major work, the five-volume *The Flora of the Malay Peninsula* written at Kew after his retirement and published between 1922 and 1925. Orchids are described in Volume 4 which deals with monocotyledons. Ridley authored 500 publications and described over 200 novel orchid species. Other taxonomists have named several orchids after Ridley but they are not medicinal.

M.V. Alvins

M. V. Alvins was stationed at Malacca by Nathaniel Cantley (superintendent of the Singapore Botanic Gardens) between 1884 and 1888 to study the composition of its forests. This hardworking, meticulous botanist sent a thousand botanical specimens to the Herbarium of the Singapore Botanic Gardens within the first year. Then in 1885, he sent 1840 specimens and appeared to have ventured into the adjacent Malay state of Negri Sembilan after systematically surveying Malacca.

Alvins numbered his specimens in the field. He also recorded the medicinal uses of several orchids in Malacca. [My attempt to locate his original reports was not successful. Burkill (1935) gave Alvins credit for assigning medicinal usage of several Malayan orchids, and such mentions will be described below.]

Isaac Henry Burkill (1870–1965) and Mohamed Haniff (d. 1930)

IH Burkill served as an economic officer in India before assuming the post of director of the Singapore Botanic Gardens in 1912. This experience inspired him to produce a two-volume *A Dictionary of the Economic Products of the Malay Peninsula*. The *Dictionary* made extensive use of data contained in the annual reports and kindred materials from the Gardens which were contributed by HJ Murton, Nathaniel Cantley, Henry N Ridley and others. Two forest officers, Frederick W Foxworthy and JG Watson, a

fisheries expert, William Birtwistle, and geologist J B Scrivenor assisted him with contributions. Work was intensified after Burkill's retirement in 1925, and the Dictionary was published in 1935 (Fig. 15.23).

Statements on orchids in the *Dictionary* give us some idea of the information that was available to researchers a century ago. This is a useful starting point. Unfortunately, although the data is not exhaustive, it remained the primary source for the majority of subsequent articles on the medicinal usage of orchids in the Malay Peninsula. Not much original work appears to have been undertaken since Burkill (1935).

Earlier on in 1930, Burkill had collaborated with Mohamed Haniff to publish a collection of Malay village medical prescriptions which were collected from two large Malay states, Perak on the West Coast and Pahang on the eastern side of the peninsula. Mohamed Haniff (d. 1930) joined the Penang Waterfall Gardens in 1890 and spent most of his time in Penang working as an overseer. He collected extensively in the Malay Peninsula and had a dozen orchids named after him, including *Bulbophyllum haniffii* and *Dendrobium haniffii*. Much admired by his European superiors, he was appointed assistant curator of the Singapore Botanic Gardens in 1925 (Fig. 15.24).

In *Malay Village Medicine* (1930), the authors appear to have studied Malay native medicine

Fig. 15.23 Isaac Henry Burkill (1870–1965). Photo: courtesy of Singapore Botanic Gardens

Fig. 15.24 Mohamed Haniff. Photo: courtesy of Singapore Botanic Gardens

practice by visiting Malay villages along the north-south road linking the two states, leaving out those towns where Chinese populations were dominant, such as Taiping and Ipoh. This is evident from their statement of the sites where their collections were made: Kuala Kangsar (*Dendrobium crumenatum, Dendrobium* sp., *Plocoglottis porphyrophylla, Aplostelis flabelliformis*), Telok Anson (*Dendrobium subulatum, Cymbidium finlaysonianum, Vanda hookeriana, Tropidia curculigoides*), Tapah (*Spathoglottis plicata*), Kuala Lipis and Bentong (*Tropidia curculigoides*). They did, however, obtain an undetermined orchid called *cheok seng* (in Cantonese) from a Chinese herbal establishment in Penang which they tentatively identified as a *Cystorchis* species, but this item did not qualify as Malay village medicine.

In the Malay Archipelago, folk medicine was practised by *bomohs*, native healers who often doubled as magicians because illness is often attributed to evil spirits; and, in lieu of

aphrodisiacs, charms were sought to entice desirable consorts. Most grandmothers, especially the learned ones, also possessed a good knowledge of folk remedies for simple conditions like headache, tummy ache, skin lesions, insect and scorpion bites, cuts and wounds, etc. They administered first-line treatment whenever the need arose. Knowledge of folk medicine was passed down vertically from mother to daughters or in the case of *bomohs* from father to son. Malay midwives known as *bidans* in addition to attending at childbirth also provided advice on what herbs to employ as *ubat meroyan*, medicine to ward off evil spirits that might attack the new mother.

Henry Ridley (1906) reported that he knew a bomoh who made his living by chanting a spell and spitting chewed betel nut over the head of any patient who sought his help. I remember being treated in like manner as child by my maternal grandmother. She was the daughter of a Chinese *kapitan* (headman) in Medan, Sumatra, and was sent to marry my English-educated grandfather in British Straits Settlement island of Penang while still a teenager. Grandmother read Romanized Malay, loved Malay poetry, read Chinese classics in Romanized Malay and related their stories to me in Hokkien during my childhood. Before leaving home she was taught the necessary culinary skills expected of a *Peranakan* lady and enough folk medicine to deal with trauma, poisonous bites and common illnesses. The latter knowledge served us well during the Japanese Occupation (1942–1945).

Indonesian and Malay folk medicine made extensive use of dried spices and culinary herbs because these were articles that grandmothers, *bomohs* and *bidans* were most familiar with. Next in popularity were plants that grew in their garden or village. The orchids which featured in local folk medicine were those commonly found on roadside trees, in the *belukar* or at the edge of primary lowland forests. *Dendrobium crumenatum, Cymbidium finlaysonianum, Acriopsis liliifolia, Spathoglottis plicata* and *Tropidia curculigoides* are examples of such species. The old healers were not trained in botany, most of them did not even have a formal education and if they wanted an *angrek*, they might not be particular about species. Nevertheless, all data collected by our early scholars were obtained from such healers. Burkill and Haniff obtained information from *bomohs* and *bidans* who brought them specimens of the herbs that they employed in their practice. These samples were numbered, taken to the Singapore Botanic Gardens, studied, identified and then preserved in its Herbarium. The scientists tried to avoid unexpected questions when they interviewed their experts, 'for they beget gusts of fancy and incorrect assertions'. Mohamed Haniff spent considerable time to obtain information from Malay-speaking *Sakais* (aborigines of northern Peninsular Malaysia), but the bulk of the knowledge was derived from Malay sources.

A total of 1675 numbered items were collected from the *bomohs* and *bidans*. On their usage, *ubat meroyan* was mentioned 194 times, taking second place only to fever (207 mentions), and far more often than other complaints for which the people sought treatment, such as headache, diarrhoea, joint pains, boils or snake bites (Table 15.1). *Nervilia aragoana* was the only orchid submitted for use as *ubat meroyan*. In contrast to the practice in Europe, the Middle East and India, no orchid was used as an aphrodisiac: Malays relied on the rhizomes of *tongkat ali* (*Smilax calophylla*) and the related *tanding* (*Smilax myosotiflora*) for sex stimulants (Burkill and Haniff 1930). They still do (Fig. 15.25).

The following medicinal orchids were described by Burkill and Haniff (1930) or by Burkill (1935):

1. *Acriopsis javanica* (now renamed *Acriopsis liliifolia*) Malay names denote its appearance (*sakat bawang*, onion orchid), habitat (*anggerek darat*, river bank orchid) and medicinal usage (*sakat ubat kepialu*, fever medicine orchid). Burkill gave credit to Alvins who collected this information on this orchid in Malacca sometime between 1884 and 1888: a decoction of the orchid was administered for unremitting high fever that the Malays called *kepialu*.

Table 15.1 Medicinal items submitted by *bomohs* and *bidans* to IH Burkill and Mohamed Haniff in 1928 in the Malay Peninsula, and the indications for their use

Indication for use number of items submitted	
Fever (including ague)	207
Childbirth (*ubat meroyan*)	194
Headache	49
Diarrhoea	47
Joint pains	41
Boils	40
Snake bites	5

2. *Anoectochilus* Bl.
   Ridley observed that formerly, when it was still abundant, *Anoectochilus* sp. (it was more likely to be the look-alike, commoner *Ludisia discolor*) was sold as pot-herb in the market. A. H. Berkhout, a Dutch botanist, reported that it was cultivated in a Chinese plantation in Parit Buntar in the northwest of the Malay Peninsula for use as medicine, but he could not discover its usage. The vegetative forms of the various species are very similar, and *Anoectochilus* has numerous medicinal uses in Taiwan (Teoh, 2016). In 2002, Christensen reported that Iban and Kelabit in Borneo made use of *Anoectochilus reinwardtii* to treat infertility (Christensen 2002) (Fig. 15.26).

3. *Aplostellis flabelliformis,* Ridl. (correct name: *Nervilia aragoana* Gaud.) Malay name: Daun sa-helai sa-tahun
   A decoction of the leaf was consumed and the remainder poured into bathwater for use by the parturient during the first 3 days of the puerperium as a preventive preparation against complications (*ubat meroyan*). In the old days, Malays believed that evil spirits responsible for causing puerperal illnesses had the strongest power over the parturient during the first 3 days following childbirth. *Ubat meroyan* is constituted by many herbal and not *Daun sa-helai sa-tahun* alone. Any pharmacological action of plants that constituted *ubat meroyan* would be purely coincidental.

Fig. 15.25 *Nervilia aragoana.* From: Blume CL, *Collection des Orchidees les plus remarquables de l'archipel et du Japon.* T. 57, Fig. 3 (1858)

POGONIA DISCOLOR. Fig.1 _ P. CONCOLOR Fig. 2 _ P. GRACILIS, Fig.3.

4. *Bromheadia finlaysoniana* Rchb. f. Malay name: *Seraman*

   Alvins reported that decoction of *Seraman* was consumed for rheumatism in Malacca. A similar usage in Sarawak has now been reported; flower stalks are chewed to extract the juice which is thought to be effective for treating asthma (Go and Hamzah 2008) and toothache (Christensen 2002) (Fig. 15.27).

5. *Bulbophyllum vaginatum* Rchb. f.

   This is a common orchid which is still to be seen on roadside trees right in the city in Singapore. Alvins reported that hot juice of

Fig. 15.26 *Ludisia discolor* (©Teoh Eng Soon 2019. All Rights Reserved.)

Fig. 15.28 *Bulbophyllum vaginatum* (©Teoh Eng Soon 2019. All Rights Reserved.)

Fig. 15.27 *Bromheadia finlaysoniana* (©Teoh Eng Soon 2019. All Rights Reserved.)

roasted fruits is dropped into the ear to treat earache in Malacca (Fig. 15.28).

6. *Cymbidium* sp.
   One unidentified species was an emetic.
7. *Cymbidium finlaysonianum* Malay name: *Sepuleh* (Restorative)
   Cymbidium finlaysonianum was employed as a talisman to ward off evil spirits from a house. Ridley specified this orchid to be the *jarang songsang* mentioned in Maxwell's *Mantra Gajah.* Chewed roots were spat on a sick elephant following recitation of a mantra (Fig. 15.29).
8. *Dendrobium crumenatum* Malay name: *Daun sepuleh tulang* (Restorative for bones)
   A ubiquitous orchid in the Malay Peninsula and in Singapore, the pigeon orchid was known as *Anggerik merpati* (dove orchid) and also as *Daun sepuleh tulang* (bone restoring leaf). Burkill did not give any explanation for its second Malay name. Instead attention

was drawn to the fact that a poultice made from leaves of *Dendrobium crumenatum* was used by Malays to treat boils and pimples. It was one of several orchids employed to treat earache. The plant was used as a besom to sprinkle magic rice to ward off evil influences alleged to be the cause of ill fortune, illness or demise (Burkill and Haniff 1930).

9. *Dendrobium planibulbe* Malay name: *Miga*
   A poultice made by pounding the plant was used to treat dermatological lesions affecting the back of the neck.

10. *Dendrobium pachyphyllum* (syn. *Dendrobium pumilum*)
    A decoction of roots of this lowland orchid was reported by Alvins to be used for dropsy in Malacca which was then part of the British Straits Settlements (Fig. 15.30).

11. *Dendrobium subulatum*
    Poultice made by pounding the leaves of this orchid which was prevalent in the north of the peninsula was applied to the forehead to relieve headache in the northwestern state of Perak.

12. *Desmotrichum pallidiflorum*
    Burkill was intrigued by the Malay name for these orchids, *Susu kubong*, which was applied to *Desmotrichum pallidiflorum* Ridl. (revised name: *Dendrobium xantholeucum* Rchb.f.) because the term means 'an extremely severe abscess'. However, he could not discern any connection between the orchid and abscesses. *Desmotrichum* is now reclassified under *Dendrobium*. *Dendrobium xantholeucum* Rchb.f. is found in Taiping, Perak.

13. *Eria pannea* (*Mycaranthes pannea*). A medicinal bath made by adding a potion of boiled *Eria pannea* (called *kura kubong*) was employed to treat ague (malaria or any severe fever).

14. *Hippeophyllum scortechinii* (Hook.f.) Schltr. Sakais living in Penjom in northern Pahang used it to treat earache. It was used in the manner of *Dendrobium crumenatum* and other orchids and went by a name which describes the process, *pokok setawar bakar perah* (Fig. 15.31).

Fig. 15.31 *Hippeo-phyllum scortechinii* [as *Oberonia scortechinii*]. From: *Naturalis Biodiversity Centre*, Wikimedia Commons

15. *Spathoglottis plicata* Bl. Sakai (aboriginal) name: *Wah*

    It was used by in Tapah, Perak, to treat rheumatism.

    The plant was decocted; a small amount was drunk and the rest employed as a foment.

16. *Tropidia curculigoides* Lindl. Malay names: *Serugat*; *Ranchang hantu*

    The plant had unrelated usages in two different localities barely 200 kilometres apart. In Telok Anson, where it is known as *Serugat*, one drank a decoction of the roots when one had diarrhoea. In Bentong where it is known as *ranchang hantu*, it was used to treat malaria. A decoction prepared using the entire orchid plant and leaves of *Ardisia* was administered during the cold stage of fever. (It should be noted that here the Malay word *hantu* is interpreted here as 'wild' and not as 'ghost'.) Telok Anson is near the coast, Bentong is at the centre of the peninsula and they are separated by 113 km as the crow flies (Fig. 15.32).

17. *Papilionanthe hookeriana* Rchb. f. (syn. *Vanda hookeriana*) Malay name: *Tulang* (bone). Common name: Kinta weed

    A decoction of the plant was used as a foment to treat joint pains in Telok Anson which was located near the natural habitat of this orchid, the swampy parts of the Kinta Valley (Fig. 15.33).

Table 10.2 provides a summary of the orchid species formerly employed as medicine in the Malay Peninsula according to the complaint or illness for which they are employed.

Fig. 15.32 *Tropidia curculigoides* [as *Tropidia assamica*]. From: Blume *Collection des Orchidees les plus remarquables de l'archipel et du Japon.* T. 41, Fig. 2 (1858) [AJ Wendel]

TROPIDIA SQUAMATA. Fig.1 _ TR. ASSAMICA. Fig.2 _ TR. GRAMINEA. Fig. 3.

Additional Sources

In the *Journal of the Bombay Historical Society*, Caius (1936) reported three additional medicinal orchids used by Malays. *Hetaeria obliqua* Bl. which is common in forests in Borneo, the

Malay Peninsula and Sumatra was employed by Malays for poulticing sores. They also used leaves of *Oberonia anceps* Lindl. which occurred from the isthmus through the Malay Peninsula to Java. *Vanilla griffithii* Rchb. f. (Malay names: *Akar punubal, Telinah kerbau bukit*) was

Fig. 15.33 *Papilionanthe hookeriana* (©Teoh Eng Soon 2019. All Rights Reserved.)

common in the Malay Peninsula and the Karimun islands. Leaves and stem were pounded and applied to hair to promote hair growth. Flowers were rubbed on the body to relieve fever (Caius 1936).

In 1929, David Hooper described the medicinal substances collected by I.H. Burkill from Chinese herbalists in Malaya when the latter travelled in the Malay Peninsula. The identities of these samples were confirmed by the Royal Botanic Gardens, Kew. Three orchids were present in the collection of 456 items:

1. *Bletia hyacinthina* (*Bletilla striata*).
   The root was exported from Hankow and Ningpo. The Singapore sample came from Canton. It was used as a demulcent for children with dyspepsia and also given for dysentery, haemorrhoids and ague.
2. *Dendrobium nobile* and other spp.
   It was mentioned that species were all lithophytic, hence the name *Shih hu*.

3. *Nervilia Fordii* Schltr. (*Pogonia fordii* Hance, *P. pulchella* Hook f.)
   The orchid occurs in Hong Kong and at Lo Fan Shan Mountains in Guangzhou. Leaves were used in medicine.
   The prominent omission was *Tianma* (*Gastrodia elata*).

Original research publications on medicinal orchids in the Malay Archipelago are few and far between. Kwan Koriba who retired from a professorship at the University of Kyoto in 1942 was made director of the Singapore Botanic Gardens when Singapore became Syonan. He permitted Eric Holttum, E.J.H. Corner, George Alphonso and other staff members formerly employed by the British administration to continue their work in the Gardens, and materials in the Gardens were well preserved. An *Illustrated Useful Plants in Malaya* was published in Japanese at the end of the Japanese Occupation of Malaya and Singapore. It described 194 plants with only one orchid—*Dendrobium crumenatum*.

Hanne Christensen who spent several years with the Iban and Kelabit of Borneo reported that they applied the sap either of *Arachnis flosaeris* or *Bromheadia finlaysoniana* on the tooth and gum to relieve toothache. The shoots of *Calanthe* or *Phaius* (species not identified) were softened by heating over a fire and eaten to cure swollen parts of the body. Kelabit ate *Cymbidium* (species unidentified) as vegetable (Christensen 2002) (Fig. 15.34).

Magic and Charms

Plants of *Dendrobium crumenatum* (Malay name: *Sepulah tulang*), other *Dendrobium* species (Malay name: *Sepulah rumah*), *Plocoglottis lowii* (syn. *P. porphyrophylla* Ridl.; Malay name: *Sepuleh dudok* or *Sepuleh dudor*) and *Cymbidium finlaysonianum* (Malay name: *Sepuleh*) were used in the state of Perak to sprinkle water in the house after a recent demise to prevent the departed spirits from haunting it

(Burkill   and   Haniff   1930).   *Dendrobium
crumenatum* was also employed to sprinkle rice
paste about a house to invite the return of beneficent spirits (Burkill 1935) (Fig. 15.35).

Across the South China Sea in Sarawak, the
Kelabit make use of orchids in rice harvest
ceremonies.   A   piece   of   *Agrostophyllum
bicuspidatum* or *Appendicula cornuta* is worn
by the Kelabit as a protective charm against
curses.   Infertile   couples   place   a   plant   of

Anoectochilus reinwardtii under their sleeping
mat so that the wife may become pregnant. It is
believed that one could predict the sex of the baby
by   deciphering   the   pattern   on   the   leaves
(Christensen 2002) (Fig. 15.36.).

Then, as though to remind one of the transient
natures of human life, it was a custom in Java to
plant the beautiful but ephemeral *Corymborkis
veratrifolia* in sacred burial places (Backer and
van den Brink 1968).

Fig. 15.35 *Plocoglottis lowii* (©Teoh Eng Soon 2019.
All Rights Reserved.)

Fig. 15.36 *Appendicula cornuta* (©Teoh Eng Soon
2019. All Rights Reserved.)

The second edition of John Desmond Gimlette's *Malay Poisons and Charm Cures* (1931), published after his retirement from an over 30-year service as Resident Surgeon in Peninsular Malaysia, reported the following old Malay charm to cast out forest spirits or demons causing disease:

Al-salam 'alikum, hai maseh di-rimba penghulu di hu-tan,
Yang Tanggong sahat humi,
Putera di –sini yang memegang da'erah bumi hutan sini,
Aku tahu sal-mu;
Nama-mu yang sal-mu-lah yang bernama Sang Ranjuna,
Jadi charang dewana, jadi gunong Sing Bima,
Jadi (?) pelana sari maha puteh, jadilaut;
Dengarkan oleh-mu perkataan-ku, aku tahu sal kejadian-mu,
Mu jadi dari-pada chahya yang kelam, aku jadi dari-pada chahaya yang cherah,
Mu jadi dari-pada tanah, yang halus,
Aku jadi dari-pada tanah yang kasar, aku jadi teleheh dahulu dari-pada-mu,
Hai sakalian Aja-aja di-gunong sini,
Aja-aja di sini, di-luwok sini,
Dengar-dengar kata-ku, kalu mu tidak dengar aku, derhaka-lah mu ka-pada aperbakala Dewa,
Yang sedia, Dewa yang lenyap,
Dewa yang ghaib pada pandangan, dan pada penguchapan, tamat.

It is interesting to note that the reciter of the charm made an appeal to *Dewa* (*Deva*) which attests to Malaysia's long relationship with Indian civilization. During a healing ritual, offerings are made to *Dewas* and their *peng* (a Hokkien Chinese word meaning 'soldiers'), with a separate placing for *Dewa Betara Kala* (leader of the Asuras who is popularly featured in Indonesian *wanyang kulit* or shadow plays). However, one charm for neutralizing poison is definitely Islamic. Dr. Gimlette spent the first 14 years of his medical career in Pahang, Selangor, Perak and then from 1903 to 1921 in Kelantan. His knowledge of Malay poisons and charms was derived from Kelantan.

Philippines

Information on medicinal usage of orchids in the Philippines is scarce. Pseudobulbs of *Geodorum densiflorum* (syn. *Geodorum nutans*) were made into a poultice for ripening boils and abscesses. A liniment prepared with pseudobulbs and rice water was also applied to chronic wounds and abscesses (Guerrera 1921; Burkill 1935). It was used as disinfectant in Luzon (Lawler 1986). Leaves of *Nervilia discolor* were chewed and then rubbed over the abdomen to relieve stomachache: women drank an aqueous extract to facilitate childbirth. Salted leaves of *Phalaenopsis aphrodite* were pounded and the poultice applied to relieve headache and also applied on the back or chest (for pain relief) (Lawler 1986). Leaves of *Phalaenopsis schilleriana* were heated and placed over centipede bites to relieve pain (Figs. 15.37 and 15.38). Decoction of *P. schilleriana* was used to treat stomach disorders and new cases of tuberculosis. In Palau, naturalized *Vanilla planifolia* was employed for hysteria, irregular menstruation and fever (Perry

Fig. 15.37 *Phalaenopsis aphrodite* (©Teoh Eng Soon 2019. All Rights Reserved.)

Fig. 15.38 *Phalaenopsis schilleriana* (©Teoh Eng Soon 2019. All Rights Reserved.)

and Metzger 1980). Juice from heated leaves of *Bulbophyllum* sp. was applied on wounds, and powdered, salted leaves of other orchid species were plastered on the head to relieve headache (Fox 1950 quoted by Lawler 1984). During the Japanese Occupation, leaves of *Spathoglottis plicata* substituted for tobacco (Lawler 1986).

Numerous name changes have occurred since the publication of the works referred to in this chapter. The accepted names of the orchids discussed are set out in Table 15.2.

Table 15.2 Accepted names of orchid species discussed, from the Malay Archipelago

Acriopsis liliifolia (J. Koenig) *Seidenf.* (syn. *Acriopsis javanica*)

Agrostophyllum bicuspidatum (J.J.Sm) Schuit.

Anoectochilus reinwardtii Bl.

Apostasia nuda R.Br.

Appendicula cornuta Bl.

Bromheadia finlaysoniana Rchb. f.

Arachnis flos-aeris (L.) Rchb.f.

Bletilla striata Rchb. f.

Bulbophyllum haniffii Carr

Calanthe rubens Ridl.

Calanthe triplicata (Willimet) Ames (syn. *Calanthe veratrifolia* R. Br. ex Ker Gaw)

Coelogyne rumphii Lindl.

Corymborkis veratrifolia (Reinw.) Bl.

Cymbidium finlaysonianum Lindl.

Dendrobium anosmum Lindl.

Dendrobium bicaudatum Reinw. ex Lindl.

Dendrobium crumenatum Sw.

Dendrobium ephemerum (J.J.Sm.) J.J.Sm.

Dendrobium faciferum J.J.Sm.

Dendrobium mirbelianum Gaud.

Dendrobium nobile Lindl.

Dendrobium pachyphyllum (Kuntze) Bakh.f. (syn. *Dendrobium pumilum* Roxb.)

Dendrobium planibulbe Lindl.

Dendrobium purpureum Roxb.

Dendrobium salaccense. (Bl.) Lindl.

Dendrobium strebloceros Rchb.f.

Dendrobium subulatum Lindl.

Dendrobium sutiknoi P. O'Byrne

Dendrobium tortile Lindl. (syn. *Dendrobium haniffii* Ridl. ex Burkill)

Dendrobium utile J.J.Sm.

Dendrobium xantholeucum Rchb.f. (syn. *Desmotrichum pallidiflorum* Ridl.)

Geodorum densiflorum (Lam) Schltr. [syn. *Geodorum nutans* (Presl.) Ames]

Grammatophyllum scriptum (L.) Bl.

Hippeophyllum scortechinii (Hook.f.) Schltr.

Hetaeria obliqua Bl.

Liparis condylobulbon Rchb. f. (syn. *Liparis treubii* J.J. Sm.)

Luisia confusa (syn. *Luisia amboinensis*)

Macodes petola (Bl.) Lindl.

Mycaranthes pannea (Lindl.) S.C.Chen & J.J.Wood *(syn. Eria pannea* LIndl.*)*

Nervilia concolor (Bl.) Schltr. (syn. *Aplostelis flabelliformis* Ridl.)

(continued)

Table 15.2 (continued)

Nervilia fordii Schltr. (syn. *Pogonia fordii* Hance, *Pogonia. pulchella* Hook f.)

Oberonia anceps Lindl.

Oxystophyllum moluccense (J.J.Sm.) M.A.Clem. (syn. *Dendrobium moluccense* J.J.Sm.)

Papilionanthe hookeriana Rchb.f.(syn. *Vanda hookeriana*)

Pectalis suzanne (L.) Rafin

Peristylis goodyroides (D.Don) Lindl.

Phaius amboinensis Bl.

Phaius tankervilleae (Banks) Bl.

Phalaenopsis amabilis (L.) Bl.

Phalaenopsis schilleriana Rchb.f.

Plocoglottis lowii Rchb.f. (syn. *Plocopglottis porphyrophylla* Ridl.)

Renanthera matutina (Bl.) Lindl.

Renanthera moluccana Bl.

Spathoglottis plicata Bl.

Tropidia curculigoides Lindl.

Vanda celebica Rolfe

Vandopsis lissochiloides (Gaud.) Lindl.

Vanilla abundiflora J.J. Sm.

Vanilla griffithii Rchb.f.

Vanilla planifolia Andrews

References

Arditti J (1989) History of several important research contribution by South East Asia scientists. Malay Orchid Rev 23:64–80

Baas P, Veldkamp JF (2013) Dutch pre-colonial botany and Rumphius's Ambonese Herbal. Allertonia 13:9–19

Backer CA, van den Brink RCB Jr (1968) Flora of Java. English book edition. P. Noordhoff, Groningen

Beekman EM (trans., ed. with annotation and introduction) (2003) Rumphius' orchids. Orchid texts from the Ambonese Herbal by Georgius Everhardus Rumphius. Yale University Press, New Haven

Beekman EM (trans. with annotation and introduction) (2011) The Ambonese Herbal of Georgius Everhardus Rumphius, vol 1. Yale University Press, New Haven

Burkill IH (1935) (1966 reprint, 2nd ed., with contributions by Birtwistle W, Foxworthy FW, Scrivenor JB, Watson IG) A dictionary of economic products of the Malay Peninsula, vol II. Ministry of Agriculture & Co-operatives, Kuala Lumpur

Burkill IH, Haniff M (1930) Malay village medicine. Gardens Bull Straits Settlements 6:165–321

Caius JF (1936) The medicinal and poisonous orchids of India. J Bombay Nat Hist Soc 38(4):791–799

Christensen H (2002) Ethnobotany of the Iban and the Kelabit. Forest Department Sarawak; NEP Con Denmark; and University of Aarhus, Denmark

de Vogel EF (1969) Monograph on the tribe Apostasiaceae (Orchidaceae). Blumea 17:313–350

de Wit HCD (1977) Orchids in Rumphius' Herbarium Amboinense. In: Arditti J (ed) Orchid biology. Reviews and perspectives, I, vol 99. Cornell University Press, New York, pp 47–94

Gimlette JD (1931) malay poisons and charm cures, 2nd edn. J & A Chirchill, London

Gimlette JD, Burkill IH (eds) (1930) The medical book of Malayan medicine by Ismail Munshi (trans.), 1886. Gardens Bull. Straits Settlements VI

Go R, Hamzah KA (2008) Orchids of peat swamp forests in Peninsular Malaysia. Peat Swamp Forest Project, UNDP/GEF Funded (MAL/99/G321) Ministry of Natural Resources & Environment, Kuala Lumpur

Guerrera LM (1921) Medicinal uses of Philippine plants. Philippine Bur Forestry Bull 22:149–246

Handoyo F (2010) Orchids of Indonesia, vol 1. Indonesian Orchid Society, Jakarta

Heyne K (1927) De Nuttige Planten van Nederlandsch Indie (The useful plants in Dutch East Indies)

Holdsworth DK (1974) A phytochemical survey of medicinal plants in Papua New Guinea Part I. Sci New Guinea 2(2):142–154

Hooper D (1929) On Chinese medicine: drugs of Chinese pharmacies in Malaya. Gardens Bull Straits Settl 6:1–163

Lawler LJ (1984) Ethnobotany of the Orchidaceae. In: Arditti J (ed) Orchid biology. Reviews and perspectives, vol III. Cornell University Press, London, pp 27–149

Lawler (1986) Orchid Ethnobotany in the Asean Area. In: Rao AN (ed) Proc 5th Asean Orchid Congress. Parks & Recreation Department, Ministry of National Development, Singapore, pp 42–45

O'Byrne P (2001) The A-Z of South East Asian orchid species. Orchid Society of South East Asia, Singapore

Perry LM, Metzger J (1980) Medicinal plants of East and Southeast Asia: attributed properties and uses. MIT, Cambridge

Piso W (1658) De Indiae utriusque re naturali et medica. Elzevier, Amsterdam

Popenoe W (1928) The mangosteen in America. J Hered 19(12):537–546

Ridley HN (1906) Malay drugs. Agric Bull (Straits Settlements and FMS) 5:245–254

Ridley H (1894) The Orchidaceae and Apostasiaceae of the Malay Peninsula. J Linn Soc 32:335–338

Ridley H (1907) Materials for a flora of the Malay Peninsula, vol 1. Methodist Publishing House, Singapore

Teoh (2016) Medicinal orchids of Asia. Springer, Cham

Van Steenis CGGJ (ed) (1948) Flora Malesiana, vol 4, Part 1. Noordhoff-Kolff N.V, Batavia

Van Steenis CGGJ (ed) (1948–1954) Short history of the phytography of Malaysian vascular plants, Flora Malesiana Series 1, vol 4. Noordhoff-Kolff N.V, Djakarta, pp I–LXIII

Veldkamp JF (2011) Georgius Everhardus Rumphius (1627–1702), the blind seer of Ambon. Proceedings of the 8th Flora Malesiana Symposium. Gard Bull 63 (1&2): 1–16

Australian Orchids as Food and Medicine

16

In respect of ethnobotany, Australian orchids are more commonly discussed as food rather than as medicine. In Australia terrestrial orchids occur in moist, sheltered, forested habitats, commonly in large communities. *Peristylis nutans* (common names, nodding greenhood, parrot's beak orchid) occur at an amazing density of 440 plants in a square metre. Not everywhere, of course, but where they do occur, the communities can be large. The species is widespread. Melbourne botanist Beth Gott discovered that 440 plants yielded 800 tiny tubers which weighed, in toto, 126 gm. Not a great amount, but considering that they were easy to find and easy to dig out, they were a readily available source of food to the aborigines. They were not the only people who ate orchids. In the late nineteenth century, Australian botanist Joseph Maiden commented that 'there is hardly a country boy who has not eaten the so-called yam which are the tubers of numerous kinds of terrestrial or ground growing orchids' (Maiden 1898) (Table 16.1).

Survival in the Australian bush depends on finding suitable food when one is not armed with hunting tools. In an early period, Hedley (1888) commented that the common Queensland tree orchid *Cymbidium canaliculatum* (common name, tiger boat-lipped orchid) 'seems to me as likely to afford the most substantial aid to a man lost in the bush'. Its pseudobulbs are large, 80–120 by 30–40 mm, crowded, each with 2–6 thick, rigid leaves 300–500 by 30–40 mm,

forming unmistakable clumps of plants that emerge from tree hollows (Jones 2006) (Fig. 16.1).

Fascinated by these accounts, Tim Low, biologist and renowned Australian writer, decided to explore South Australia for edible orchids. He tasted over 20 species of Australian terrestrial orchids which belonged to 12 genera. They did not make him sick, although some were quite unpalatable, and *Pterostylis longifolia* (syn. *Bunochilus longifolius*; common names, tall greenhood, common leafy greenhood) was bitter. However, a few were exceptionally tasty, especially the walnut-sized, juicy and fragrant, waxy potatoes of *Lyperanthus suaveolens* (brown beaks) and the fragrant *Petalochilus carneus* (horned orchid). Low also found the paired, peanut-sized, white tubers of *Petalochilus carneus* (syn. *Caladenia carnea*; pink fingers) to be sweet and juicy (Low 1987). The orchids that Tim Low sampled are still, either locally common (e.g. *Pterostylis longifolia*; *Petalochilus carneus*) or widespread and common (e.g. *Lyperanthus suaveolens*; *Petalochilus carneus*) [Jones 2006] (Figs. 16.2, 16.3, and 16.4).

For a substantial meal, one would need the tubers of the saprophytic orchid, *Dipodium punctatum* (blotched hyacinth orchid), which flourishes in the bush near Brisbane, under certain species of eucalyptus. Expanded roots of the blotched hyacinth orchid are 7–8 mm thick and longer than a man's fingers. They are fibrous and

© Springer Nature Switzerland AG 2019
E. S. Teoh, *Orchids as Aphrodisiac, Medicine or Food*, https://doi.org/10.1007/978-3-030-18255-7_16

Table 16.1 Australian orchid species that are eaten or employed as medicine by natives

Pterostylis longifolia R.Br. (syn. *Bunochilus longifolius* (R.Br.) D.L.Jones & M.A.Clem.)

Ceratostylis latifolia Bl.

Cryptostylis erecta

Cymbidium canaliculatum R.Br.

Cymbidium madidum Lindl.

Cymbidium suave R.Br.

Dendrobium affine

Dendrobium canaliculatum R.Br.

Dendrobium discolor Lindl.

Dendrobium speciosum Sm.

Dendrobium teretifolium R.Br. [syn. *Dockrillia teretifolia* (R.Br.) Brieger]

Dipodium squamatum (G.Forst.) R.Br (syn. *Dipodium punctatum* (Sm.) R.Br.

Diuris maculata Sm.

Eriochilus cucullatus (Labill.) Rchb.f.

Gastrodia sesamoides R.Br.

Geodorum densiflorum (Lam) Schltr. (syn. *Geodorum pictum* Lindl.)

Habenaria multipartita Blume ex Ktraenzl.

Habenaria rumphii (Brongn.) Lindl.

Lyperanthus suaveolens R.Br.

Microtis unifolia (G.Forst.) Rchb.f.,

Pterostylis nutans R.Br.

Caladenia carnea R.Br. [syn. *Petalochilus carneus* (R. Br.) D.L.Jones & M.A.Clem.]

Renanthera moluccana Blume

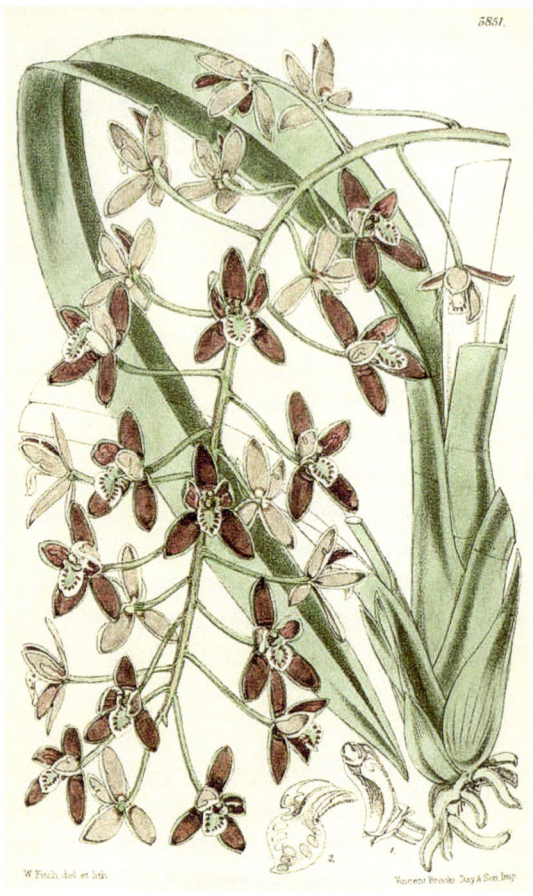

Fig. 16.1 *Cymbidium canaliculatum*. From: *Curtis Botanical Magazine* vol. 96 [ser. 3, vol. 26] t. 5851 (1870) [WH Fitch]

need to be cooked before they are eaten (Low 1987). Another edible orchid is the saprophytic *Gastrodia sesamoides* (native potato, potato orchid, cinnamon bells, bell orchid) which occurs in eastern Australia (from Queensland to South Australia and Tasmania) and New Zealand, in forests and coastal scrub and in sandy or clay loam. Its tubers are up to 15 cm long and 4.5 cm thick. Roasted tubers are eaten by Tasmanian natives. They are also a traditional food of aborigines in New South Wales (Biodiversity Conservation Unit, Adelaide Region 2012). The taste has been described as resembling beetroot, though rather insipid and watery (Lawler and Slaytor 1970). The vertical inflorescence of *Gastrodia sesamoides* is 30–60 cm tall, and it carries 2–20, nodding, bell-shaped, scented flowers with fused petals and sepals of light

brown. Flowering which occurs between November and December is enhanced by summer fires (Jones 2006). However, when looking for 'native potato', aborigines generally do not search for the flowers: they look for tell-tale signs in known habitats where bandicoots have scratched the ground to search for these orchids. These animals are drawn to its scent. Although *Gastrodia sesamoides* is naturalized in South Africa, their natives have yet to discover that its tubers can be eaten (Figs. 16.5 and 16.6).

Diuris maculata (spotted doubletail), known for its sweet tasting 'yams', is another food orchid. South Australian aboriginals ate tubers of *Microtis unifolia* (common onion orchid), *Eriochilus cucullatus* (leafless parson's bands,

PTEROSTYLIS Longifolia

Fig. 16.2 *Peristylis longifolia* From: Fitzgerald RD: *Australian Orchids* vol. 1: t. 1 (1875–1882) [RD Fitzgerald]

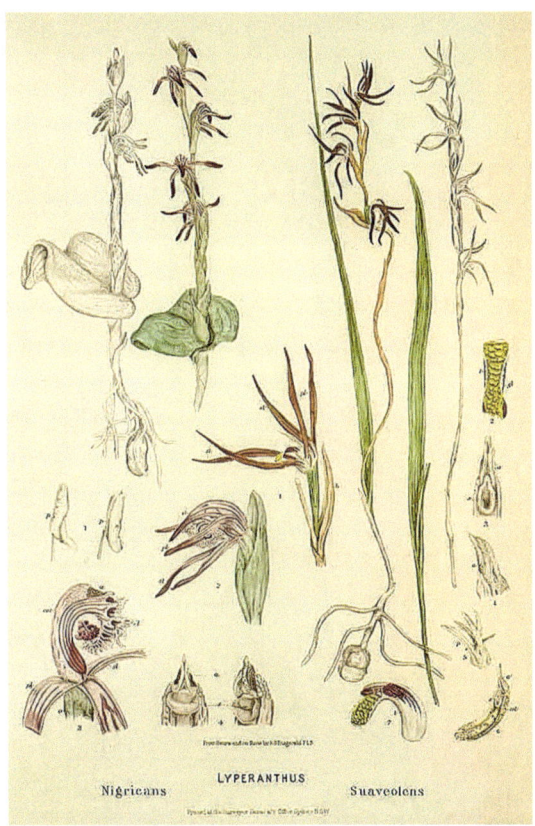

LYPERANTHUS
Nigricans   Suaveolens

Fig. 16.3 *Lyperanthes suaveolens*. From: Fitzgerald RD: *Australian Orchids* vol. 1: t. 36, Fig. 2 (1875–1882) [RD Fitzgerald]

pink autumn orchid) and *Caladenia carnea* (syn. *Petalochilus carneus*). Aboriginals in Victoria were reported to use tubers of several terrestrial genera as food: *Acianthus*, *Caladenia*, *Cryptostylis*, *Dipodium*, *Diuris*, *Glossodia*, *Lyperanthus*, *Microtis*, *Prasophyllum*, *Pterostylis* and *Thelymitra*. New South Wales aborigines regarded *Cryptostylis erecta* (bonnet orchid), *Cymbidium canaliculatum* and *Cymbidium suave* (grassy boat-lipped orchid) as food (Lawler 1984). Tubers of *Geodorum densiflorum* (Lam) Schltr. (formerly known as *Geodorum pictum* Lindl., pink nodding orchid, shepherd's crook orchid) were eaten by the aboriginals of Gladstone and Rockhampton in Queensland. They

were known to the aborigines of Gladstone as *Yeenga*. Rockhampton aboriginals called it *Uine* (Hedley 1888). The species is distributed in an arc from Western Australia across the tip of Northern Territory to Queensland and northern New South Wales. It is also widely distributed from India to Sri Lanka and across Southeast Asia to Japan and New Caledonia (Figs. 16.7, 16.8, 16.9, and 16.10).

Pseudobulbs of *Cymbidium madidum* and two *Dendrobium* species, *D. canaliculatum* and *D. speciosum*, were eaten as food by Australian aboriginals although they do not contain much nutritive matter. Pseudobulbs were chewed raw or rendered into powder much like sago (White 1938) (Figs. 16.11 and 16.12).

Australian aboriginals and her early settlers were not the only people who ate orchid tubers

Fig. 16.4 *Petalochilus carneus* [as *Caladenia carnea*] From: *Curtis Botanical Magazine* vol. 124: t. 7630 (1898) [M.Smith]

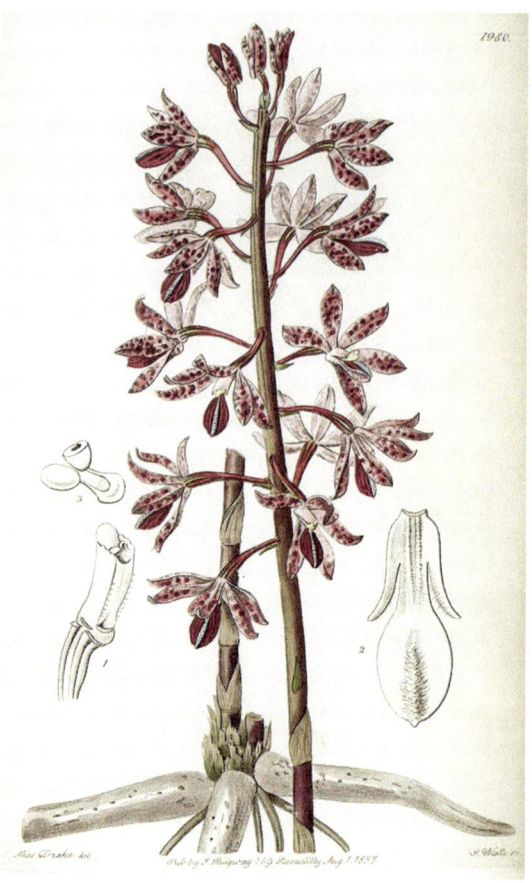

Fig. 16.5 *Dipodium squamatum* [as *Dipodium punctatum*] From: *Edward's Botanical Register* vol. 23: t. 1980 (1837) [SA Drake]

as food. Corms of the beautiful pink *Calypso bulbosa* (common name, fairy slipper) which contain a large amount of mucilaginous substance were eaten by Indians in British Columbia (Correll 1978). Orchids had a high place in the menu of hill tribes in the Himalayas where they were considered a delicacy. New shoots of *Cymbidium* were ground into a semi-liquid paste, and with the addition of species it becomes a sauce to improve the bland cereal diet. Curried pseudobulbs only needed a pinch of salt (Pemphahishey 1974). In Amboin (Moluccas), pickled leaves of *Renanthera moluccana* (*anggrik merah, bunga karang*) were a delicacy

(Rumphius 1741–1755). They tasted like capers but were very fibrous. Candy was made from tubers of *Habenaria rumphii*, preferably picked before the flowering season. During a field trip around 1975 to South Sukami in West Java, Rafai discovered that the local people ate young leaves of *Ceratostylis latifolia* (*ki pahit*), either raw as salad or cooked. Tubers of *Habenaria multipartita* (*uwi-uwi*) were eaten in Central Java (Rafai 1975).

On a long, lone track, African natives chew roots of plants for nourishment and, therefore, it is not surprising that orchids with tubers would be among the items of diet. Numerous orchids are eaten in Africa and a few are even traded as food

Fig. 16.6 *Gastrodia sesamoides*. From: Hooker JD, *The botany of the Antarctic voyage of H.M. Discovery ships Erebus and Terror in the Years 1839–1843, under the command of Captain Sir James Clark Ross*, vol. 3 (2): t. 126 (1866) [W Archer]

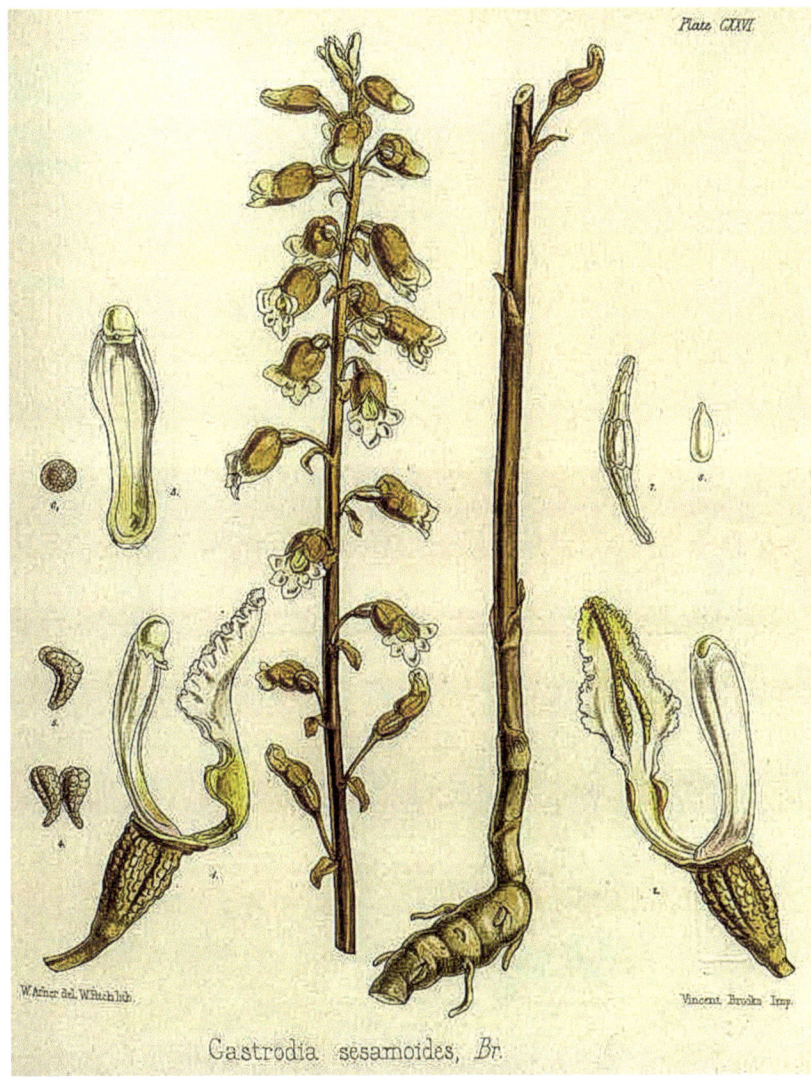

Plate CXXVI.

Gastrodia sesamoides, Br.

items. *Chikanda*, a popular cake sold in Tanzanian markets, is prepared from tubers of several orchid species. There is no agreement regarding the extent to which this forms a threat to existing species. The Darwin Initiative is making an effort to promote sustainable harvest of these orchids to provide a livelihood for women and children many of whom have become destitute when the family's breadwinners perish from AIDs (Challe et al. 2011; Kim 2016) (Figs. 16.13 and 16.14).

Medicinal Usage of Australian Orchids

Many indigenous Australian orchids commemorate the names of physicians and surgeons who lived and worked in Australia. Being schooled in both botany and medicine during their undergraduate years, these doctors continued to maintain an interest in botany despite their busy medical practice. Robert Brown (1773–1858), a Scottish military surgeon, was exceptionally prodigious in his

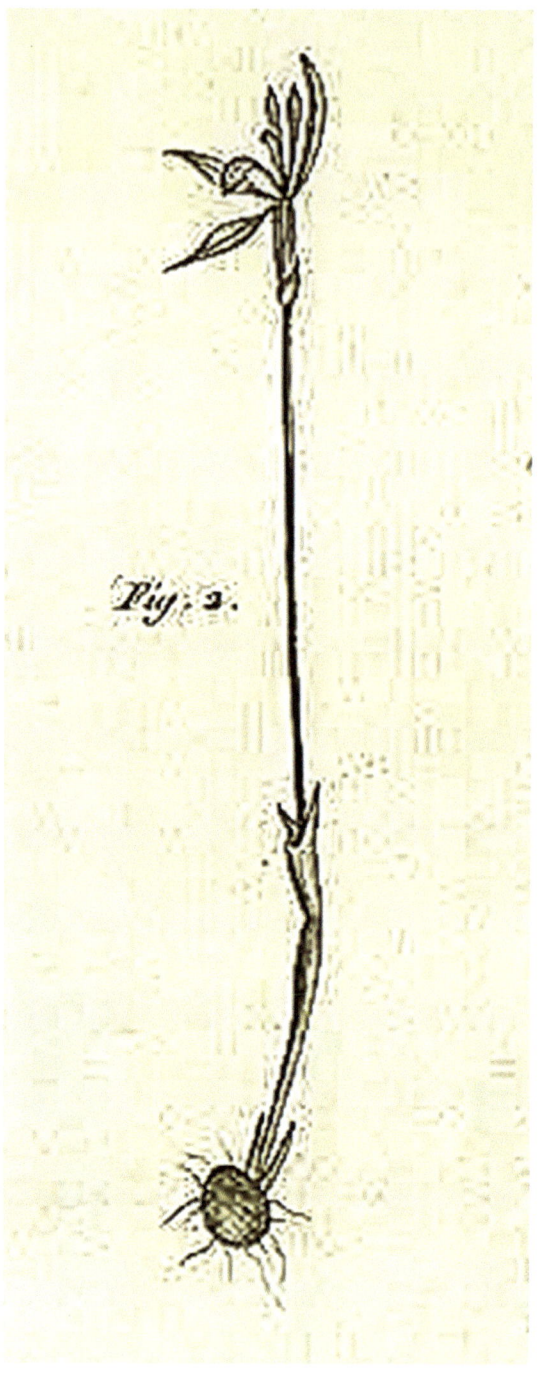

Fig. 16.7 *Diuris maculata.* From: Hooker JD. *The botany of the Antarctic voyage of H.M. Discovery ships Erebus and Terror in the Years 1839–1843, under the command of Captain Sir James Clark Ross*, vol. 3(2): t. 104 (1860) [W Archer & WH Fitch]

Fig. 16.8 *Eriochilus cucullatus* [as *Epipactis cucullata*] From: La BillardiereJJ, Houlton de, *Novae Hollandiae plantarum specimen*, vol. 2, t. 211 (1804)

Fig. 16.9 *Cryptostylis erecta*. From: Fitzgerald RD: *Australian Orchids* vol. 1: t. 124, Fig. 1 (1875–1882) [RD Fitzgerald]

CRYPTOSTYLIS

Erecta   Leptochila

Sydney, N.S.W. Thomas Richards, Government Printer.

botanical effort and is well remembered as a botanist. He described 120 new Australian orchid species, and his description of the movement of pollen grains in water is commemorated by the term 'Brownian motion'. An additional 15 Australian doctors also have their names immortalized by orchid species (Pearn 2013). However, none of them paid any attention to the use of orchids as medicine.

Australian aboriginals apparently do not think much of orchids as medicinal plants. Very few orchids are employed and they are only used to

Fig. 16.10 *Cymbidium suave* From: Bauer F, *Australian botanical drawings*. p.163: t. 163

treat minor illnesses. *Dendrobium discolor* which occurs in Queensland and is abundant in Papua New Guinea was made into a poultice or liniment to draw a boil and to treat ringworm. Mature canes were pulverized with a hammer to extract the pulp for making the liniment. In northern Queensland, *Cymbidium canaliculatum* and *Cymbidium madidum* were used to treat dysentery, and seeds of the latter species were employed as contraceptive (Lawler and Slaytor 1970). Shepherds in Queensland chewed on pseudobulbs of *Cymbidium suave* to arrest dysentery (Lawler 1984) (Fig. 16.15).

In the Northern Territory of Australia, aboriginals employ the sticky sap obtained by crushing pseudobulbs of *Cymbidium canaliculatum* (vernacular name in Ngankikurungkurr, *Tjalamarinj*) to apply on boils, sores, wounds and dry patches. It adheres to the skin and acts as an emollient. It has antiseptic properties but does not contain any alkaloid. An alternative usage involves applying pounded pseudobulbs on fractures. *Cymbidium canaliculatum* grows on the trunk and branches of hallbark and bloodwood, *Eucalyptus* spp.

Fig. 16.11 *Dendrobium canaliculatum* (©Teoh Eng Soon 2019. All Rights Reserved)

Fig. 16.12 *Dendrobium speciosum*. From: *Curtis Botanical Magazine* t. 3039–3122, vol. 58 [ser. 2., vol. 5]: t. 3074 (1831) [WJ Hooker]

Fig. 16.13 *Renanthera*
moluccana. From: Blume
CL, *Rumphia*, vol. 4: t. 193,
Fig. 2 (1848)

Dendrobium affine (vernacular names in Ngankikurungkurr *Tjalamarinj*; Burarra and Djinang *Marndaja*; Djambarrpuyngu *Djalkur*rk) and *Dendrobium canaliculatum* (Burarra and Djinang *Marndaja*) which are commonly found on *Melaleuca* spp. in paperbark swamps and other species of *Dendrobium* are employed in a similar manner to treat skin lesions. The paste is applied many times. Viscous sap from pseudobulbs of *Dendrobium affine* is used as a dressing for minor burns (Fig. 16.16). Alkaloid is absent in this species (Aboriginal Committee of the Northern Territory 1993). Australian aboriginals only employ a generic name for epiphytic orchids and make no effort to provide different names for individual species.

Fig. 16.14 *Habenaria rumphii*. Adapted from: Duperrey LI, *Voyage autour du monde sur la corvette de S.M. La Coquille, pendant les annees* 1822–1825, Atlas, t. 38 * (1826) [as *Platanthera rumphii*]

Fig. 16.15 *Dendrobium discolor* (©Teoh Eng Soon 2019. All Rights Reserved)

Fig. 16.16 *Dendrobium affine* From: Lindemann E von, *Lindenia, Iconography of Orchids* vol. 2: t. 291 (1891) [A Goossens]

Some orchids occurring in Australia and elsewhere are not employed by Australian aborigines to treat disease although they enjoy widespread medicinal uses in other countries. *Geodorum densiflorum* is an example (Burkill 1935; Dash et al. 2008; Davies and Steiner 1982; Musharof Hossain 2009; Rao 2007; Reddy et al. 2005). Mashed leaves of *Dockrillia teretifolia* (syn. *Dendrobium teretifolium*; thin pencil orchid, bridal veil orchid), a widespread and common epiphytic or saxicolous species in Queensland and New South Wales, were applied on the head to relieve headache in Tahiti (Lawler 1984), but it is not employed by Australian aboriginals (Fig. 16.17).

Fig. 16.17 *Dendrobium teretifolium*. From: *Curtis Botanical Magazine* [t. 4689–4757, vol. 79 [ser. 3, vol. 9] t. 4711 (1853) [WH Fitch]

Conservation Issues

Aboriginal consumption of orchids as food do not pose a threat to native orchids of Australia. In his editorial in the September 1937 issue of the *Australian Orchid Review*, G. H. Slade mentioned that the spread of settlement and the timber industry took a toll on orchids, but these were matters beyond control. He called instead for a restriction on the sale of native orchids collected from private lands: such orchids were frequently offered at the Sydney municipal markets. The (Australian) Wild Flowers and Native Plant Protection Act 1927–1931 only forbad the collection of wild orchids from crown lands.

References

Aboriginal Committee of the Northern Territory (1993): Traditional aboriginal medicines in the Northern Territory of Australia. Darwin: Conservation Commission of the Northern Territory of Australia

Biodiversity Conservation Unit, Adelaide Region (2012) Adelaide and Mount Lofty Ranges, South Australia.

Threatened species profile. http://www.environment. sa.au/

Burkill IH (1935) (1966 reprint, 2nd ed., with contributions by Birtwistle W, Foxworthy FW, Scrivenor JB, Watson IG) A dictionary of economic products of the Malay Peninsula, vol II. Ministry of Agriculture & Co-operatives, Kuala Lumpur

Challe JF, Niehof A, Struik PC (2011) The significance of gathering wild orchid tubers for orphan household livelihoods in a context of HIV/AIDS in Tanzania. Afr J AIDS Res 10(3):207–218

Correll DS (1978) Native orchids of North America north of Mexico. Stanford University Press, Stanford

Dash PK, Sahon S, Bal S (2008) Ethnobotanical studies on orchids of Niyamgiri Hill Ranges, Orissa, India. Ethnobot Lett 12:70–78

Davis RS, Steiner ML (1982) Philippine orchids. A detailed treatment of some 100 native species. M & L Lucidine Enterprises (reprint), Manila

Hedley G (1888) Uses of some Queensland plants. Proc R Soc Qld 5:10–15

Jones DL (2006) A complete guide to native orchids of Australia including the island territories. Reed New Holland, Sydney

Kim SL (2016) see http://www.facebook.com/ chikandaorchidconservation

Lawler LJ (1984) Ethnobotany of the Orchidaceae. In: Arditti J (ed) Orchid biology. Reviews and perspectives III. Cornell University Press, Ithaca, pp 27–149

Lawler LJ, Slaytor M (1970) Uses of Australian orchids by aborigines and early settlers. Med J Aust 2:1259

Low T (1987) Australian wild foods. Aust Nat Hist 22(5): 2202–2203

Maiden JH (1898) Some plant foods of the aborigines. Agric Gaz NSW 9:349–354

Musharof Hossain M (2009) Traditional therapeutic uses of some indigenous orchids of Bangladesh. Med Aromat Plant Sci Biotechnol 3:100–106

Pearn JH (2013) Australian orchids and the doctors they commemorate. Med J Aust 198(1):52–54

Pemphahishey KT (1974) Orchid eaters of Shangri-la. Am Orchid Soc Bull 43(8):716–725

Rafai MA (1975) Extraordinary uses of orchids in Indonesia. Report of the First Asean Orchid Congress. Ministry of Agriculture, Ministry of Foreign Affairs and Kasetsart University, Bangkok

Rao TA (2007) Ethno Botanical data on wild orchids of medicinal value as practised by tribals at Kudremukh National Park in Karnataka. Orchid Newslett 2(2):1–7

Reddy KN, Reddy CS, Raju VS (2005) Ethno-orchidology of orchids of Eastern Ghats of Andhra Pradesh. EPRTI Newslett 11(3):5–9

Rumphius GE (1741–1755) Herbarium amoinense. M. Uytwerf, Amsterdam

White CT (1938) Some economic use of orchids. Orchidol Zeylan 5:74–77

Northern African countries belong to the Mediterranean region, and having once formed the outer reaches of the Ottoman Empire during the last millennia, current traditional herbal practices are closely linked with that of the Middle East and ancient Greece. Information on the use of orchids in this region is scarce. A rare reference to medicinal orchids from northern Africa describes *Habenaria* species as *etse yihayu* (restorative plants) which are used to overcome impotence in Ethiopia. In the traditional health practice of that country, plants are the primary source of drugs (Dawit 1987; Dawit and Estifanos 1986). In Tunisia, *Ophrys lutea* and *Orchis anthropophora* (syn. *Aceras anthropophora* R. Br.) are served as appetizers.

African salep makes use of *Dactylorhiza sambucina* (L) Soo (syn. *Orchis lutea* Dulac) and *Orchis provincialis*. In Beni-Suef, Upper Egypt, salep prepared with tubers of *Orchis hircine* Crantz. which is imported is boiled in milk to treat peptic ulcer (AbouZid and Mohamed 2011). However, Egyptian herbal medicine has existed for seven millennia, if not much longer, and its usage of herbs could have strongly influenced Greek medicine. The Ebers Papyrus (c. 1552 BC) which predates the Herbal of Dioscorides (c. 40–90 CE) by 1600 years describes over 700 different substances (von Klein 1905). Whether orchids feature in ancient Egyptian medicine will only be known when the full texts of all extant papyri are translated and made readily available to the public. The rest of this chapter will therefore focus on usage of orchids as medicine or food only in Central and South Africa (Figs. 17.1 and 17.2).

Africans enjoy freedom from brittle bones because in addition to their genetic constitution, they eat root crops and roots of wild plants when foraging in forests. In the southern part of Africa, tubers of many orchid species are eaten as food, mostly as a relish or snack. Edible orchids, especially *Satyrium carsonii*, are preferred to fish, meat and other vegetables. There exists a belief in Malawi that eating edible orchids (*chinaka*; *chikanda*, a cake or meatless sausage prepared with edible orchids and peanuts) confers protection against illness and in one survey over two-thirds of the 117 people interviewed said that they ate orchids at least once a week. The popular edible orchids were *Disa engleriana*, *D. robusta*, *D. zombica*, *Habenaria clavata*, *Satyrium ambylosaccos*, *S. buchananii* and *S. carsonii* (Kasulo et al. 2009). Edible orchids contain 5.36% protein and 2.2% minerals and sufficient vitamin C and beta-carotene in 10 g of the tuber to meet minimum daily requirement. There is a large amount of calcium [22.120–33.574 mg/100 g (Lalika et al. 2013) or 48 mg/100 g (Kasulo et al. 2009)] even when

© Springer Nature Switzerland AG 2019
E. S. Teoh, *Orchids as Aphrodisiac, Medicine or Food*, https://doi.org/10.1007/978-3-030-18255-7_17

Fig. 17.1 *Orchis anthropophora* [as *Aceras anthropoph*ora]. From: Barla JB, *Flora illustre de Nice et des Alpes-maritimes. Iconographie des Orchidees*, t. 23, Fig. 1–13 (1868)

Fig. 17.2 *Orchis provincialis* Balb. Ex Lam. & D.C. From: Barla JB, *Flore illustre de Nice et des Alpes-maritimes. Iconographie des Orchidees*, t. 38 (1868)

compared with other root crops like potato
(12 mg/100 g), sweet potato (14 mg/100 g),
greater yam (24 mg/100 g), carrot (34 mg/100 g)
and tapioca root (46 mg/100 g) (Leong and Morris 1947; Teoh and Teoh 1999).

Chikanda has been a village delicacy for
hundreds of years, but the custom did not pose a
threat to the edible orchid species in sub-Saharan
Africa until the last decade of the twentieth century (Davenport and Ndangalasi 2003). In
Malawi edible orchid tubers are known as
Chanaka, Chikande or *Nyama Yapans*, the last
term meaning 'underground meat'. Women collect tubers of 15 species of terrestrial orchids
(*Disa, Habenaria, Satyrium*) from the mountains,
wash and sun-dry them to keep for later use.
When boiled in potash, they produce a jelly-like
substance which is eaten as a relish (Morris
1996). An investigation in Malawi found that all
respondents reported eating orchid tubers as
food. Their survey involved interviewing
147 individuals and 74 households in 18 villages
in 2 districts where edible orchids occurred
(Kasulo et al. 2009). High demand in some places
like Zambia, a big consumer of *chikanda*, has
given rise to cross-border trade and questions
over its sustainability. It is estimated that between
2.2 and 4.1 million, orchid plants consumed in
Zambia originate from Tanzania (Challe and
Price 2009). They are comprised almost entirely
of species of *Disa, Habenaria* and *Satyrium*
which are called by their vernacular name,
Chikanda. The government of Tanzania has
designated 135 square kilometres of the Kitulo
Plateau as a new National Park to protect the
local flora and fauna, and perhaps this may help to
protect some orchid species.

There is also an active cross-border trade of
orchid tubers from Malawi to Zambia (Kasulo
et al. 2009), and more recently supplies have
arrived from Angola, the Democratic Republic of
the Congo and Mozambique (Veldman et al.
2017). Edible orchids in the originating country
include *Satyrium carsonii, Disa zombica,
Satyrium buchananii, Satyrium ambylosaccos,
Disa engleriana, Disa robusta*, other unidentified
Disa species, *Habenaria clavata* and *Habenaria
walleri*, all found predominantly in the highlands
with *Disa* occurring in montane bogs. *Chikanda*

*t*ubers are varied and consist of a mixture of many
species of terrestrial orchid tubers, some collected
from neighbouring countries like Congo and
Tanzania, and depending on their origin, the species would be different. They are traded in bulk at
Zambia's Soweto Market in Lusaka, with
middlemen selling *myala* tubers at 150 ZMW
(approximately US$15) per tin of 41 kg. A large
bag contains 410 kg. On the other hand, vendors at
Nakadoli Market in Kitwe sell tubers in small
amounts, with portions costing only US 50 cents.
Girls learn their trade from their mothers, and some
continue working in the market for many years. To
prevent loss from rot, tubers may be sun-dried and
then grounded and sold in powder form. *Chikanda*
cakes are also sold in these markets (Kim 2016)
(Figs. 17.3, 17.4, 17.5, and 17.6).

Thirteen species belonging to the three genera
that were identified during a survey conducted by
Nyomora in February to March when plants were
in bloom, but the author submits that this may
represent only 15% of all orchid species collected
for *chikanda* (Nyomora 2005). An early study
reported that in excess of four million tubers of
85 species of the three genera were collected
annually (Davenport and Ndangalasi 2003).
Most of these orchids are gathered by AIDS-/
HIV-affected households, often headed by
orphaned children who rely almost entirely on
the sale of edible orchids for their livelihood.
They are living from hand to mouth (Challe and
Price 2009; Challe et al. 2011). They also eat the
tubers in lieu of vegetable and meat (Nyomora
2005). Up to 80% of households in Tanzania are
involved in the gathering of orchid tubers (Bone
2016) (Figs. 17.7 and 17.8). Table 17.1 lists the
edible orchid species, mostly employed to prepare
chikanda which have been positively identified.

Roots (or pseudobulbs) of *Disa barbata* (syn.
Herschelia barbata), *Disa venusta* (syn.
Herschelia venusta), *Satyrium bicorne, Satyrium
candidum, Satyrium carneum* and *Satyrium
erectum* were reported to contain a large quantity
of sweet, mucilaginous and nutritious juice (Watt
and Breyer-Brandwijk 1962). Gatherers are able
to distinguish between edible and nonedible species within a genus. Edible tubers are sweet and
contain sugar crystals, and they are referred to as
manseke, vinsake, lidala or *sidala*, i.e. 'female

442 Microspermae. — Orchidaceae.

Fig. 313. *A—F* Disa Engleriana Krzl. *A* Ganze Pflanze; *B* Blüte; *C* Labellum; *D* Blumenblatt; *E* Gynostemium von der Seite gesehen; *F* dasselbe von vorn. *G—J* Disa Carsoni N. E. Brown. *G* Habitus; *H* Blüte; *J* Pollinien. — Original.

Fig. 17.3 *Disa engleriana* (**a–f**) and *Disa erubescens* (**g–j**). From: Engler A, Drude O, *Die Vegetation Erde*: *Sammlung pflanzengeographischer Monographien* vol. 9 (2): p 442 (1896–1928)

orchids', whereas inedible tubers are bitter, watery and non-marketable and they are referred to as 'male' (Challe and Struik 2008; Lalika et al. 2013). *Disa erubescens*, *Disa robusta* and *Satyrium atherstonei* are additional examples of sweet tubers, and *Satyrium crassicaule* and

Habenaria praestans var. *praestans* are examples of orchids with bitter tubers. Due to over-collection, in the areas studied, over a short period, populations of the sweet species were overtaken by the bitter species (Challe and Struik 2008). Several studies found that villagers prefer to collect tubers of *Disa* over those of *Satyrium* and *Habenaria* (Davenport and Ndangalaasi 2003; Nyomora 2005; Hamisy 2007; Challe and Price 2009). In Cameroon, tubers of *Habenaria keayi* and *Habenaria zambesina* are eaten as food (Kasulo et al. 2009). Three orchids (*Habenaria cirrhata*, *Eulophia horsfallii*, *Nervilia bicarinata*) are eaten in Benin (Assese et al. 2017).

As popular species become less easy to collect, alternative species are sought (Bone 2016). DNA barcoding employed by the team of Veldman, de Boer and their colleagues showed that many more species of orchids are being used for *chikanda* in Zambia than was previously discovered (Veldman et al. 2018). A recent investigation which examined six prepared *chikanda* cakes by DNA barcoding found *Disa* to be present in all samples, *Satyrium* in five out of six and *Habenaria* only in one (Veldman et al. 2017) (Figs. 17.9, 17.10, 17.11 and 17.12).

The Darwin Initiative recently funded a project to manage the wild edible orchid trade in Zambia that will provide security and good livelihood for poor rural women and girls while at the same time preventing over-exploitation and cross-border trade of such orchids. The project is undertaken by the Royal Botanic Gardens, Kew, and supervised by Ruth Bone.

An Uppsala University team of researchers working with the Darwin Initiative has started a programme to identify the exact orchid species that are consumed in *chikanda* (Kim 2016; Veldman et al. 2017). They noted that a vernacular name may refer to several different plants because native plant traders are not familiar with their scientific names. For instance, *iphamba* refers to 12 different species, namely, *Cyrtorchis arcuata*, *Diaphananthe millarii*, *D. xanthopollinia*, *Eulophia ensata*, *E. ovalis*, *E. leontoglossa*, *Microcoelia exilis*, *Mystacidium capense*, *M. venosum*, *Polystachya transvaalensis*, *Tridactyle bicaudata* and *T. tridentata*.

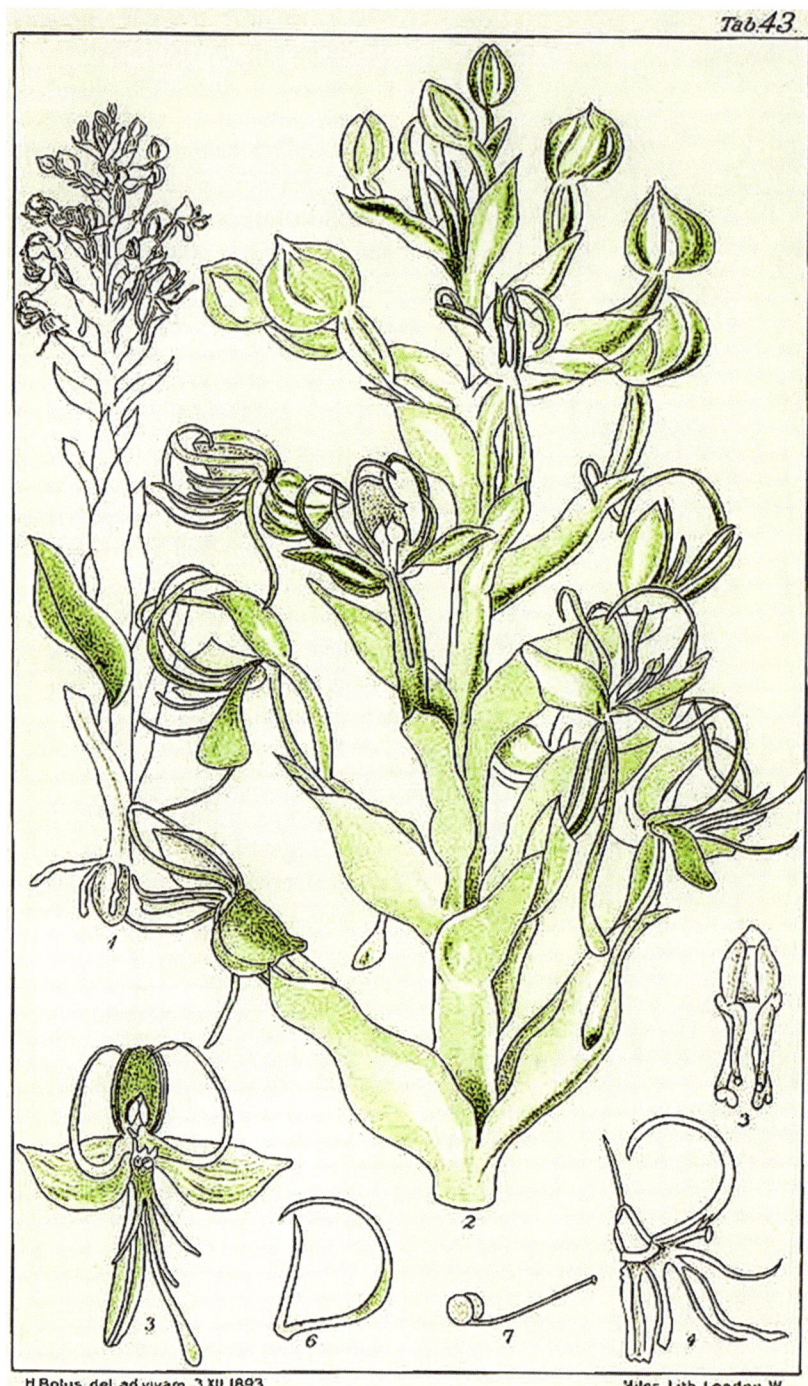

Fig. 17.4 *Habenaria clavata* (Lindl.) Rchb.f. From: Bolus, H, *Icones orchidearum Austro-Africanarum extratropicarum* (*Orchids of South Arica*), vol 2: t. 43 (1896–1913) [H Bolus]

Fig. 17.5 *Chikanda*, a popular Central African cake made with orchid tubers. Photo: Soel-Jong Kim

Fig. 17.6 *Chikanda* tubers from varied species on sale at a local African market. Photo: Soel-Jong Kim

Lourance Njopilai David Mapunda related that he met an old lady from Ilindiwe Village in Tanzania who informed him that she managed to grow orchids in the field by broadcasting a mature flower bunch mixed with soil in a farm. Some farmers reported obtaining larger tubers when they grew the orchid instead of collecting it from the wild. In the past, villagers only collected edible orchids from plants that did not flower (ironically referred to as female plants, *Lidala, Sidala*), whereas male plants that flowered (*Ligosi, Sigosi, Likose*) were not collected because they were responsible for seeding the next generation. This traditional practice is now ignored, and male plants are also collected.

In addition to species in the three traditional edible genera of *Satyrium, Disa* and *Habenaria*, Mapunda reported that *Brachycorythis pleistophylla* and *Eulophia schweinfurthii* were also collected for food (Mapunda 2007). In South Africa and Swaziland, tubers of *Neobolusia tysonii* and two varieties of *Satyrium longicauda* are also eaten, albeit not to a great extent (Long 2005) (Figs. 17.13 and 17.14). Tubers of *Eulophia livingstoneana* (Madagascar local name, *Felatrandraka*) and leaves and fruits of *Eulophia reticulata* (Madagascar names, *Kamasina, Tandrokondrylahy, Kitandrokondrilahy*) are eaten in Madagascar (Cribb and Hermans 2009).

The edible orchid, *Bulbophyllum scaberulum*, is moderately effective in killing roundworms. Another orchid, *Cyrtorchis arcuata*, is more effective, but it is only employed as a charm in South Africa (Chinsamy 2012) (Fig. 17.15).

People in Zaire were advised not to eat *Eulophia kalende* which is poisonous.

Orchids as Medicine

The role of orchids in Central and South Africa's folk medicine takes a unique form that is closely bound to local cultural beliefs. It is influenced by their notion that illnesses, unease, even difficult situations and natural catastrophes are caused by spirits, witchcraft or charms employed by enemies and opportunists. Brain Morris who lived, worked and studied medicinal plants in

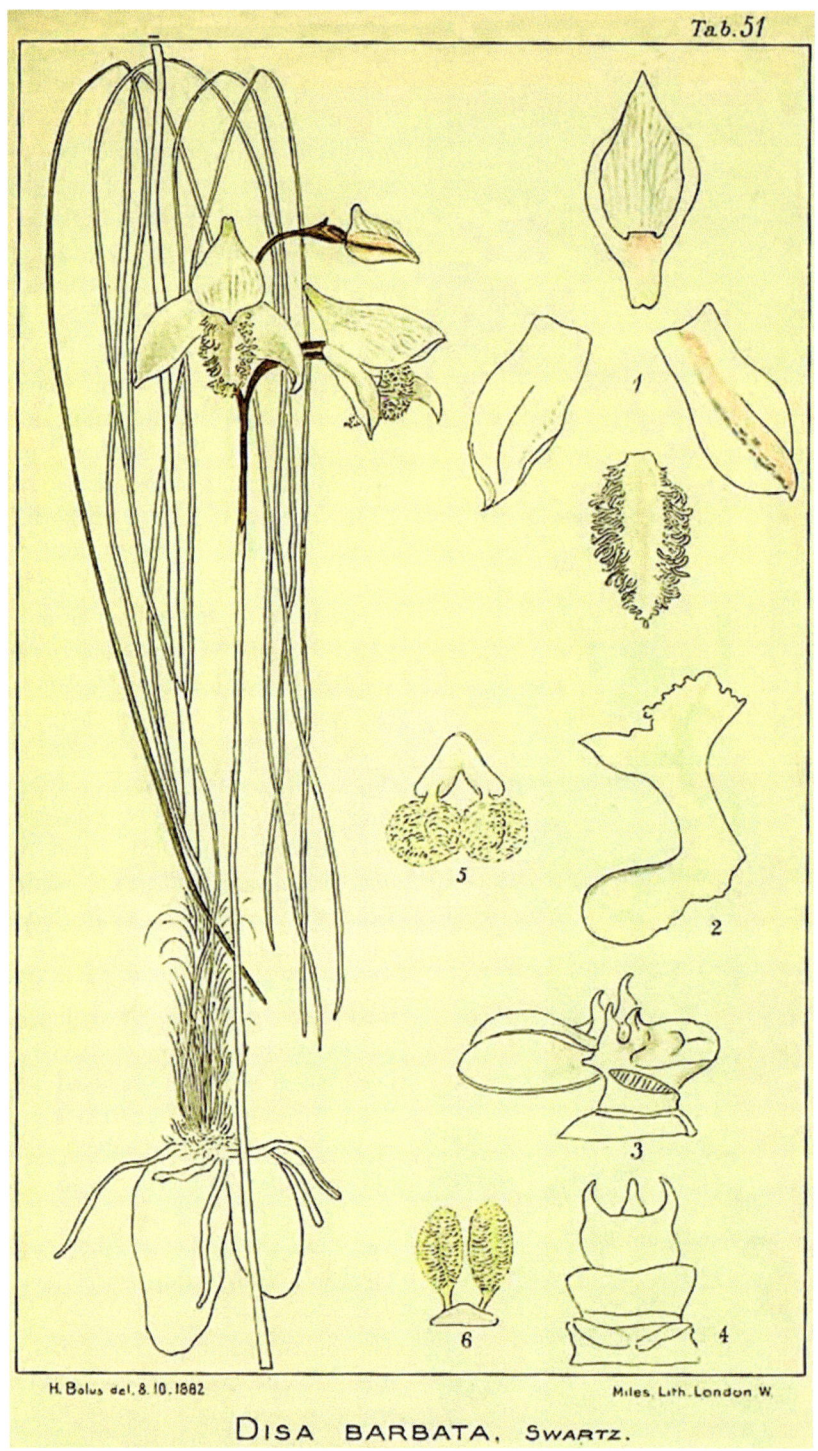

Fig. 17.7 *Disa barbata*. From: Bolus H, *Icones orchidearum Austro-Africanarum extra-tropicarum* vol 3: t. 51 (1896–1913) [H Bolus]

Fig. 17.8 *Satyrium bicorne* [as *Satyrium cucullatum*]. From: *Edward's Botanical Register* vol. 5: t. 416 (1819) [S Edwards]

Table 17.1 African edible orchid species. Many species are employed in the making of *chikanda*

Brachycorythis ovata Lindl.

Brachycorythis pleistophylla Rchb. f.

Bulbophyllum sandersonii (Hook. f.) Rchb.f.

Bulbophyllum scaberulum (Rolfe) Bolus

Disa baurii Bolus (syn. *Disa hametopetala* Rendle)

Disa engleriana Kraenzl.

Disa barbata (L.f.) Sw.

Disa caffra Bolus

Disa celata Summerh.

Disa engleriana Kraenzl.

Disa equiloba Summerh.

Disa erubescens Rendle

Disa leucostachys Rolfe

Disa miniata Summerh.

Disa ochrostachya Rchb.f.

Disa robusta N.E.Br.

Disa satyriopsis Kraenzl.

Disa tanganyikensis Summerh.

Disa venusta Bolus

Disa walleri Rchb.f

Dis welwitschii Rchb. f.

Disa zombica N.E.Br.

Eulophia horsfallii (Bateman) Summerh.

Eulophia livingstoneana (Rchb.f.) Summerh.

Eulophia reticulata Ridl.

Eulophia schweifurthii Kraenzl.

Habenaria adolphii Schltr.

Habenaria cirrhata (Lindl.) Rchb.f.

Habenaria clavata (Lindl.) Rchb.f.

Habenaria cornuta Lindl.

Habenaria humilior Rchb.f.

Habenaria keayi Summerh.

Habenaria praestans Rendle

Habenaria praestans var. *praestans*

Habenaria sochensis Rchb. f.

Habenaria walleri Rchb.f.

Habenaria xanthochlora Schltr.

Habenaria zambesina Rchb.f.

Himantoglossum hircinum (L.) Spreng. (syn. *Orchis hircine* (L.) Krantz

Neobolusia tysonii

Nervilia bicarinata (Bl.) Schltr.

Oeceoclades beravensis (Rchb.f.) R. Bone & Buerki (syn. *Eulophia beravensis* Rchb.f.)

Platycoryne crocea Rolfe

Roeperochian wentzeliana Kraenzl.

Satyrium ambylosaccos Rchltr.

Satyrium atherstonei Rchb. f.

(continued)

Table 17.1 (continued)

Satyrium bicorne (L.) Thunb.

Satyrium breve Rolfe

Satyrium buchananii Schtr.

Satyrium candidum Lindl.

Satyrium carneum (Alton) Sims

Satyrium carsonii Rolfe

Satyrium chlorocorys Rolfe

Satyrium crassicaule Rendle

Satyrium erectum Sw.

Satyrium kitimboense Kraenzl.

Satyrium longicauda var. jacottetianum

Satyrium longicauda var. longicauda

Satyrium macrophyllum

Satyrium sacculatum (Rendle) Rolfe

Satyrium sceptrum Schltr. (syn. *Satyrium acutirostrum* Summerh.)

Eulophia schweinfurthii Kraenzl.

Satyrium shirense Rolfe

Satyrium trinerve Lindl.

Satyrium volkensii Schltr.

Data from Assese et al. (2017), Bingham (2002), Morris (1996), Davenport and Ndangalasi (2003), Nyomora (2005), Hamisy (2007), Challe and Price (2009), Kasulo et al. (2009), and Veldman et al. (2016, 2018)

Malawi for over 20 years stated that the term for medicine in Malawi is *Mankhwala*, cognate of the widespread term *Bwanga*. They cover a wide variety of substances—charms, amulets, protective medicine, medicine in the usual sense and modern pharmaceuticals—that are believed to possess an inherent ability to prevent or cure illnesses. One might say their usage is holistic: it extends beyond the specific therapeutic need. *Mankhwala* is also a good luck charm that helps a person to be liked by friends and employers, to get married, to establish a family, and to achieve success in hunts and agriculture. It protects a person against witchcraft. It promotes health and potency and has a role in life cycle passage (Morris 1996).

Emetics or compounds that induce a person to vomit are widely used in African folk medicine because there is a belief that this facilitates removal of the cause of the ailment. Orchids are generally employed in this manner, be it to treat cough in children, induce fertility in women and cure madness, or even as a love charm either

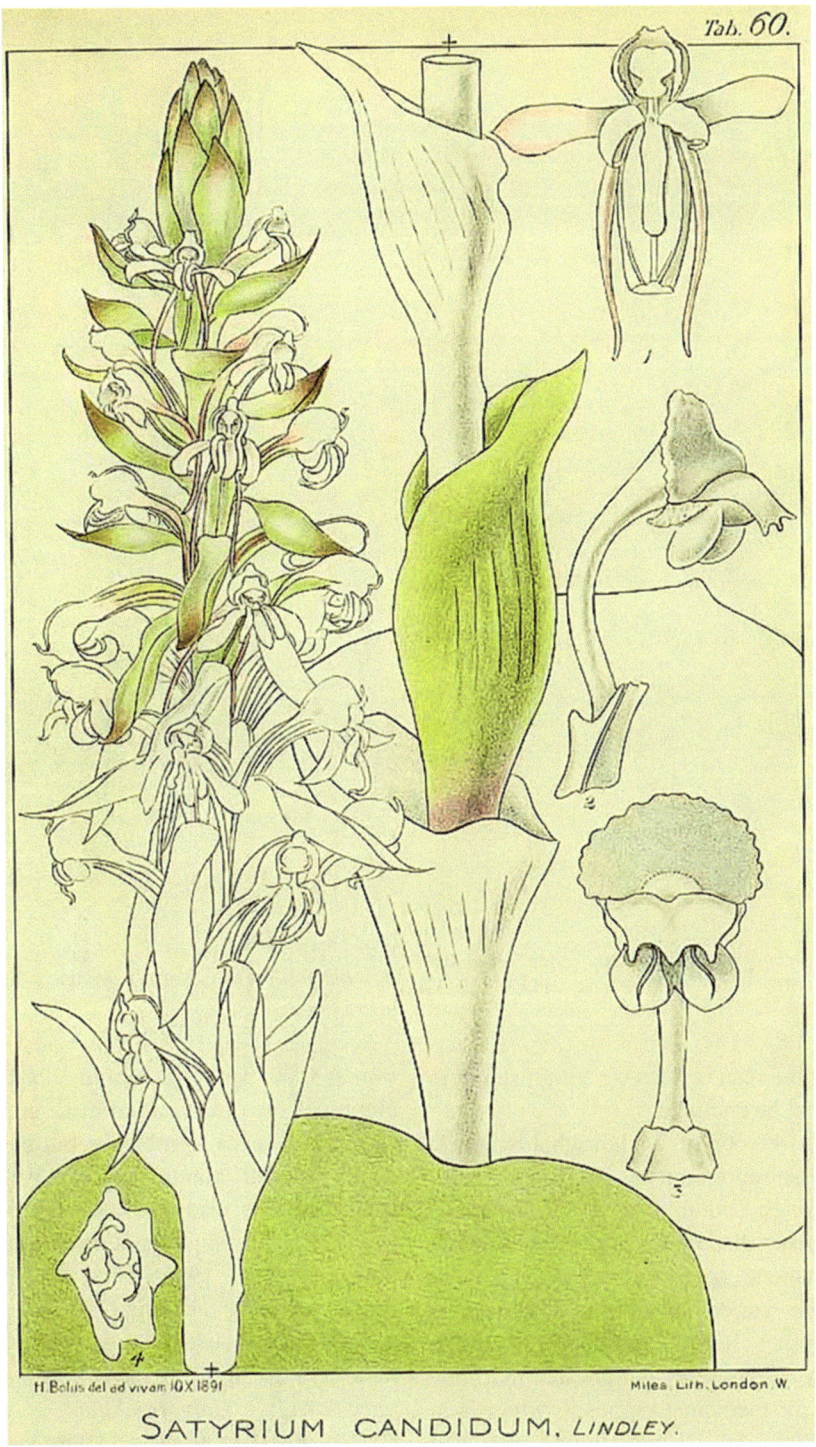

Fig. 17.9 *Satyrium candidum*. From: Bolus H, *Icones orchidearum Austro-Africanarum extra-tropicarum* vol 2: t. 60 (1896–1913) [H Bolus]

Fig. 17.10 *Satyrium carneum*. From: *Curtis Botanical Magazine* t. 1502–1547, vol. 37; t. 1512 (1813) [ST Edwards]

Fig. **17.11** *Satyrium erectum* [as *Satyrium pustulatum*]. From: *Edward's Botanical Register* vol. 26: t. 18 (1840)

consumed by the man or secretly administered to the woman that he cherishes.

In Malawi, 45 out of 74 households (61%) interviewed also reported using orchids as medicine to treat cough, abdominal and chest pain, eye infection, urinary disorders, ringworm, rheumatism and 'fortanelle' (presumably meaning headache?). Usage varied significantly from one district to another. It was common in Kasungu District where 38 households indicated that they used orchids for medicinal purposes, whereas in Mzimba district, only 7 households indicated such usage (Kasulo et al. 2009). In an extensive study of herbs employed in Southern Malawi, Brian Morris discovered that *Bulbophyllum*

sandersonii and *Eulophia cucullata* were employed to treat infertility, impotence and other reproductive problems, but they were only 2 out of 80 plants so employed. *Eulophia cucullata* was also used as a goodwill charm (*mwayi*, *chimwemwe*) and in friendship or love potions (Morris 1996) (Figs. 17.16 and 17.17). Seven years later, Morris (1996) added ten more species to his list of medicinal orchids. The majority (*Acampe pachyglossa*, *Angraecopsis parviflora*, *Calyptrochilum christyanum*, *Cyrtorchis arcuata*, *Microcoelia exilis*, *Polystachya tessellata*, *Tridactyle bicaudata* and also *Bulbophyllum sandersonii*) were employed to treat stomach ailments (*Mwanawamphepo*)

Fig. 17.12 *Habenaria cornuta* [as *Habenaria ceratopetala*]. From: Richard A, *Tentamen florae Abyssinicae, Atlas*, t. 88 (1851) [AC Vauthier]

(Figs. 17.18, 17.19, 17.20, and 17.21). *Bulbophyllum fuscum* was used to treat heart ailments. *Bulbophyllum maximum* provided protection against sorcery, and roots of *Tridactyle tricuspis* was used to treat insanity. The orchids were used in infusion: additionally, stems of *Tridactyle tricuspis* were employed as a wash for mad people (Morris 1996). *Disa polygonoides*

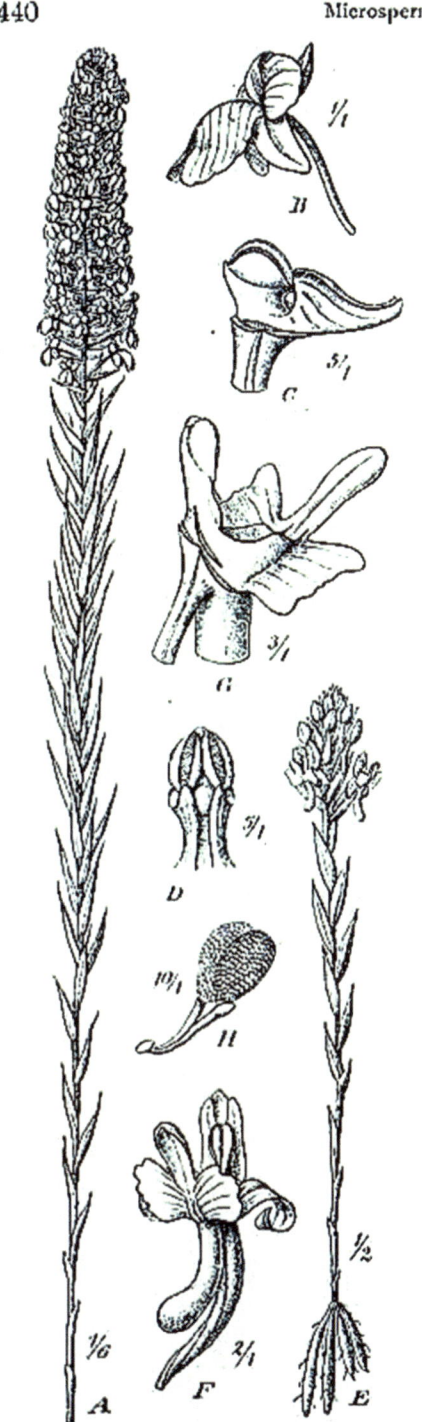

Fig. 17.13 *Brachycorythis pleistophylla*. From: A Engler, O Drude, *Die Vegetation der Erde: Sammlung pflanzengeographischer Monographien* vol. 9(2): p 440, Fig. 31 A–D. (1896–1928). Fig. E illustrates plant of *Brachycorythis tenuior* showing its tubers

Fig. 17.14 *Satyrium longicauda* Lindl. From: Bolus H, *Icones orchidearum Austro-Aficaranum extratropicarum* vol. 1: t. 70 (1896–1913) [H Bolus]

was employed to restore a person's voice after an illness, and *Polystachya ottoniana* (*iphamba lehlathi*) was a medication for diarrhoea in Swaziland (Long 2005) (Fig. 17.22, 17.23, 17.24, and 17.25).

Orchids are also employed as medicine in Benin, West Africa, but here the situation is complex with four predominant tribes having different names (peculiar for their dialect) as well as different usages for the same species. *Calyptochilum christyanum* was employed by ethnic Gourmantche to handle dysmenorrhea and swollen feet. The Berba used it to treat malaria and snake bite, and the Waama employed it to treat 'faith disease'. All three ethnic groups used it to get their babies to start walking early. *Eulophia guineensis* was employed by the three tribes to treat (1) cough, stomachache, (2) fever or (3) stomachache, respectively (Fig. 17.26). Berba used another *Eulophia* species (unidentified) as a laxative and to treat cough. A fourth tribe, Fulani, employed *Habenaria cirrhata* for bracing and *Habenaria schimperiana* to improve their eyesight. Middle-aged Fulani women who had more knowledge of orchids than the men used the root and fruit of *Eulophia guineensis* to treat a variety of diseases (not specified). Fulani also used *Nervilia kostchyi* to treat cough and stomachache, whereas *Gourmantche* employed it to relieve painful menstruation (Fig. 17.27). *Nervilia bicarinata* was employed by Gourmantche to treat fever, by Berba to relieve muscle ache, by

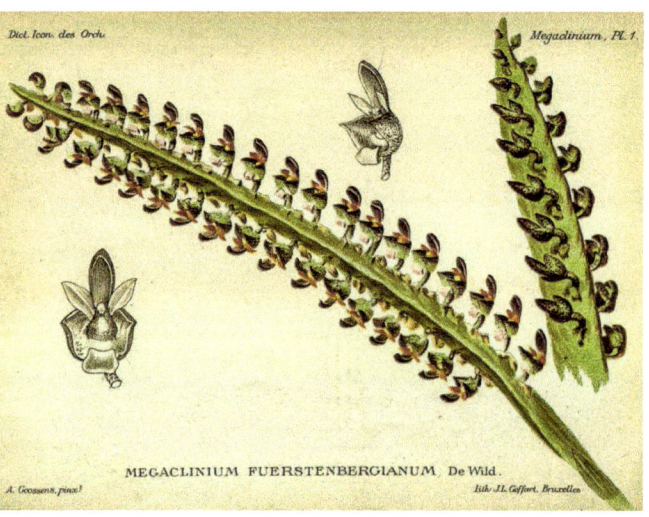

Fig. 17.15 *Bulbophyllum scaberulum* [as *Megaclinium fuerstenbergianum*]. From: Cogniax A, Goossens A, *Dictionaire Iconographique des orchidees, Maxillaria*, vol. 11: Fasicle Megaclinium, t. 1 (1896–1907) [A Goossens]

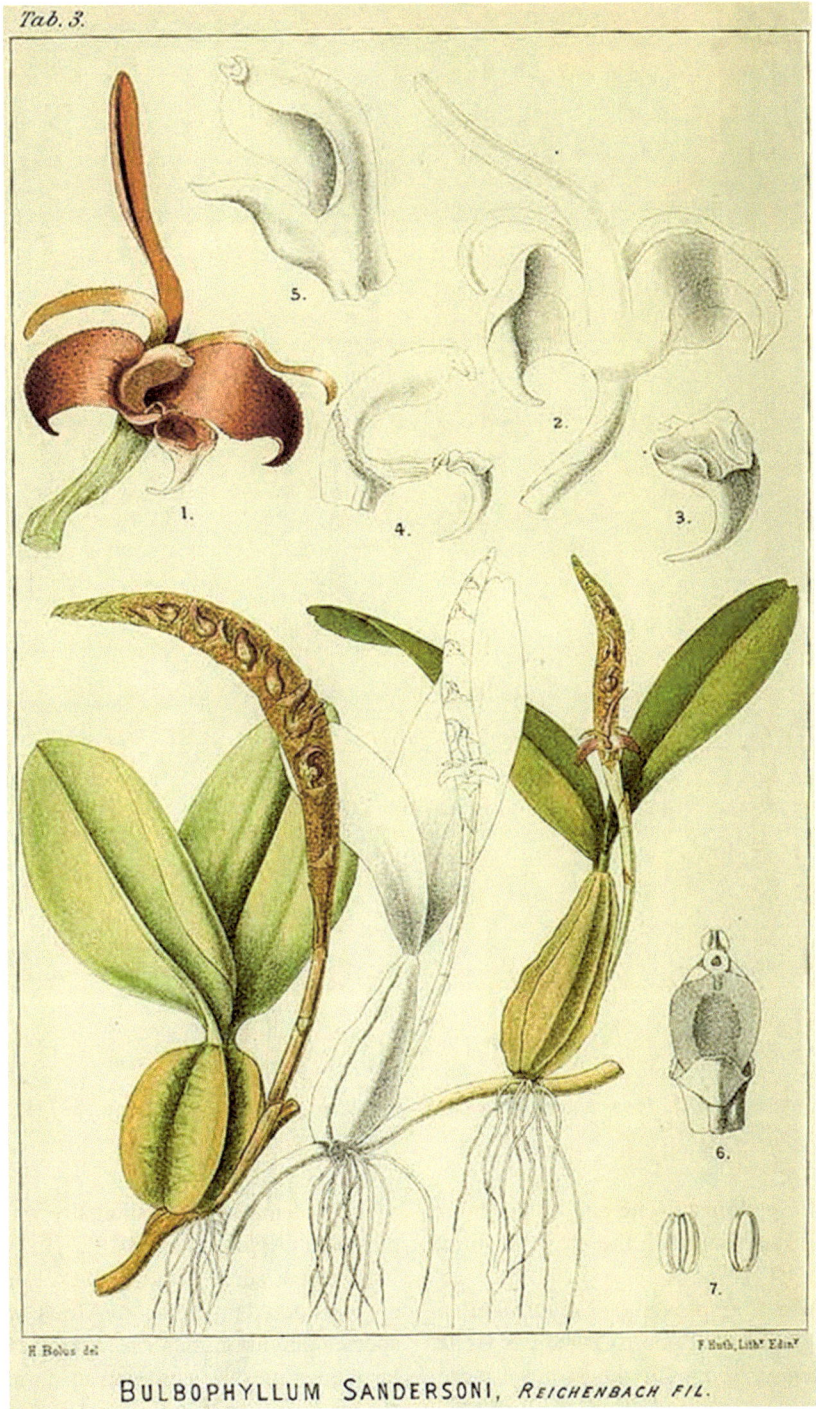

Fig. 17.16 *Bulbophyllum sandersonii*. From: Bolus H, *Icones orchidearum Austro-Africanarum extra-tropicarum* vol 1: t. 3 (1896–1913) [H Bolus]

Fig. 17.17 *Eulophia cucullata*. From: *Curtis Botanical Magazine* vol. 137 [ser. 34, vol. 7] t. 8397 (1911) [M Smith]

Waama to relieve stomachache and by Fulani to treat jaundice and improve the flow of urine (Assese et al. 2017).

Gourmantche employed *Calyptrochilum christyanum* to treat sick poultry, whereas Berba used *Habenaria cirrhata* for the same purpose. Three orchid species (*Calyptrochilum christyanum*, *Eulophia guineensis* and *Nervilia bicarinata*) also had spiritual usage. The authors noted that abundance of a species gave it more cultural significance and increased its likelihood

of being employed medicinally or for spiritual purposes (Assese et al. 2017).

A recent survey found that 49 orchid species are considered medicinal in Africa, with *Eulophia* species playing a major role. However, vernacular names may be common to numerous plants; for instance, *iphamba* refers to 12 different species, namely, *Cyrtorchis arcuata*, *Diaphananthe millarii*, *D. xanthopollinia*, *Eulophia ensata*, *E. ovalis*, *E. leontoglossa*, *Microcoelia exilis*, *Mystacidium capense*, *M. venosum*, *Polystachya*

Fig. 17.18 *Acampe pachyglossa*. From: Pole-Evans IB, Phillips EP, Dyer RA, Codd LE, *The flowering plants of South Africa*. Vol. 30: t. 1175 (1954–1955) [C. Letty]

transvaalensis, *Tridactyle bicaudata* and *T. tridentata*. Native plant traders are not familiar with scientific names. Only 15 species were handled in the usual manner of medicinal herbs, whereas three quarters of over 50 species were employed as love charms or as charms for protection (Chinsamy 2012). Many orchids are also eaten as food (Chinsamy et al. 2008, 2011). Table 17.2 lists the African orchid species that have been employed as medicine. Orchid species employed as charms are excluded from this list and presented separately in Table 17.3.

Ansellia

Ansellia is now regarded as a single variable species (www.theplantlist.org, 2010) with flowers of different sizes, colours and patterns. It is the largest epiphytic orchid in Africa, growing conspicuously on the lower branches of lofty trees. In the typical form, *Ansellia africana*, petals are broader than the sepals, and they are characterized by broad brown bars over a yellow background. These plants are widely distributed across tropical East Africa. The so-called *Ansellia*

Fig. 17.19 *Angraecopsis parviflora* (Thouars) Schltr. [as *Angraecum parviflorum* Thouars]. From: Du Petit-Thouars LMA, *Histoire paryiculiere des plantes Orchidees*, t. 60 (1822)

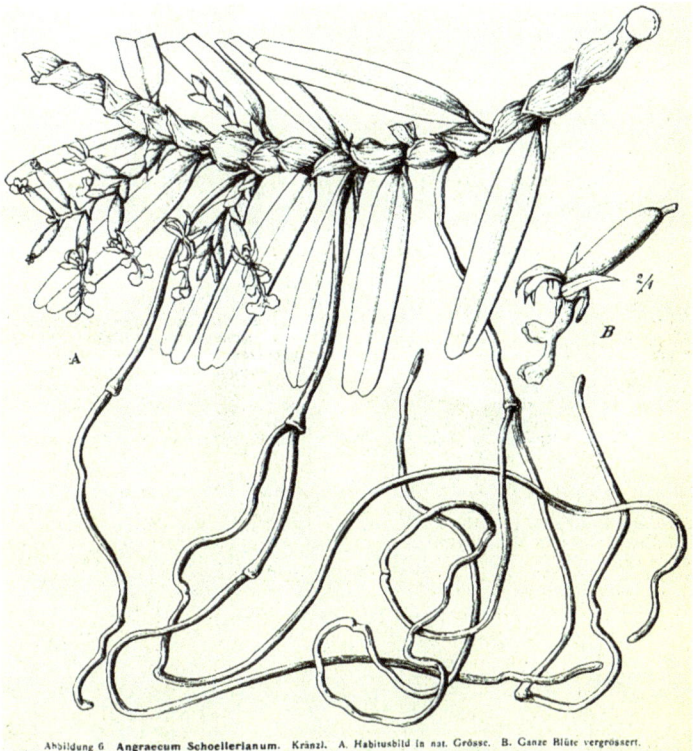

Fig. 17.20 *Calyptrochilum christyanum* (Rchb.f.) Summerh. [as *Angraecum schoellerianum* Kraenzl. Ex Schweinf.]. From: *Orchis. Monatsschrift für Orchideenkunde*, vol. 2: p. 41, Fig. 6 (1908)

Fig. **17.21** *Cyrtorchis arcuata* [as *Angraecum sedenii*]. From: von Lindemann E, *Lindenia* vol. 3: t. 135 (1887) [P. de Pannemaeker]

gigantea (now considered to be a variant of *Ansellia africana*) occurs in South Africa. Flowers are smaller and of a pale yellow, with a small spots on the tepals. Sepals and petals are of the same width. Many intermediate colour forms exist (Stewart 1981). The amount of *Ansellia africana* sold in herbal markets exceeds that of all other orchids because of its alleged aphrodisiac properties (Chinsamy 2012) (Fig. 17.28).

Stem of *Ansellia gigantea* being regarded as an aphrodisiac in the Acornhoek area of South Africa, Zulus would administer decoction of the stem to women with salacious intent. On the other hand, Zulu youths employed root of *Ansellia* (variety *humilis*) as a charm to prevent an unmarried woman from getting pregnant. They might also wear a leaf underneath their bangle when out courting. The way to be rid of bad dreams was to immerse one's head in the smoke of burning roots of *Ansellia* or to partake of its stem infusion (Watt and Breyer-Brandwijk 1962).

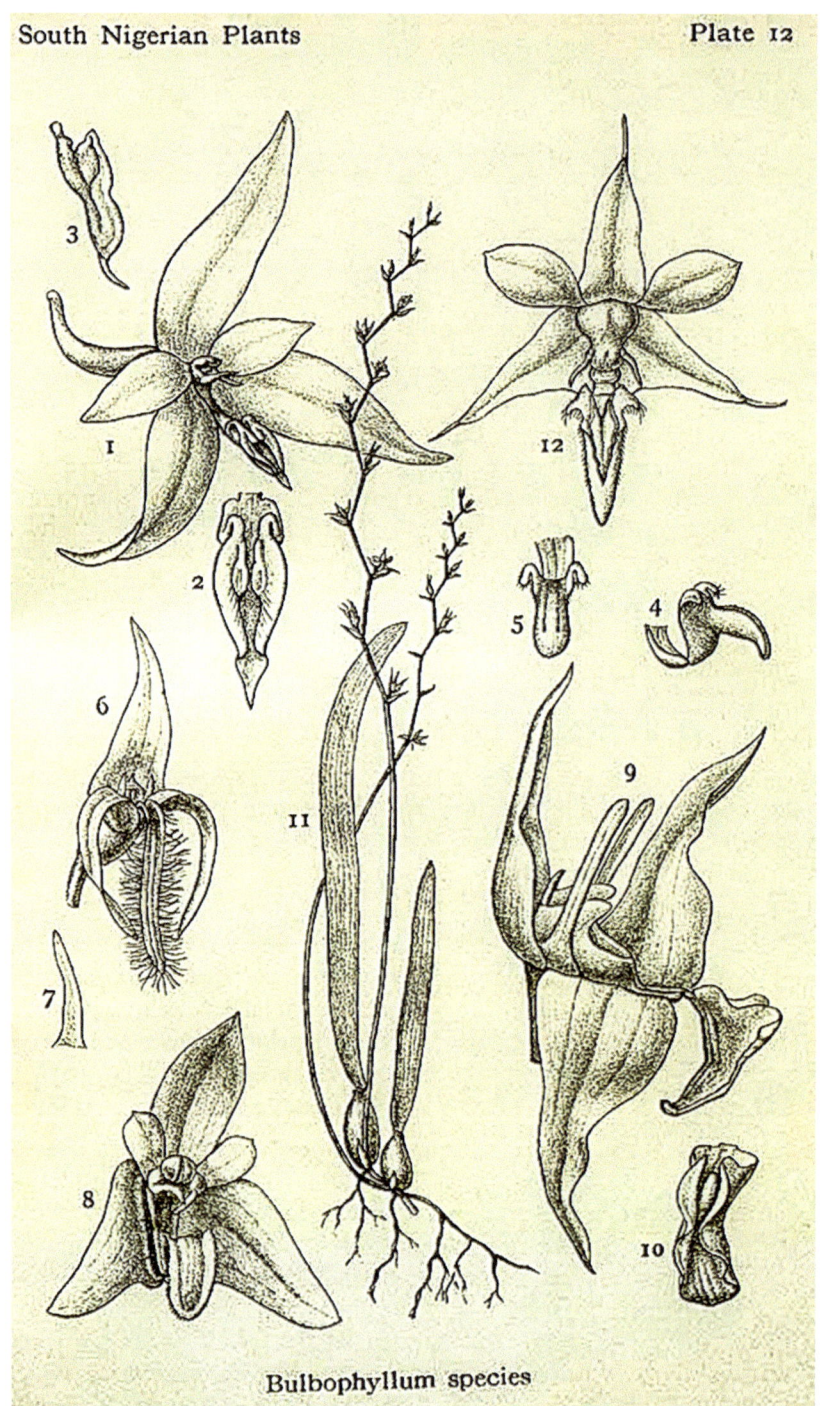

Fig. 17.22 *Bulbophyllum fuscum* Lindl. [as *Bulbophyllum obanense* Rendle]. From: Rendle AB, Baker EG, *Catalogue of the plants collected by Mr. & Mrs. P.A.Talbot in Oban district, South Nigeria*, t. 12 (1913) [Talbot]

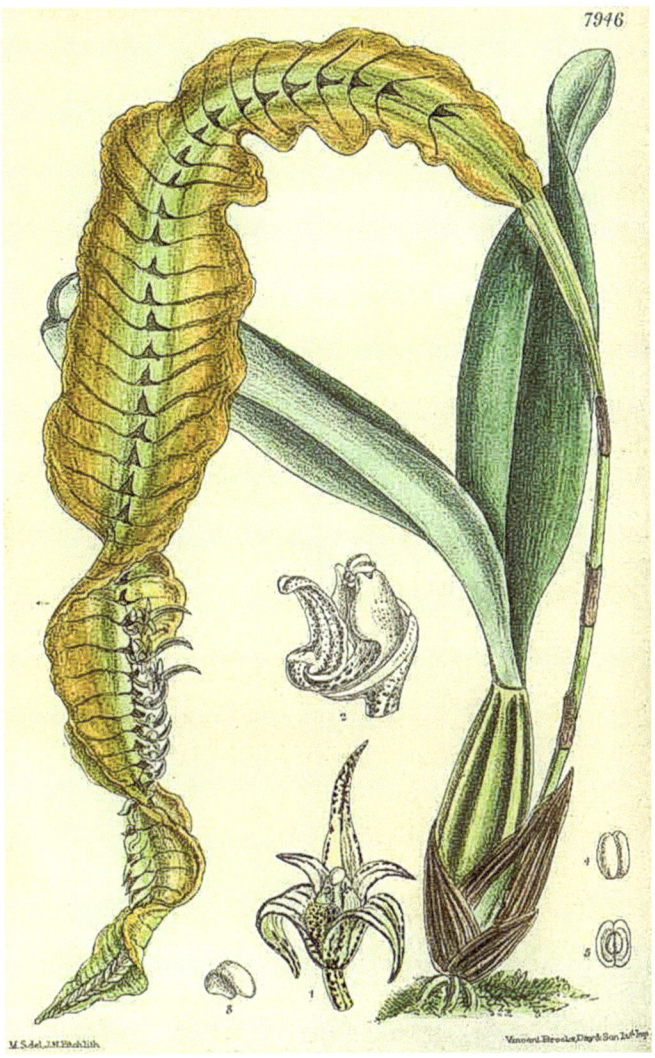

7946

Fig. 17.23 *Bulbophyllum maximum* (Ridley) Ridley [as *Megaclinium platyrhachis* Rolfe]. From: *Curtis Botanical Magazine* vol. 130 [ser. 3, vol. 60]: t. 7946 (1904) [M Smith]

In Zambia (formerly Northern Rhodesia), infusion of the leaf and stem was a folk remedy for madness (Watt and Breyer-Brandwijk 1962). Other orchid remedies for psychiatric conditions include Xhosa using roots of *Brachycorythis ovata* Lindl. (*Imfeyamasele yentaba*) in decoction to treat insanity; seeds of *Polystachya ottoniana* (*Amabelejongosi*) as a snuff in to produce a psychoactive or hallucinogenic effect which is an essential feature in certain African cultures; and an emetic prepared from *Eulophia* species to treat hysteria (Chinsamy et al. 2011) (Fig. 17.29).

The whole plant of *Ansellia africana* is employed to treat respiratory disease, mainly asthma in southern Mozambique (Bandeira et al. 2001). Juice of *Ansellia africana* and roots of the plant (vernacular name *Imfe-nkawu*) were used to treat malaria in Senegal. In this country, sick children were bathed with a decoction of the plant. Infusion of the stem and leaves was employed for madness in Zambia. Zulu herbalists produced an emetic by boiling any part of the plant, whereas the Pedi in Transvaal employed an infusion of the plant to treat children with

Fig. 17.24 *Tridactyle tricuspis* (Bolus) Schltr. [as *Angraecum tricuspis* Bolus]. From: Bolus H, *Icones orchidearum Austro-Afixaranum extratropicarum*, vol. 3: t. 13 (1896–1913) [H Bolus]

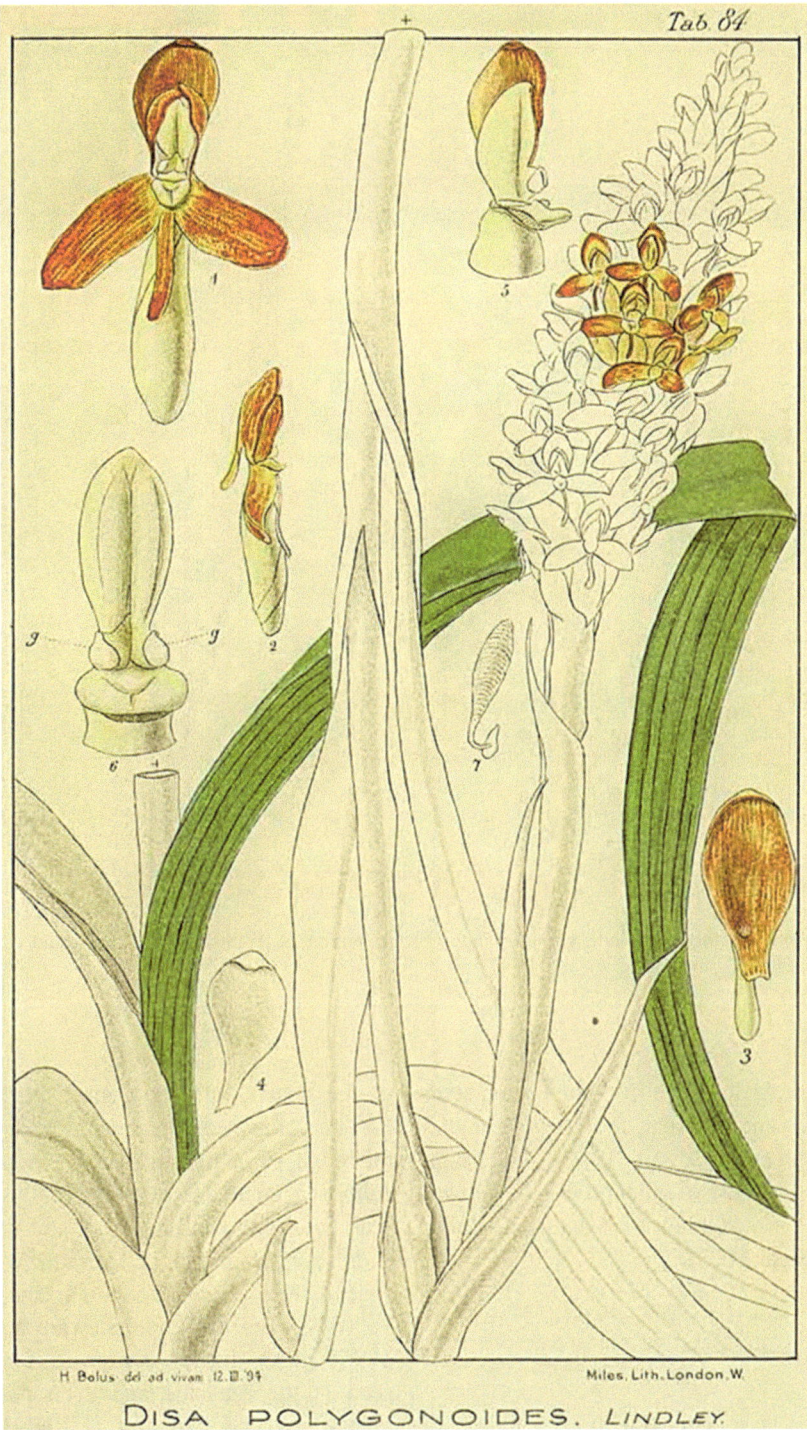

Fig. 17.25 *Disa polygonoides*. From: Bolus H, *Icones orchidearum Austro-Africanarum extra-tropicarum* vol 2: t. 84 (1896–1913) [H Bolus]

Fig. 17.26 *Eulophia guineensis* Lindl. var. *purpurata*. From: Warner R, Williams BS, The Orchid album vol. 1883: t. 89 (1883) [JN Fitch]

coughs. In East Africa juice from the heated stem was squeezed into the ear to relieve earache (Watt and Breyer-Brandwijk 1962), this usage being similar to the use of other orchids employed for earache in Southeast Asia and India (Dagar and Dagar 2003).

Eulophia Species

Eulophia is a large genus with 120 members distributed throughout the tropics but with a concentration in Africa. They are mostly terrestrial, growing in open scrub or grassland. The giant within the genus, *Eulophia petersii*, even thrives

in the desert, and the Asian species *Eulophia herbacea* has been discovered popping out from the edge of a macadamized road. In Malawi, *Eulophia cucullata*, a metre tall, showy plant, with rose pink petals and lip, and the even taller *Eulophia streptopetala* (1.5 m) whose widely spaced flowers are dark green and brown with yellow petals and lip (Stewart and Campbell 1996) grow almost everywhere: they can be seen at the roadside and even persist in gum plantations (La Croix et al. (1991). Plants of *Eulophia* have robust, clustered pseudobulbs and narrow, plicate leaves. Long-lasting flowers are borne on an erect inflorescence which arises from the base of the pseudobulb. The generic

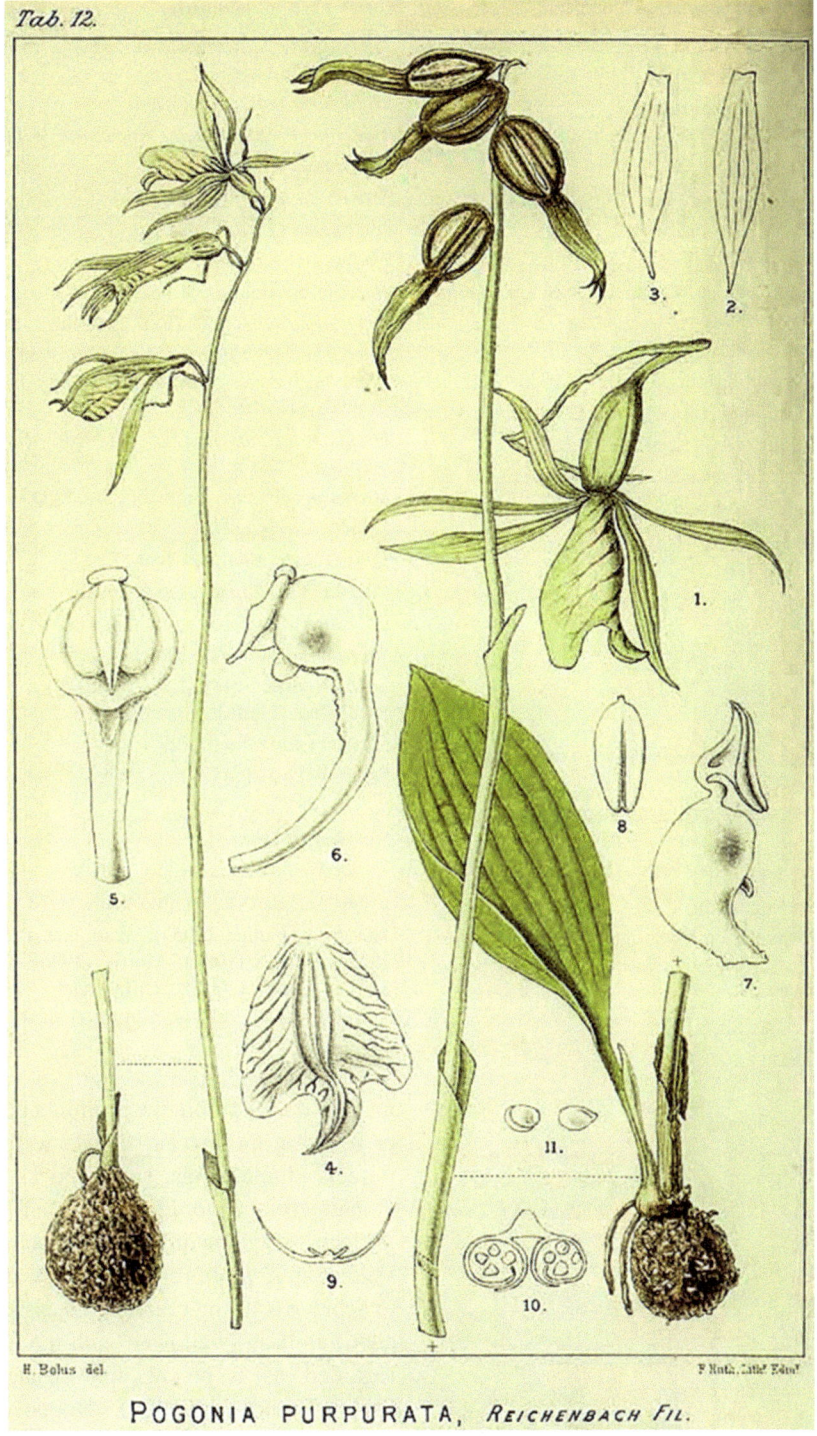

Fig. 17.27 *Nervilia kotschyi* (Rchb.f) Schltr. [as *Pogonia purpurata* Rchb. f. & Sonder]. From: Bolus H, *Icones orchidearum Austro-Africaranum extratropicarum,* vol. 1: t. 12 (1896–1913) [H Bolus]

Table 17.2 African medicinal orchid species

Acampe pachyglossa Rchb. f.

Aerangis biloba (Lindl.) Schltr.

Aerangis thomsonii (Rolfe) Schltr.

Angraecopsis parviflora (Thouars) Schltr.

Angraecum dives Rolfe

Angraecum moandense De Wild. (syn. *A. chevalieri* Summerh.)

Ansellia africana Lindl.

Bonatea steudneri Rchb. f.) T. Durand & Schinz. (syn. *Habenaria steudneri* Rchb. f)

Brachycorythis ovata fc v Lindl.

Bulbophyllum fuscum Lindl.

Bulbophyllum maximum (Lindl.) Rchb.f.

Bulbophyllum sandersonii (Hook.f.) Rchb.f.

Calyptrochilum christyanum (Rchb.f.) Summerh.

Ceratandra grandiflora Lindl.

Corymborkis corymbis Thouars.

Cynorkis spp.

Cyrtorchis arcuata (Lindl.) Schltr.

Cyrtorchis sp.

Dactylorhiza sambucina (L) Soo (syn. *Orchis lutea* Dulac)

Disa aconitoides Sond.

Disa barbata (L.f.) Sw. (syn. *Herschelia barbata*)

Disa polygonoides Lindl.

Disa venusta Bolus (syn. *Herschelia venusta*)

Eulophia angolensis (Rchb. f.) Summerh. (syn. *Eulophia lindleyana* Rchb. f.)

Eulophia barteri Summerh.

Eulophia cucullata (Afzel. Ex Sw.) Steud. [syn. *Eulophia dilecta* (Rchb. f.) Schltr.; *Lissochilus dilectus* Rchb.f.]

Eulophia ensata Lindl.

Eulophia galeoloides Kraenzl.

Eulophia guineensis Lindl.

Eulophia hians Lindl.

Eulophia herbacea Lindl.

Eulophia livingstoneana (Rchb.f.) Summerh. (syn. *E. robusta*)

Eulophia nuda Lindl. (syn. *spectabilis* Suresh., *Eulophia campestris* Wall.)

Eulophia ovalis Lindl.

Eulophia petersii (Rchb.f.) Rchb.f.

Eulophia pulchra (Thouars.) Lindl. (syn. *Eulophidium silvaticum, Oeceoclades pulchra*)

Eulophia reticulata Ridl.

Eulophia speciosa (R. Br ex Lindl.) Bolus

Eulophia streptopetala Lindl.

Eulophia tenella Rchb. f. (syn. *E. flaccida* Schltr.)

Habenaria cirrhata (Lindl.) Rchb. f.

(continued)

Table 17.2 (continued)

Habenaria epipactidea Rchb. f [syn. *Habenaria foliosa* (Sw.) Rchb. f.]

Habenaria macrandra Lindl.

Habenaria schimperiana Hochst. ex A. Rich

Habenaria walleri Rchb. f.

Manniella gustavi Rchb. f.

Microcoelia exilis Lindl.

Neobolusia tysonii (Bolus) Schltr.

Nervilia bicarinata (Bl.) Schltr.

Nervilia kotschyi (Bl.) (Rchb.) Schltr.

Oeceoclades beravensis (Rchb.f.) R. Bone & Buerki [syn. *Eulophia beravensis* Rchb. f., *Lissochilus beravensis* (Rchb. f.) H. Perr.]

Orchis anthropophora (L.) All. (syn. *Aceras anthropophora* R. Br.)

Orchis provincialis Balb. ex Lam. & D.C.

Ophrys lutea Cav.

Polystachya ottoniana Rchb. f.

Polystachya stauroglossa Kraenzl.

Polystachya concreta (syn. *Polystachya tessellata* Lindl.)

Satyrium bicorne (L.) Thunb.

Satyrium bracteatum (L.f.) Thunb. (syn. *Satyrium cordifolium* Lindl.)

Satyrium candidum Lindl.

Satyrium carneum (Alton) Sims

Satyrium erectum Sw.

Stenoglottis fimbriata Lindl.

Tridactyle bicaudata (Lindl.) Schltr.

Tridactyle tricuspis (Bolus) Schltr.

Data from Hulme (1954), Watt and Breyer-Brandwijk (1962), Lawler (1984), Morris (1996), Bandeira et al. (2001), Long (2005), Cribb and Hermans (2009), Chinsamy et al. (2011), Chinsamy (2012), and Assese et al. (2017)

name is derived from the prominent crested ridges or keels on the lip: *eu* (Greek, well) and *lophos* (Greek, plume) (Figs. 17.30 and 17.31).

Numerous *Eulophia* species are employed in African village medicine. Male Lobedu chew the stem of an *Eulophia* species and swallow the juice to achieve a strong erection. Occasionally cow or goat's milk is swallowed with the juice which is regarded as a potent aphrodisiac (Watt and Breyer-Brandwijk 1962). Roots of *Eulophia reticulata* were employed as an aphrodisiac in Madagascar (Cribb and Hermans 2009). Juice extracted by crushing the roots of *Eulophia angolensis* (syn. *E. lindleyana*) was instilled into painful ears, and shredded roots of the species were applied to syphilitic ulcers (Lawler 1984). *Eulophia barteri* was rendered into an astringent paste to set fractures and treat sprains. The whole plant of *Eulophia beravensis* [syn. *Lissochilus beravensis*], a lowland orchid found in deciduous woodland, sand and dunes on the western side of Madagascar (Cribb and Hermans 2009), was used to treat nervous disorders on the island (Figs. 17.32 and 17.33). Pseudobulb of *Eulophia cucullata* (syn. *Eulophia dilecta*, *Lissochilus dilectus*; vernacular names *Amabelejongosi*, *Uhlamvu*, *Lwabafazi*, *Umabelejongosi*, *Undwendweni*) was employed to treat scabies and skin disorders in Congo, Gabon and Zaire. Cooked roots of *Eulophia cucullata* was employed by *Nyanja of Nyasaland* to make a poultice. In Malawi, pseudobulbs of *Eulophia cucullata* (syn. *Lissochilus arenarius*) were boiled to make a poultice. Root infusion is consumed in small amounts by couple who were unable to have children (Hulme 1954; Watt and Breyer-Brandwijk 1962; Chinsamy 2012). Boiled pseudobulbs or roots of *Eulophia galeoloides* were fed to infants suffering from colic: *Eulophia ensata* Lindl. (*iphamba yentaba, mahlane*) and *Eulophia streptopetala* var. *streptopetala* (syn. *Lissochilus krebsii*) were also administered to sick babies (ailments not specified). Roots were employed as sedative by some African tribes (Watt and Breyer-Brandwijk 1962). Heated pseudobulbs of *Eulophia tenella* (syn. *E. flaccida*) were applied on sore limbs for pain relief in northern Lesotho (Watt and Breyer-Brandwijk 1962; Lawler 1984) (Figs. 17.34 and 17.35). *Eulophia ovalis* (local name, *iphamba*) was also employed to treat sore limbs (Chinsamy et al. 2011). *Eulophia hians* (syn. *Eulophia clavicornis*, *E. madagascariensis*, *E. vaginata*, *E. bathiei*) (local name in Madagascar, *Tongolomboalavo*), which grows in grassland, dry deciduous scrub and rocky outcrops at 1500–2500 m, is employed to treat boils in Madagascar. It is an unassuming plant with white and purple flowers that appear from December to March (Cribb and Hermans 2009). In South Africa it is employed to treat infertility (Chinsamy 2012) (Figs. 17.36 and 17.37). Root decoction of *Eulophia petersii* (*sahalisaka*), a common plant in Kenya and Natal which thrives despite being exposed to drying winds, strong sunlight and low rainfall (Stewart and Campbell 1970), served as a purgative in East Africa (Lawler 1984), and it is used to treat dropsy or heart failure in South Africa (Chinsamy 2012). Nursing women ate salted stems of *Eulophia pulchra* (syn. *Eulophidium silvaticum*, *Oeceoclades pulchra*) to encourage milk production (Lawler 1984). *Eulophia speciosa* R.Br. ex Lindl. (*Amabelejongosi*, *Umlunge omhlope*) was employed as an emetic (Chimsamy et al. 2011). In the Democratic Republic of the Congo, a 30 cm tall *Eulophia* species (unidentified) was employed as a wash to treat a form of madness known as *hehetolu*. The native name of the *Eulophia* is *torepi* meaning plant of the Efe's ancestors (tore): Efe are hunter-gatherers in the Ituri forests (Terashima and Ichikawa 2003) (Figs. 17.38 and 17.39).

Medicinal Orchids in Other Genera

Stomach disorders appear to have been common complaints. Boiled roots of *Habenaria cirrhata*, a montane species widespread from Sudan to Zambia and Madagascar at 2500–2700 m (Stewart and Campbell 1996), were administered for indigestion. In East Africa, boiled roots of *Habenaria macranda* were used as purgative, roots of *Habenaria walleri* from tropical swampy grassland for stomach disorders and decoction of roots of *Bonatea steudneri* (syn. *Habenaria steudneri*) from mountainous bushland and scrub (Stewart and Campbell 1996) for undiagnosed stomach disorders and influenza. Leaves or roots of *Corymborkis corymbis* which thrives in dense shade of lowland and evergreen forests in southern Africa were used as a purgative in Tanzania and Sierra Leone. Zulus employed root decoction of *Habenaria epipactidea* Rchb. f (syn. *H. foliosa*; vernacular name, *umabelebuca omkhulu*) to induce vomiting (Watt and Breyer-Brandwijk 1962): the orchid is widely distributed at medium altitudes in eastern Africa (Stewart and Campbell 1996). *Aerangis biloba* was used for this purpose in Gabon. South African tribes employed *Satyrium bracteatum* (syn. *Satyrium cordifolium*) as a vermifuge and *Stenoglottis fimbriata* root as enema and cure for flatulence (Figs. 17.40, 17.41, and 17.42).

Fig. 17.28 *Ansellia africana* [as *Ansellia congoensis*]. From: von Lindemann E, *Lindenia* vol. 2: t. 64 (1886) [P de Pannemaeker]

Cynorkis is a large, terrestrial genus with 125 members most of which occur in the Indian Ocean islands off Africa (Hennessey 1981). One or more of its species is employed to treat burns in Madagascar (Lawler 1984). In East Africa, sores were treated with leaf infusion of *Angraecum dives*, sore eyes and sprains with *Angraecum* species and abscesses and hernia with *Aerangis thomsonii*: the last species is distributed in Kenya, Natal (South Africa) and Madagascar. *Polystachya ottoniana* (*Amabelejongosi*) is used to sooth teething pain in children and to treat diarrhoea and *Satyrium bracteatum* (*Ubani lwenkangala*) to treat intestinal worms (Chinsamy et al. 2008) (Fig. 17.43).

Coming to more serious illnesses, plant extracts of *Acampe pachyglossa* and of a *Cyrtorchis* species were employed as antimalarials. *Angraecum moandense* (syn. *A. chevalieri*) was used to treat people who vomited or coughed up blood.

Disa is a striking species when seen in the field because most of the species have brightly coloured flowers and grow in grassland. The name of the genus was derived by Swedish botanist Bergius from his native folklore: *Disa* was a beautiful lady who appeared wrapped in a fishing net when invited to appear before the king, 'neither naked nor clothed' (Stewart and Campbell 1996). In southern Africa people who have lost their ability to speak are fed an infusion of the pseudobulbs of *Disa polygonoides*. In Swaziland, *Brachycorythis ovata* subsp. *ovata* is employed to treat insanity (Long 2005).

Fig. 17.29 *Brachycorythis ovata*. From: Bolus H, *Icones orchidearum Austro-Africanarum extra-tropicarum* vol 3: t. 51 (1896–1913) [H Bolus]

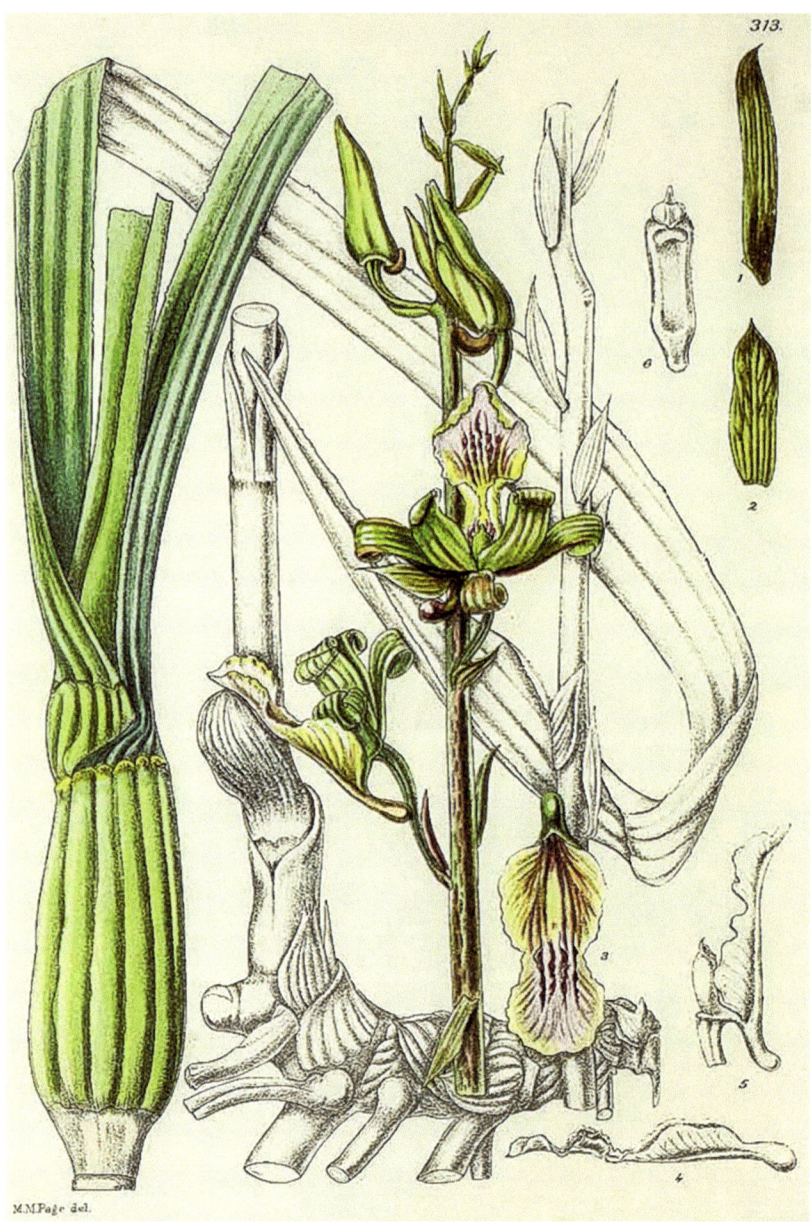

Fig. 17.30 *Eulophia petersii* (Rchb.f.) Rchb.f. [as *Eulophia caffra* Rchb.f.]. From: Pole Evans IB, *Flowering plants of (South) Africa*, vol. 8: t. 313 (1828) [MM Page]

Fig. 17.31 *Eulophia streptopetala*. From: *Edward's Botanical Register* vol. 12: t. 1002 (1826)

Human Reproduction

At least nine orchid species were used at various stages in human reproduction. Plant juice of *Manniella gustavi* served not only as a purgative, it was a counter-poison in the Congo. Barren women used it to cleanse the belly (Lawler 1984). Emetic consisting of *Disa aconitoides* root infusion was administered to women to help them conceive (Chinsamy et al. 2011) (Fig. 17.44). In Swaziland, roots of *Disa aconitoides* subspecies *aconitoides* (*ihlamvu*, *umashushu*) were prescribed for women to help them conceive (Long 2005).

In Basutoland, barren women drank daily infusions made with pseudobulbs of *Eulophia tenella* (syn. *E. flaccida*), *Eulophia hians*. or *Eulophia livingstoneana* (syn. *E. robusta*), the treatment continuing over a period of 4 months if she conceived. *Eulophia tenella* and *Eulophia hians* may be employed together to prepare the infusion (Watt and Breyer-Brandwijk 1962) (Fig. 17.45).

Occasionally, the infusion was made with *Eulophia livingstoneana*, a species which is widespread throughout continental Africa, Madagascar and the Comoro Islands, in grassland and deciduous woodland, from sea level to 2350 m (Stewart and Campbell 1996). Cooked root of *Eulophia*

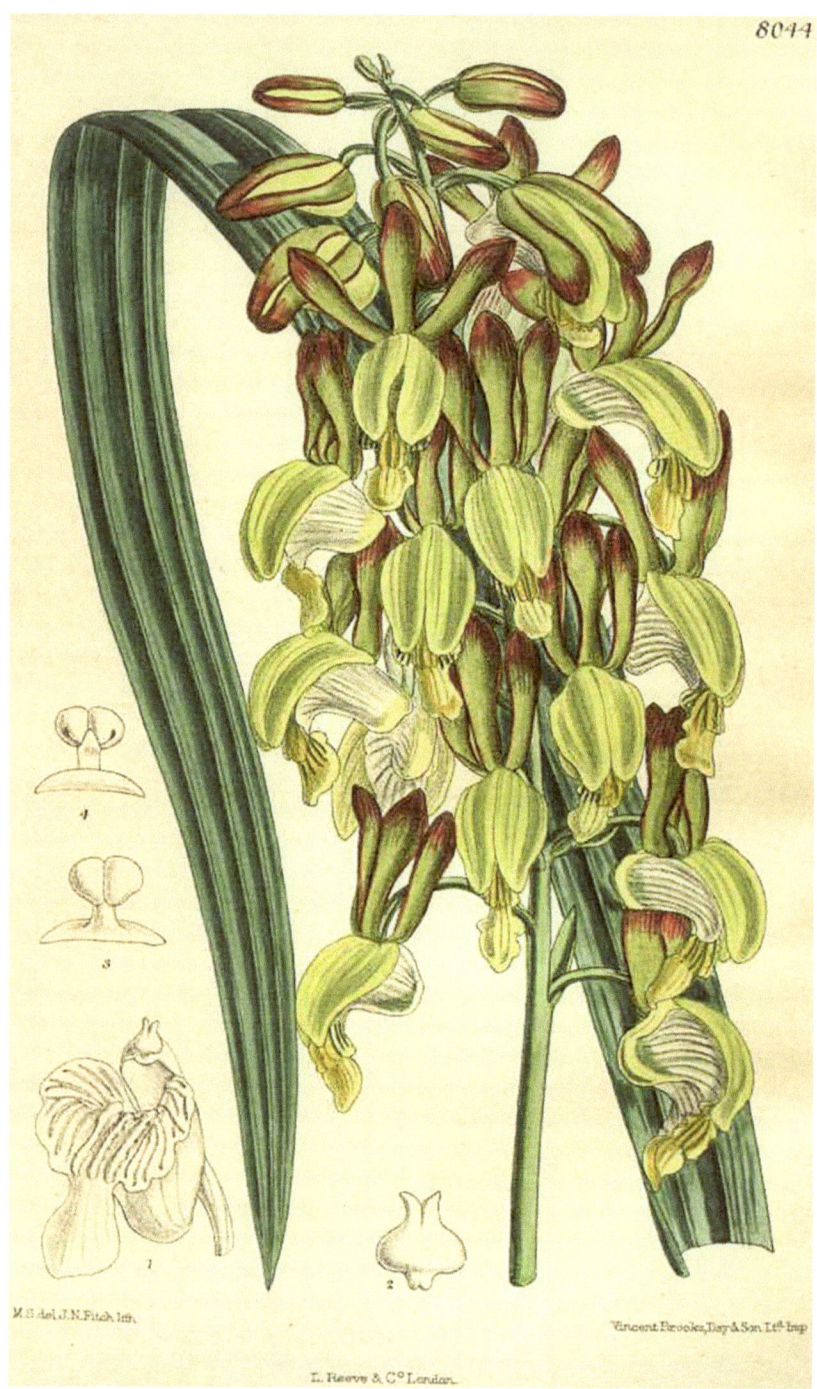

Fig. 17.32 *Eulophia angolensis* [as *Lissochilus ugandae*]. From: *Curtis Botanical Magazine* t. 7992–8057, vol. 14 [ser. 4, vol. 1] t. 8044 (1901) [M Smith]

Fig. 17.33 *Eulophia beravensis* Rchb.f. [as *Lissochilus beravensis* (Rchb.f.) Perrier]. From: *Flora de Madagascar et des Comores, Orchidees*, vol. 49 (2): Orchidees, p. 45 (1941) [MJ Vesque]

Fig. 17.34 *Eulophia ensata*. From: *Edward's Botanical Register* vol. 14: t. 1147 (1828) [M Hart]

cucullata (syn. *Lissochilus arenarius*) was a Zulu remedy for impotence and infertility (Watt and Breyer-Brandwijk 1962). Leaf decoction of *Polystachya stauroglossa* was administered to women in labour in Kenya. In East Africa, salted stems of *Eulophia pulchra* (syn. *Eulophidium silvaticum*) was employed as a galactogogue. Evening colic which is common in babies was tackled by feeding them with boiled pseudobulbs of *Eulophia galeoloides*.

Two medicinal orchids, *Corymborkis corymbis* and *Disa chrysostachya* (*D. gracilis*), are described as being harmful to humans or animals (Arnold et al. 2002).

Pharmacological Studies

Anti-inflammatory, antioxidant, anticholinesterase activity and mutagenicity of South African medicinal orchids were investigated by Chinsamy et al. (2014). More than a third of the evaluated orchid extracts showed anti-inflammatory activity with root extract of *Ansellia africana* being the most potent. *Eulophia hereroensis* was the only extract to significantly inhibit both COX-1 and COX-2 enzymes. *Bulbophyllum scaberulum* organic root extract exhibited both COX-2 selective inhibitory activity and anticholinesterase

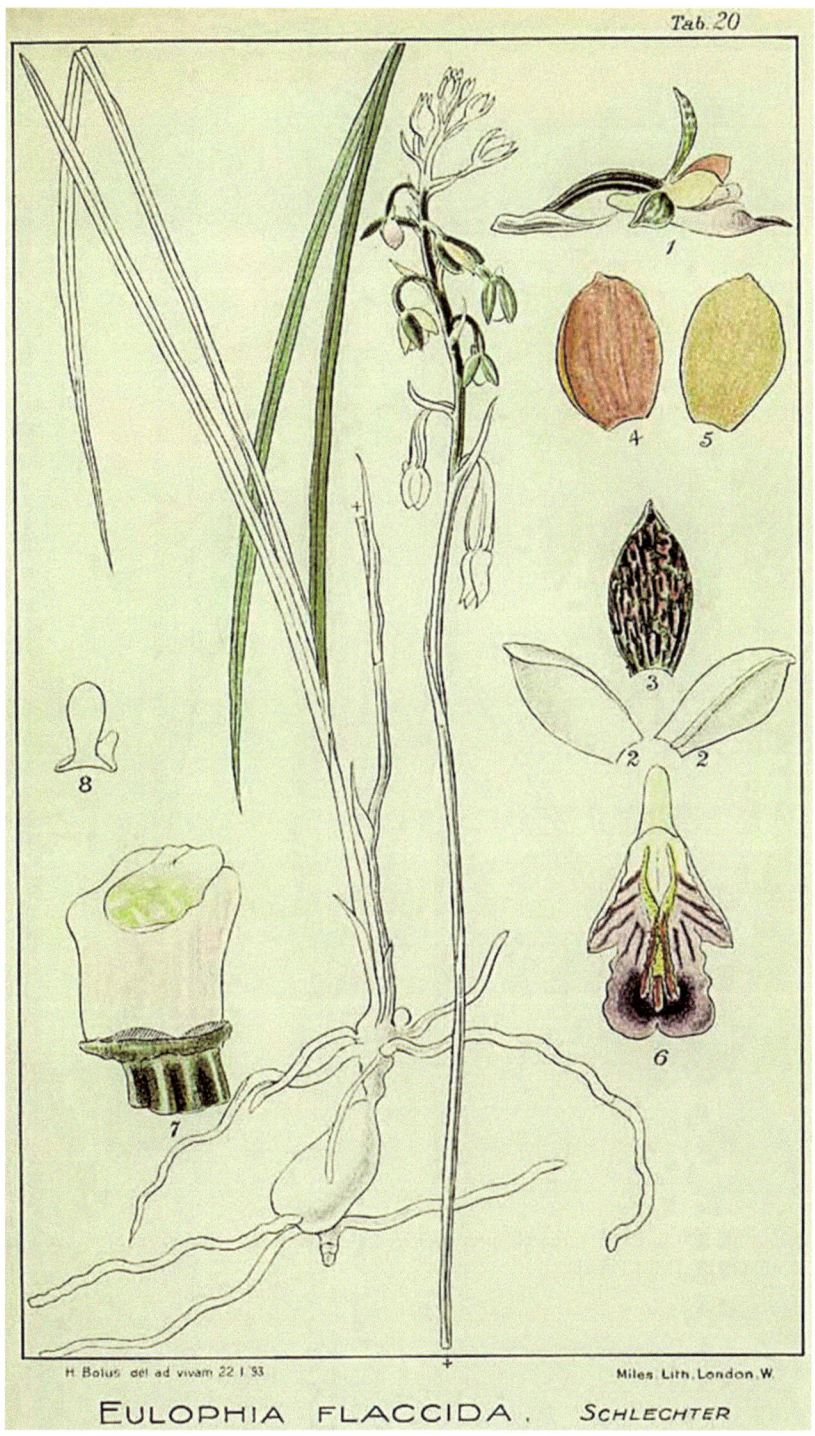

Fig. 17.35 *Eulophia tenella*. From: Bolus H, *Icones orchidearum Austro-Africanarum extra-tropicarum* vol 2: t. 20 (1896–1913) [H Bolus]

Fig. 17.36 *Eulophia ovalis* Lindl. [as *Eulophia pretoriensis* L. Bolus]. From: Bolus H, *Flowering plants of (South) Africa*, vol. 13: t. 500 (1933) [MM Page]

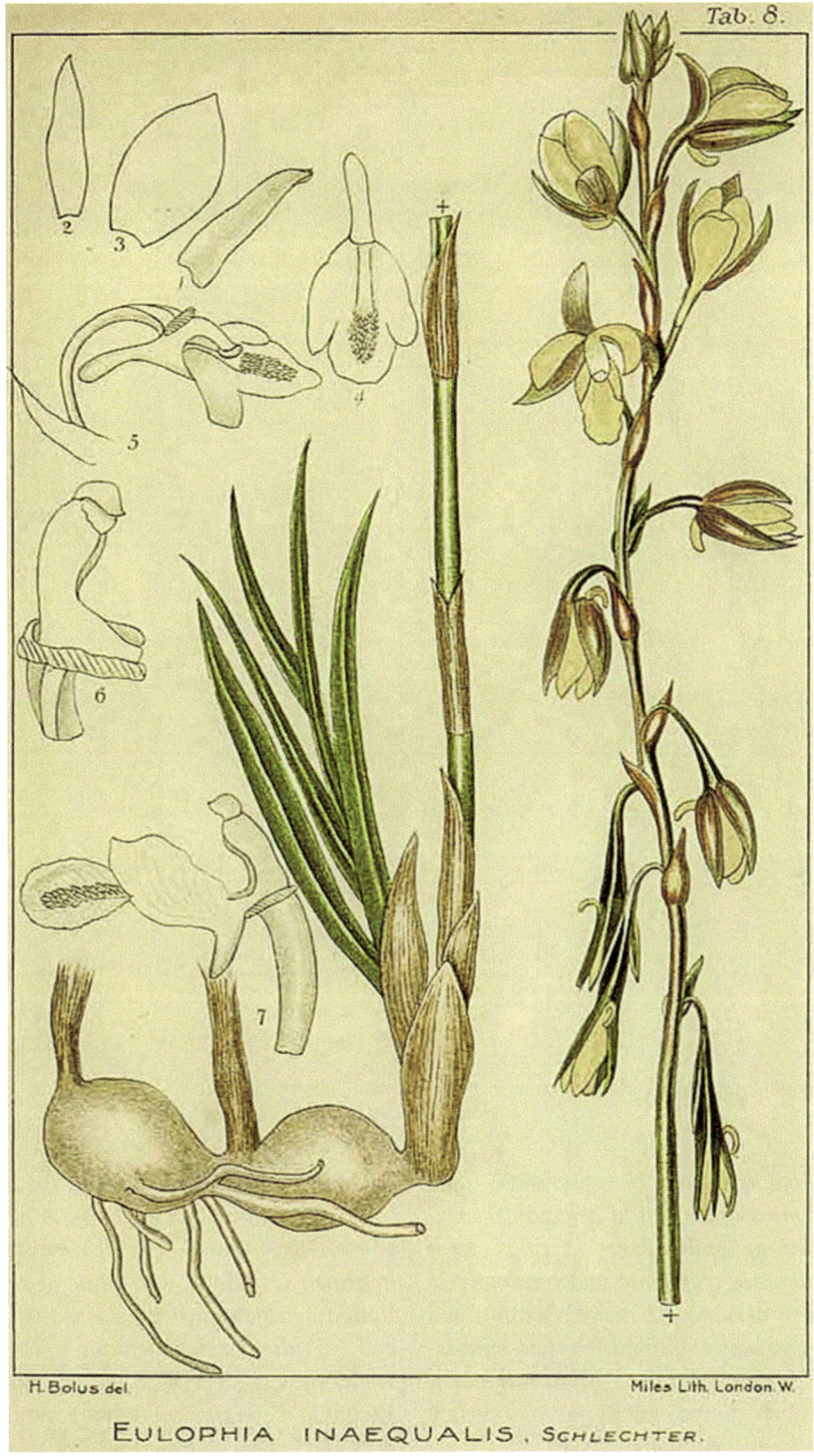

Fig. 17.37 *Eulophia hians* [as *Eulophia inequalis*]. From: Bolus H, *Icones orchidearum Austro-Africanarum extra-tropicarum* vol. 3: t. 8 (1913)

L'ILLUSTRATION HORTICOLE   PL. CLXXXI

EULOPHIA PULCHRA LINDL.

Fig. 17.38 *Eulophia pulchra.* From: *l'Illustration horticole* vol. 40: t. 181 (1893)

activity. *Eulophia petersii* pseudobulb and *Ansellia africana* root possessed free radical scavenging and strong antioxidant activities. Leaf extract of *Tridactyle tridentata* and root extracts of *Cyrtorchis arcuata* (Lindl.) Schltr. and *Eulophia hereroensis* exhibited the best antioxidant effects (Chinsamy et al. 2014). *Ansellia africana* exerts effects on the central nervous system (Bhattacharyya and van Staden 2016).

Seven South African medicinal orchids commonly traded in KwaZulu-Natal herbal markets were investigated for their antimicrobial activity by Chinsamy (2012), namely, *Ansellia africana* Lindl., *Bulbophyllum scaberulum* (Rolfe) Bolus, *Cyrtorchis arcuata* (Lindl.) Schltr., *Eulophia hereroensis* Schltr., *Eulophia petersii* (Rchb.f.) Rchb.f., *Polystachya pubescens* (Lindl.) Rchb.f. and *Tridactyle tridentata* (Harv.) Schltr. Dichloromethane (DCM) pseudobulb extract of *Eulophia petersii* pseudobulb inhibited all four bacterial strains tested [0.39 mg/ml against *Staphylococcus aureus* (which commonly infects the skin) and 0.78 mg/ml against *Bacillus subtilis*, *Escherichia coli* and *Klebsiella*

Fig. 17.39 *Eulophia speciosa* [as *Lissochilus speciosus*]. From: *Edward's Botanical Register* vol. 7: t. 573 (1822) [M Hart]

Fig. 17.40 *Corymborkis corymbis*. From: Bolus H, *Icones orchidearum Austro-Africanarum extra-tropicarum* vol 2: t. 98 (1896–1913) [H Bolus]

pneumoniae]. Lusianthridin, a phenanthrene present in *Eulophia petersii*, is active against Grampositive bacteria (Kovacs et al. 2007). Aqueous extract of *Tridactyle tridentata* root was effective against *Staphylococcus aureus*, but aqueous extracts of the other six orchids were ineffective. Apart from *Polystachya pubescens*, the orchids did not yield aqueous extracts which were effective against the common fungus, *Candida albicans* (Chinsamy 2012).

Orchids produce phytoalexins to ward off predators so all medicinal orchids would show some degree of antimicrobial activity against bacteria and fungi. The need is to demonstrate that they are effective in treating infections in humans and animals. Whereas organic root extracts of *Ansellia africana*, *Bulbophyllum scaberulum*, *Eulophia petersii* and *Tridactyle tridentata* were moderately effective against the roundworm, *Caenorhabditis elegans*, leaf and root extracts of *Cyrtorchis arcuata* were the most effective (Fig. 17.46). Its aqueous extract was the single

aqueous extract that was effective. *Caenorhabditis elegans* is a tiny free-living, non-parasitic South African roundworm made famous by the studies of Nobel Prize winner Sydney Brenner who demonstrated that humans do employ similar chemical messengers essential for species survival as those employed by the primitive roundworm, namely, oxytocin which is released during sex, childbirth and breast-feeding in humans and egg laying in the roundworm.

Various explanations have been suggested to explain the significant antimicrobial, anthelmintic, anti-inflammatory and antioxidant activities of *Eulophia petersii* pseudobulb extracts and

Fig. 17.41 *Aerangis biloba* [as *Angraecum bilobum*]. From: *Edward's Botanical Register* vol. 27: t. 35 (1841) [SA Drake]

Eulophia hereroensis tuber and root extracts (Fig. 17.47). This could be due to their high total phenolic content. Alternatively, the significant levels of gallotannin content in *E. hereroensis* may have contributed to the bioactivity. The flavonoid content of *Bulbophyllum scaberulum* and *Tridactyle tridentata* may explain the potent activity observed in the anti-inflammatory, antioxidant and acetylcholinesterase in South Africa (Chinsamy 2012) to the potent anthelmintic and antioxidant activities. The significantly higher

levels of gallotannin content may explain the significant anti-inflammatory and anthelmintic activity of *Ansellia africana*.

Charms and Talismans

Numerous cultural beliefs pertaining to sex, love, fertility and death involved the use of orchids in rural areas of South Africa. Protective charms top the list and employ any of 25 orchid species.

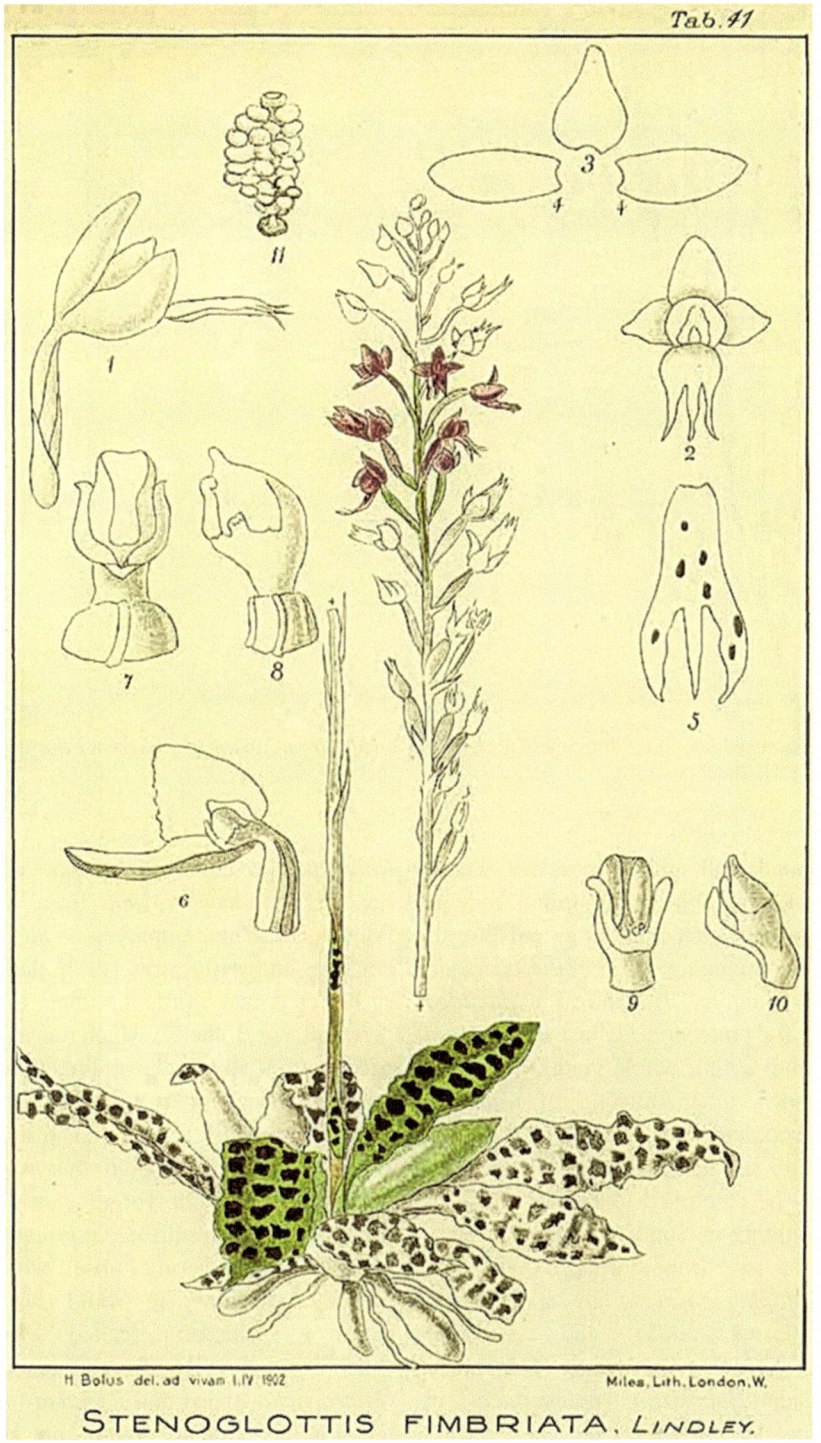

Fig. 17.42 *Stenoglottis fimbriata*. From: Bolus H, *Icones orchidearum Austro-Africanarum extra-tropicarum* vol 2: t. 41 (1896–1913) [H Bolus]

Fig. 17.43 *Polystachya ottoniana*. From: Bolus H, *Icones orchidearum Austro-Africanarum extra-tropicarum* vol 2: t. 32 (1896–1913) [H Bolus]

Charms were made with orchids to repel evil or to protect against being hit by lightning. Indeed, among natural disasters, lightning and storms appear to be the top concerns. *Ansellia africana*, *Habenaria dregeana* and *Habenaria epipactidea* are employed for protection against lightning in Swaziland (Long 2005); whole plant infusion of *Disa stachyoi*des, tuber infusion of *Eulophia leontoglossa,* root and stem infusion of *Eulophia speciosa* or bulb and leaf infusion of *Habenaria dregeana* may be employed to establish protection against lighting in South Africa (Chinsamy 2012). For protection from storms, *Disa stachyoides* and *Eulophia speciosa* are specified in Swaziland (Long 2005) and *Eulophia leontoglossa, Eulophia speciosa, Habenaria dregeana* and *Habenaria epipactidea* in South Africa (Chinsamy et al. 2011) (Fig. 17.39) (Table 17.3).

An interesting account of how a personal protective charm was prepared and employed was provided by Grant in 2016. *Polystachya pubescens* (*iphamba*) is employed to cope with situations believed to have arisen from witchcraft in Pondoland. When employed as *intelezi*, *iphamba* confuses the evil spirits, or it diverts lightning strikes. The two words are interesting: *intelezi* is derived from Buthelezi which translates it as 'slipperiness' referring to the ability of the medicine to make its user slippery and evade or get out of trouble. *Iphamba* is derived from *ukuphamb*a which translates as dodge or outwit (Grant 2016).

This was how a young man from Malawi might prepare himself before going out to socialize. Instead of covering himself with deodorizing power or perfume, he would use a good luck charm for friendship, called *konda* (meaning 'like' or 'love'). Roots of *Paliberkanto* (*Microcoelia exilis*) and *chanasa* (*Myrothamnus flabellifoliu*s, not an orchid) are soaked in oil which is then employed to anoint the body (Morris 1996) (Fig. 17.48).

Charms were employed to protect a home, the orchid being made into an infusion to sprinkle

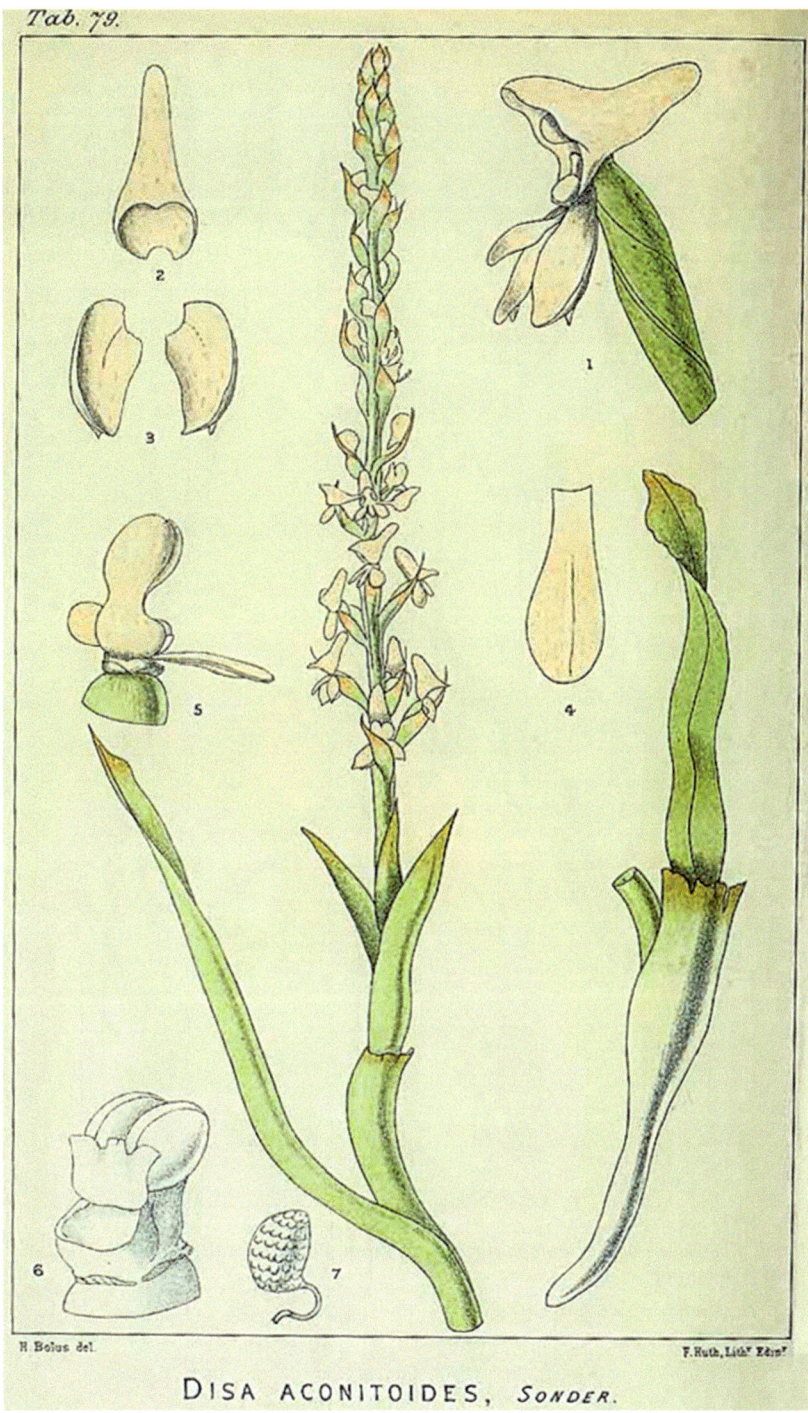

Fig. 17.44 *Disa aconitoides*. From: Bolus H, *Icones orchidearum Austro-Africanarum extra-tropicarum* vol 1: t. 79 (1896–1913) [H Bolus]

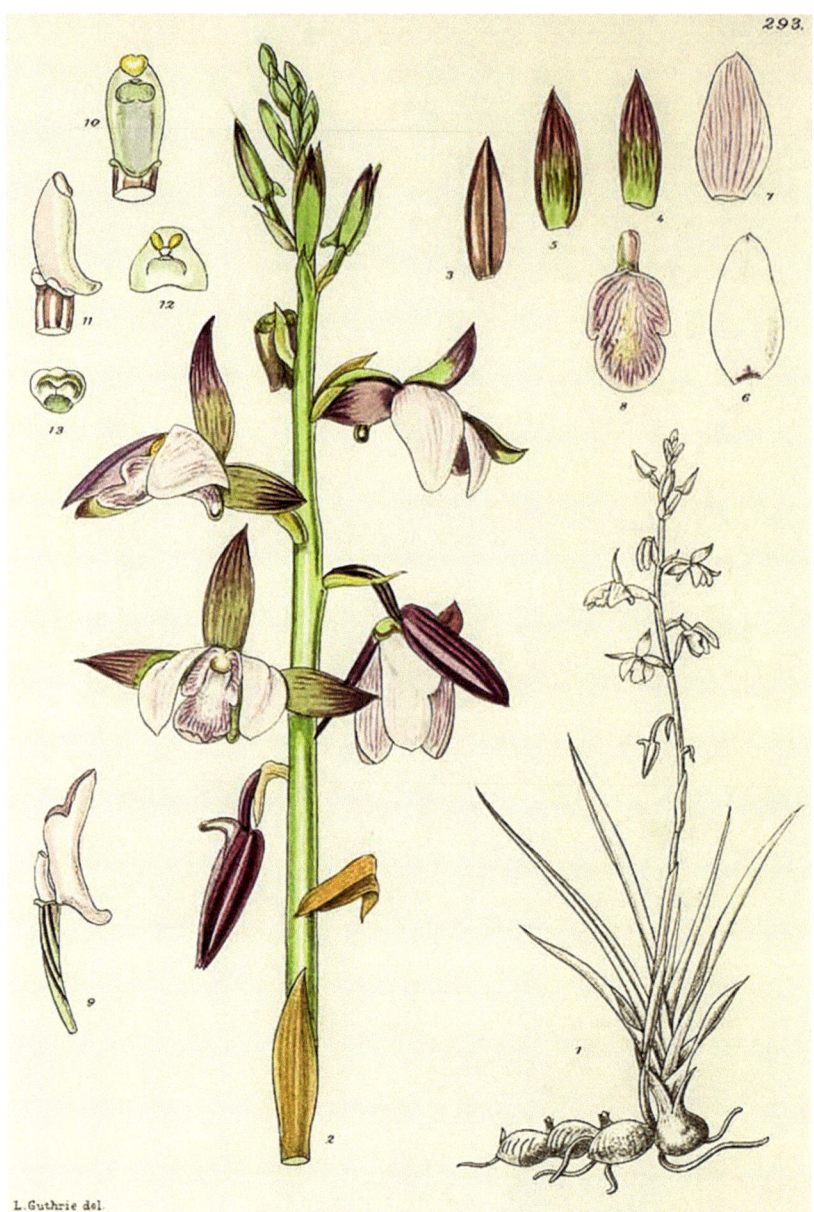

Fig. 17.45 *Eulophia livingstoneana* [as *Eulophia robusta*]. From: Pole Evans IB, *Flowering plants of (South) Africa*, vol. 8: t. 292 (1928) [L Guthrie]

Fig. 17.46 *Tridactyle tridentata* [as *Angraecum tridentatum*]. From: Bolus H, *Icones orchidearum Austro-Africanarum extra-tropicarum* vol 1: t. 53 (1896–1913) [H Bolus]

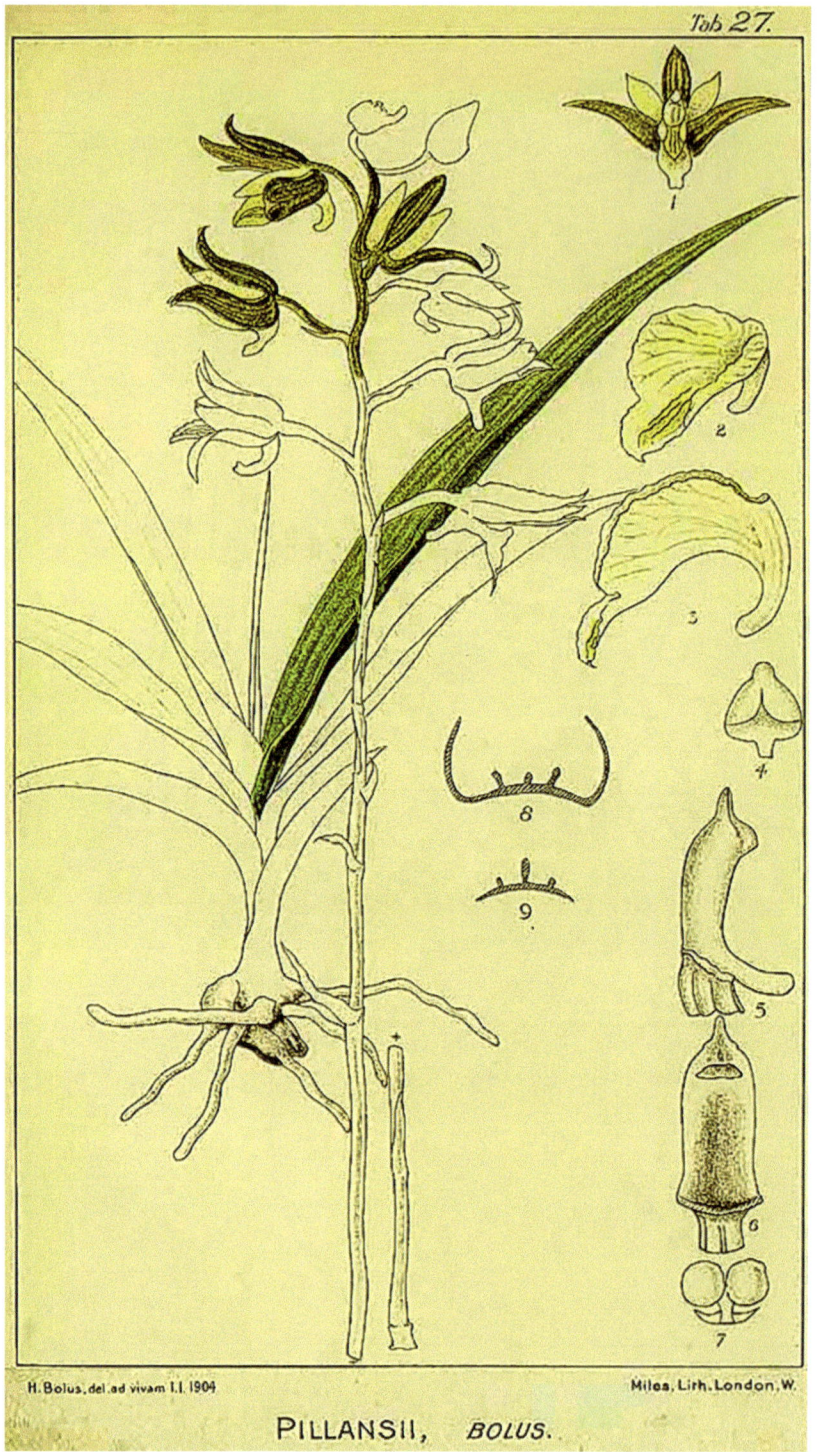

Fig. 17.47 *Eulophia hereroensis* Schltr. [as *Eulophia pillansii* Bolus]. From Bolus H, *Icones orchidearum Austro-Aficanarum extratropicarum*, vol. 2: t. 27 (1896–1913) [H Bolus]

Table 17.3 African orchid species employed as charms

Acrolophia cochlearis (Lindl.) Schltr. & Bolus

Aerangis mystacidii (Rchb. f.) Schltr.

Ansellia africana Lindl. (forma *A. gigantean* Rchb. f.)

Brachycorythis ovata Lindl.

Corycium nigrescens Sond.

Cyrtorchis arcuata (Lindl.) Schltr.

Diaphananthe millarii (Bolus) H.P. Linder

Diaphananthe xanthopollinia (Rchb. f.) Summerh.

Disa stachyoides Rchb. f.

Disa versicolor Rchb. f.

Eulophia angolensis (Rchb. f) Summerh.

Eulophia clavicornis Lindl. var. *clavicornis*

E. clitellifera (Rchb. f.) Bolus

E. cucullata (Afzel. ex Sw.) Steud.

E. ensata Lindl.

E. hians Spreng. [including var. *hians*, var. *inaequalis* (Schltr.) S. Thomas; var. *nutans* (Sond.) S. Thomas, but excluding var. *nutans*]

E. leontoglossa Rchb. f.

E. ovalis Lindl., subsp. *ovalis*

E. parviflora (Lindl.) A, V. Hall

E. petersii Rchb. f.

E. robusta Rolfe

E. speciosa (R. Br. Ex Lindl) Bolus

E. streptopetala Lindl.

E. tenella Rchb. f.

E. welwitschii (Rchb. f.) Rolfe

Habenaria dives Rchb. f.

H. dregeana Lindl.

H. epipactidea Rchb. f.

Liparis bowkeri Harv.

L. remota J.L. Stewart & Schelpe

Microcoelia exilis Lindl.

Mystacidium capense (L.f.) Schltr.

Mystacidium venosum Harv. Ex Rolfe

Oeceoclades mackenii (Rolfe ex Hemsl.) Garay & Taylor

Polystachya modesta Rchb.

Polystachya ottoniana Rchb.f.

Polystachya pubescens (Lindl.) Rchb.f.

Polystachya sandersonii Harv.

Polystachya transvaalensis Schltr.

Rangeris muscicola (Rchb.f.) Summerh.

Satyrium longicauda Lindl. hybrid [probably *Satyrium longicauda* var. *longicauda* x *Satyrium neglectum* Schltr. subsp. *woodii* (Schltr.) A.V.Hall]

Satyrium parviflorum Sw.

S. rhodanthum Schltr.

Stenoglottis fimbriata Lindl.

Tridactyle bicaudata (Lindl.) Schltr.

T. tridentata (Harv.) Schltr.

Vanilla roscheri Rchb.f.

Data from: Arnold et al. (2002), Morris (1996), Long (2005), Chinsamy (2012), and Grant (2016)

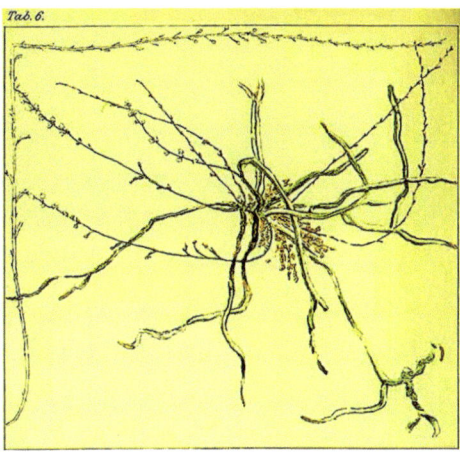

Fig. 17.48 *Microcoelia exilis* [as *Angraecum exilis*], a leafless orchid. Adapted from: Bolus H, *Icones orchidearum Austro-Africanarum extra-tropicarum* vol 1: t. 6 (1896–1913) [H Bolus]

around the house. Powder prepared by burning the whole plant of *Habenaria epipactidea* (local name, *umabelebuca omkhulu*) was mixed with sheep fat and applied to pegs placed in the ground around a new homestead (Chinsamy et al. 2011). Orchids documented for usage in a magical sense in southern Africa consist of around 48 species (Table 17.3) which include the following: *Acrolophia cochlearis* (Lindl.) Schltr. & Bolus, *Aerangis mystacidii* (Rchb. f.) Schltr., *Ansellia africana* Lindl. (forma *A. gigantean* Rchb. f.), *Brachycorythis ovata*, *Corycium nigrescens* Sond., *Cyrtorchis arcuata* (Lindl.) Schltr., *Diaphananthe millarii* (Bolus) H.P. Linder, *D. xanthopollinia* (Rchb. f.) Summerh., *Disa polygonoides* Lindl., *D. stachyoides* Rchb. f., *D. versicolor* Rchb. f., *Eulophia angolensis* (Rchb. f) Summerh., *E. clitellifera* (Rchb. f.) Bolus, *E. cucullata* (Afzel. Ex Sw.) Steud., *E. ensata* Lindl., *E. hians* Spreng. [including var. *hians*, var. *inaequalis* (Schltr.) S. Thomas; var. *nutans* (Sond.) S. Thomas, but excluding var. *nutans*], *E. leontoglossa* Rchb. f., *E. ovalis* Lindl. subsp. *ovalis*, *E. parviflora* (Lindl.) A. V. Hall, *E. petersii* Rchb. f., *E. robusta* Rolfe, *E. speciosa* (R. Br. Ex Lindl.) Bolus, *E. streptopetala* Lindl., *E. tenella* Rchb. f., *E. welwitschii* (Rchb. f.) Rolfe, *Habenaria dives* Rchb. f., *H. dregeana* Lindl., *H. epipactidea* Rchb. f., *Liparis bowkeri* Harv., *L. remota* J.L. Stewart & Schelpe, *Microcoelia*

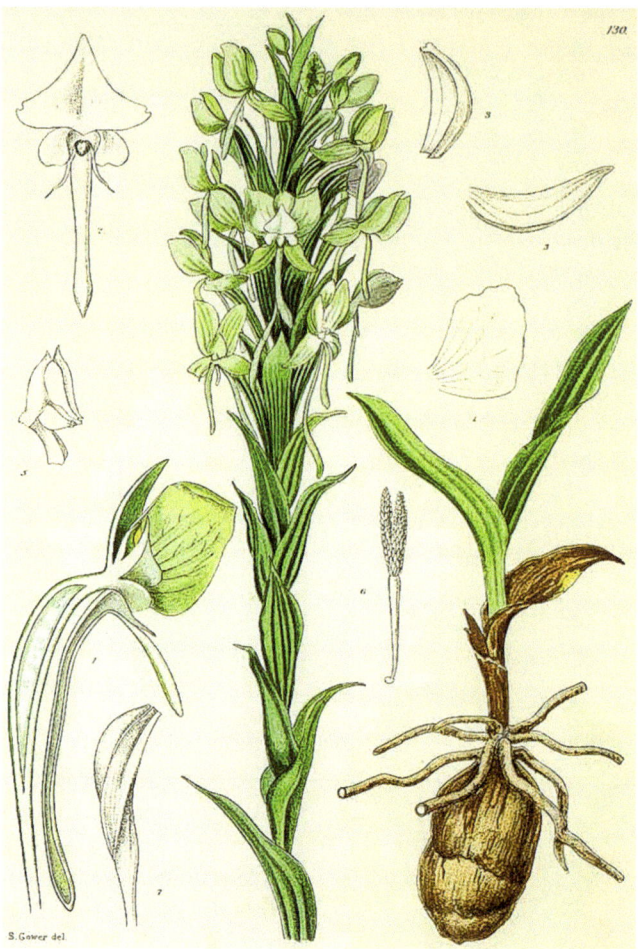

Fig. 17.49 *Habenaria epipactidea* Rchb.f. [as *Habenaria foliosa* (Sw.) Rchb.f.]. From Pole-Evans IB, *Flowering plants of* (South) *Africa*, vol. 4: t. 130 (1824) [S Gower]

exilis Lindl., *Mystacidium capense* (L.f.) Schltr., *M. venosum* Harv. Ex Rolfe, *Oeceoclades mackenii* (Rolfe ex Hemsl.) Garay & Taylor, *Polystachya modesta* Rchb., *Rangeris muscicola* (Rchb.f.) Summerh., *Satyrium parviflorum* Sw., *S. rhodanthum* Schltr., *Stenoglottis fimbriata* Lindl., *Tridactyle bicaudata* (Lindl.) Schltr. and *T. tridentata* (Harv.) Schltr. (Arnold et al. 2002). Of these, 35 (over 72.92%) were also employed medicinally (Figs. 17.49, 17.50, 17.51, 17.52, 17.53, 17.54, and 17.55).

It is not clear whether women have the knowledge or the right to use a protective (orchid) charm to defend themselves against the love charm. However, an enema-type tuber infusion could be used to neutralize evil charms put into food and drinks. This is prepared with *Eulophia tenella* (local name, *untongazibomvana*). There are also charms to get rid of bad dreams (Chinsamy et al. 2011). Water in which tubers of *Eulophia clavicornis* have been boiled is sprinkled to drive away evil. The tubers are made into dolls that represent fertility (Chinsamy 2012).

Next in popularity are love charms, prepared with any of 19 orchid species. Love charms appear to be popular in Swaziland as it makes use of 12 species (*Liparis bowkeri, Aerangis mystacidii, Ansellia africana, Eulophia angolensis, E. ensata, Eulophia petersii, Microcoelia exilis, Mystacidium capense, Polystachya pubescens, P. transvaalensis, Satyrium longicauda, S. parviflorum*). Even not counting the use of three species whose tubers in

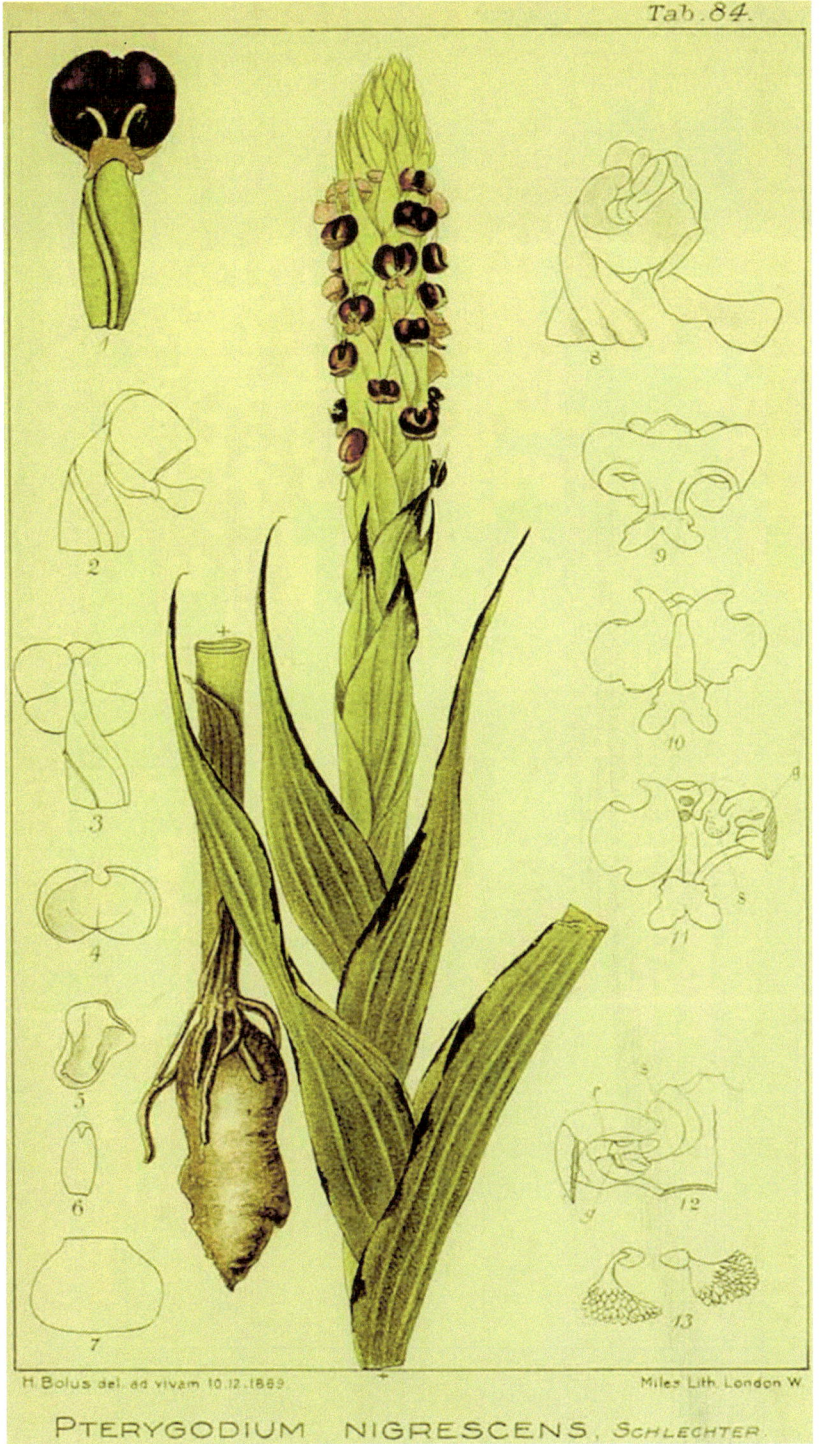

Fig. 17.50 *Corycium nigrescens* [as *Pterygodium nigrescens*]. From: Bolus H, *Icones orchidearum Austro-Africanarum extra-tropicarum* vol 1: t. 6 (1896–1913) [H Bolus]

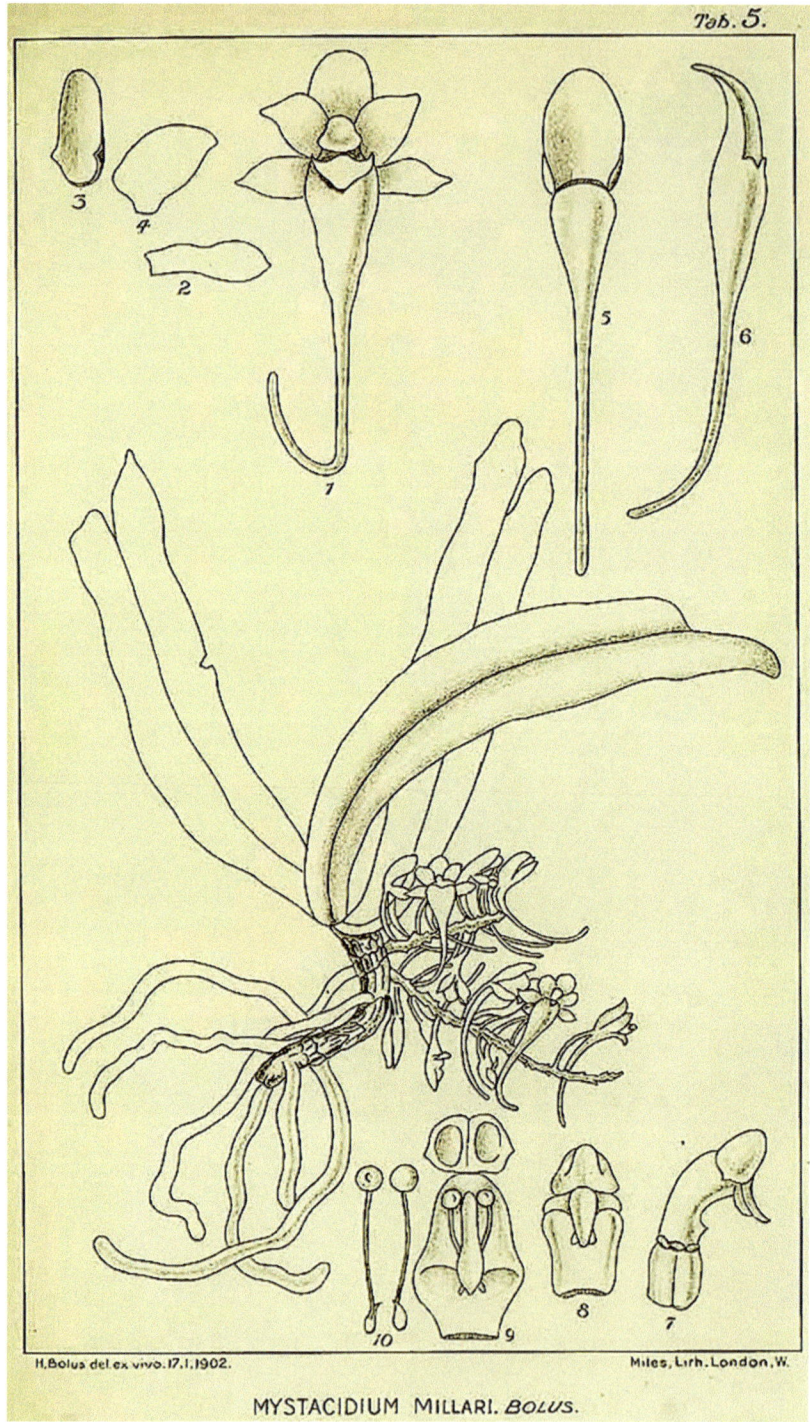

Fig. 17.51 *Diaphananthe millarii* [as *Mystacidium millarii*]. From: Bolus H, *Icones orchidearum Austro-Africanarum extra-tropicarum* vol 2: t. 5 (1896–1913) [H Bolus]

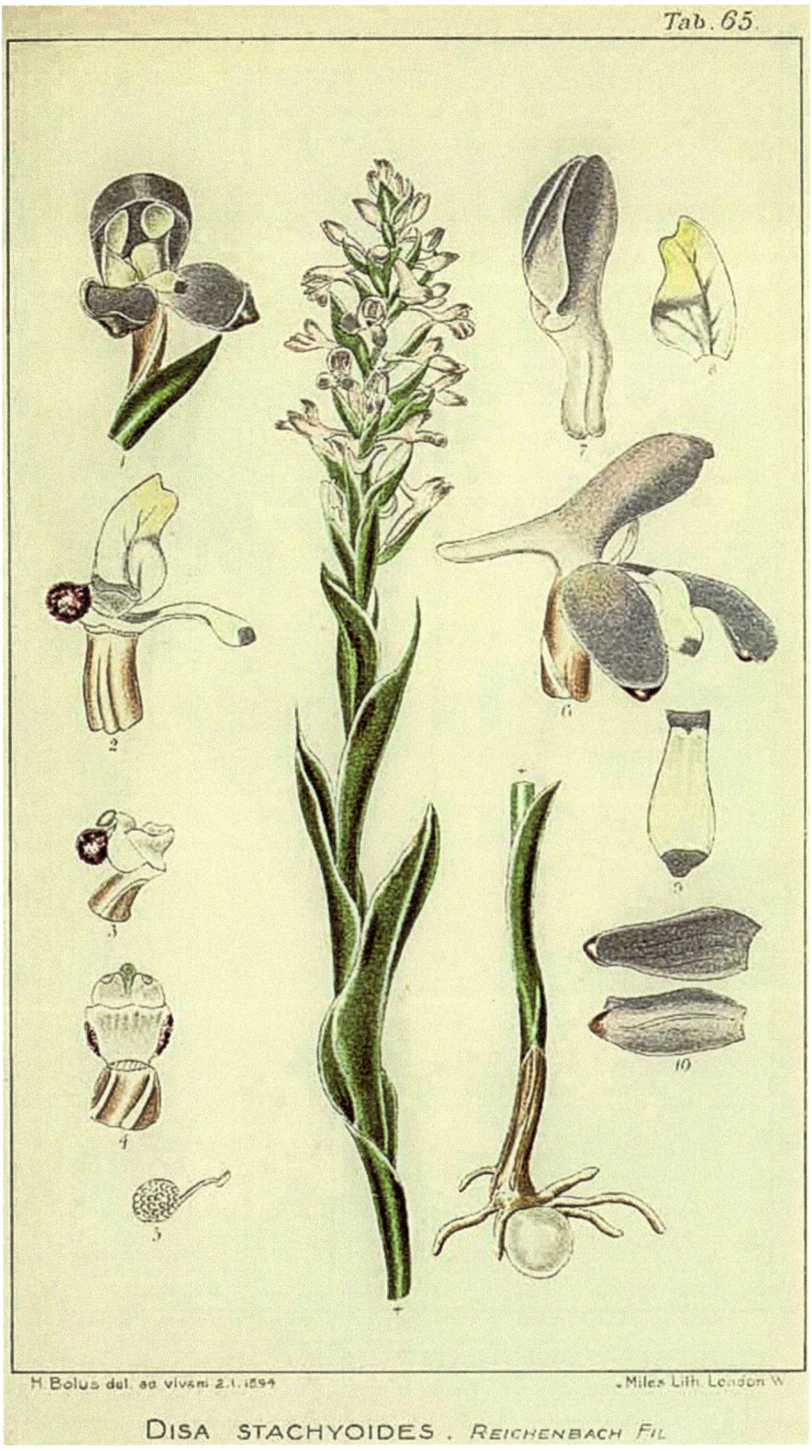

Fig. 17.52 *Disa stachyoides.* From: Bolus H, *Icones orchidearum Austro-Africanarum extra-tropicarum* vol 3: t. 65 (1896–1913) [H Bolus]

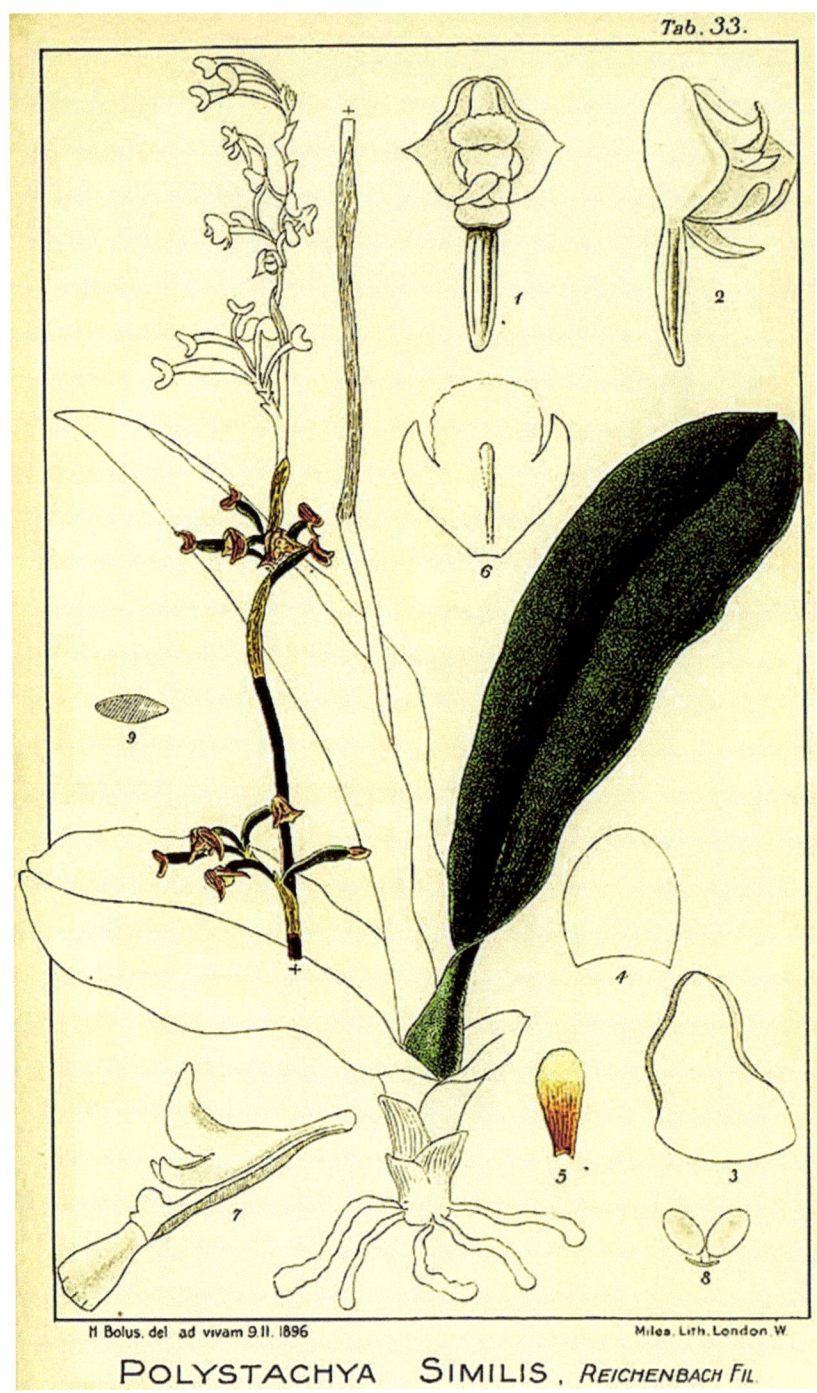

Fig. 17.53 *Polystachya modesta* [as *Polystachya similis*]. From: Bolus H, *Icones orchidearum Austro-Africanarum extra-tropicarum* vol 2: t. 33 (1896–1913) [H Bolus]

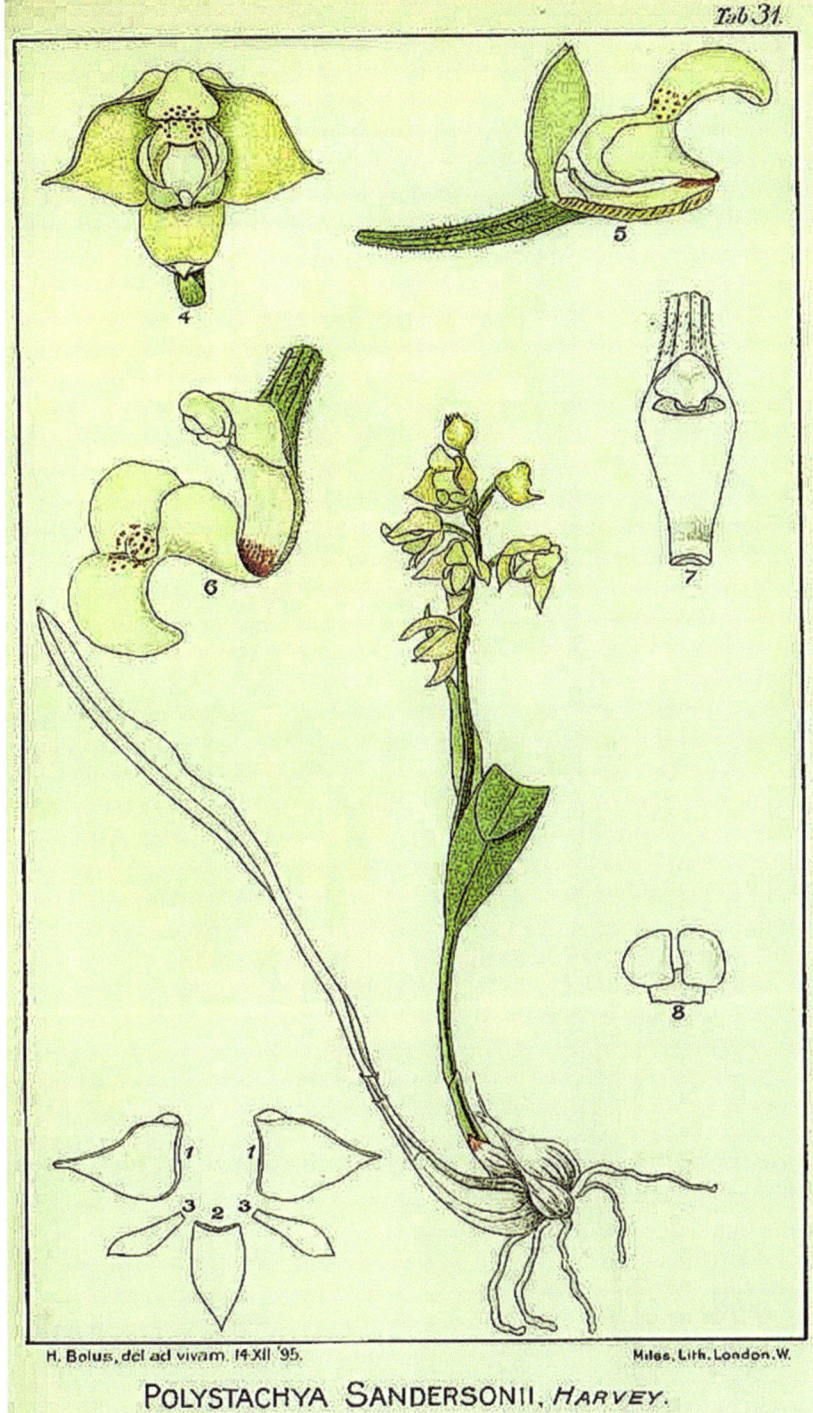

Fig. 17.54 *Polystachya sandersonii*. From: Bolus H, *Icones orchidearum Austro-Africanarum extra-tropicarum* vol 2: t. 31 (1896–1913) [H Bolus]

Fig. 17.55 *Tridactyle bicaudata* [as *Angraecum bicaudatum*]. From: Harvey WH, *Thesaurus capensis, or Illustrations of South African Flora*, col. 2: t. 108 (1863) [WH Harvey]

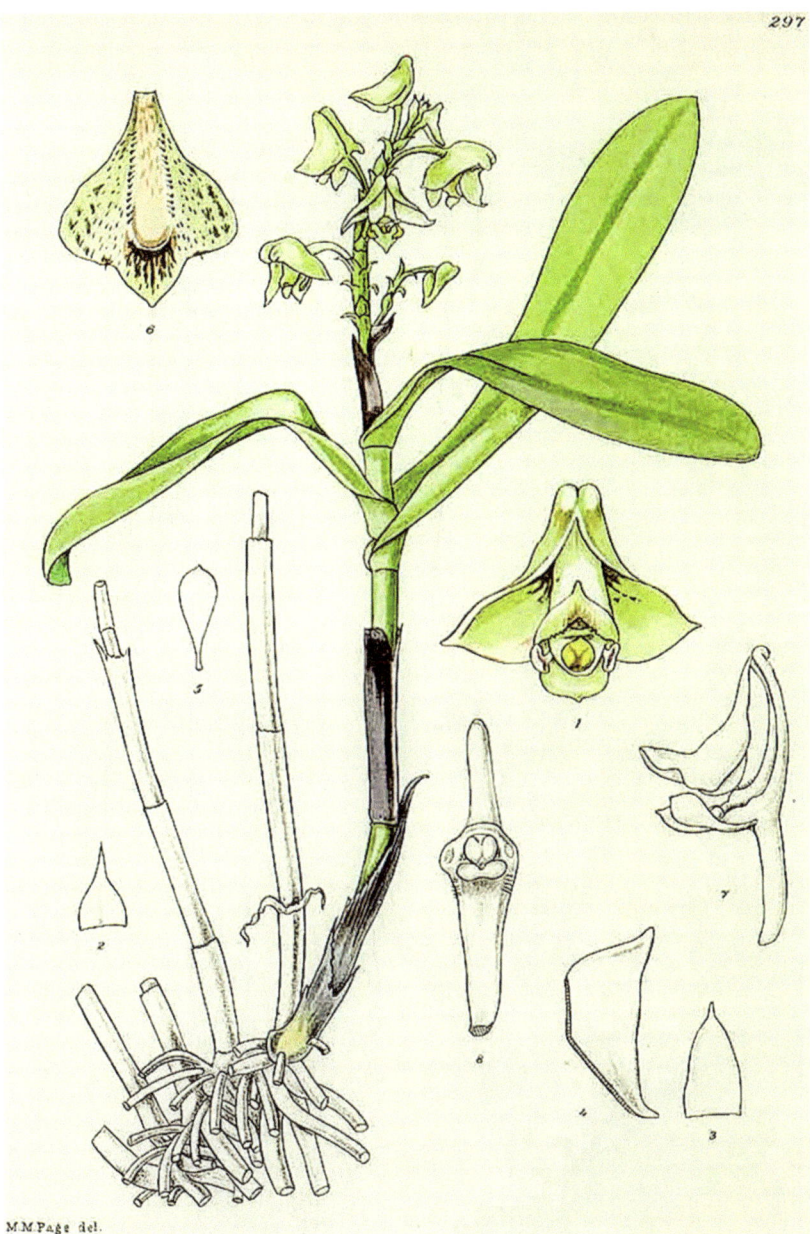

Fig. 17.56 *Polystachya transvaalensis.* From: Pole Evans IB, *Flowering plants of (South) Africa* vol. 8: t. 297 (1928) [MM Page]

infusion are consumed when young men go courting, this is more than a third of the 33 species employed as medicine, food or charms in Swaziland (Long 2005) (Fig. 17.56).

When an African native male is rejected, he might use an infertility-causing charm on the woman. An even more evil intent is embodied in the use of the 'death charm' which is prepared from dried powder of *Habenaria dives* (local name, *inhluthi yotshani*) tubers mixed with other herbs (Chinsamy et al. 2011) (Fig. 17.57). Large quantities of *Eulophia streptopetala* (*Amabelejongosi*) are

Fig. 17.57 *Habenaria dives*. From: Bolus H, *Icones orchidearum Austro-Africanarum extra-tropicarum* vol 8: t. 22 (1896–1913) [H Bolus]

traded in the herb market, being exceeded in amount only by *Ansellia africana*, because of its usage as a love/protective charm (Chinsamy 2012).

(Other orchid species have also been reported to have medicinal uses in Africa, e.g. *Bolusiella maudiae*, *Bulbophyllum nutans*, *Cheirostylis lepida* and *Disa chrysostachya*, but since their usage were not specified, they are omitted in this short discussion.)

References

AbouZid SF, Mohamed AA (2011) Survey on medicinal plants and spices used in Beni-Sueif, Upper Egypt. J Ethnobiol Ethnomed 7:18–23

Arnold TH, Prentice CA, Hawker LC et al (2002) Medicinal and magical plants of southern Africa: an annotated checklist. National Botanical Institute, Pretoria

Assese ESP, Djahoun CAMS, Azihou AF et al (2017) Folk perceptions and patterns of use of orchid species in Benin, West Africa. Flora et Vegetatio Sudano-Sambesica 20:26–36

Bandeira SO, Gaspar F, Pagula FP (2001) Ethnobotany and health care in Mozambique. Pharm Biol 39(Suppl): 70–73

Bhattacharyya P, van Staden J (2016) Ansellia africana (Leopard orchid): a medicinal orchid species with untapped reserves of important biomolecules—a mini review. S Afr J Bot 108:181–185

Bingham MG (2002) Biology of the terrestrial orchid Habenaria sochensis Rchb.f. in Zambia. Kirkia 18(1): 111–116

Bone R (2016) Chikanda Zambia: wild edible orchids, a Darwin Initiative project 2016–2019. eulophiinae.e-monocot.org

Challe JF, Price LL (2009) Endangered edible orchids and vulnerable gatherers in the context of HIV/AIDS in the southern highlands of Tanzania. J Ethnobiol Ethnomed 5:41. https://doi.org/10.1186/1746-4269-5-41

Challe JFX, Struik PC (2008) The impact on orchid species abundance of gathering their edible tubers by HIV/AIDS orphans: a case of three villages in the Southern Highlands of Tanzania. NJAS 56(3):261–279

Challe JF, Niehof A, Struik PC (2011) The significance of gathering wild orchid tubers for orphan household livelihoods in a context of HIV/AIDS in Tanzania. Afr J AIDS Res 10(3):207–218

Chinsamy M (2012) South African medicinal orchids: a pharmacological and phytochemical evaluation. Ph.D Thesis, University of KwaZulu-Natal, Pietermaritzburg

Chinsamy M, Finnie J, van Staden J (2008) The potential of South African medicinal orchids. Abstracts, World Congress on Medicinal and Aromatic Plants, Capetown

Chinsamy M, Finnie J, van Staden J (2011) The ethnobotany of South African medicinal orchids. S Afr J Bot 77:2–9

Chinsamy M, Finnie J, van Staden J (2014) Anti-inflammatory, antioxidant, anti-cholinesterase activity and mutagenicity of South African medicinal orchids. S Afr J Bot 91:88–98

Cribb P, Hermans J (2009) Field guide to the orchids of Madagascar. Royal Botanic Gardens, Kew

Dagar HS, Dagar JC (2003) Plants used in ethnomedicine by the Nicobarese of Islands in Bay of Bengal, India. In: Singh V, Jain AP (eds) Ethnobotany and medicinal plants of India and Nepal. Scientific Publishers (India), Jodhpur, pp 773–778

Davenport TRB, Ndangalasi HJ (2003) An escalating trade in orchid tubers across Tanzania's Southern Highlands: assessment, dynamics and conservation implications. Oryx 37(1):55–61

Dawit A (1987) Plants in the health care delivery system of Africa. In: Leeuwenberg AJM (ed) Medicinal and poisonous plants of the tropics. Den Haag, CIP-gegevens Koninklijke Bibliotheck

Dawit A and Estifanos H (1986): Plants as primary source of drugs in traditional health practices in Ethiopia. In: Engles Proc 1st Intern Symposium on Conservation and Utilization of Ethiopian Germplasm

Grant K (2016) Medicinal plants in Pondoland. Africa Geographic. https://africageographic.com/blog/medicinal-plants-pondoland/

Hamisy WC (2007) Development of conservation strategies for the wild edible orchid in Tanzania. Orchid Conservation Project. Progress report for the Rufford Foundation (RSGNC)

Hennessey EF (1981) *Cynorkis kassnerana* Kraenzlin. Am Orchid Soc Bull 50(7):788–790

Hulme MM (1954) Wild flowers of Natal. Shuter and Shooter, Pietermaritzburg

Kasulo V, Mwabumba L, Cry M (2009) A review of edible orchids in Malawi. J Hort For 1(7):133–139

Kim SL (2016) see http://www.facebook.com/chikanda orchidconservation

Kovacs A, Vasas A, Hohmann J (2007) Natural phenanthrenes and their biological activity. Phytochemistry 69:1084–1110

La Croix IF, La Croix EAS, La Croix TM (1991) Orchids of Malawi. The epiphytic and terrestrial of South and East central Africa. AA Balkema, Rotterdam

Lalika MCS, Mende DH, Urio P et al (2013) Domestication potential and nutrient composition of wild orchids from two southern regions in Tanzania. Time J Biol Sci Technol 1(1):1–11

Lawler LJ (1984) Ethnobotany of the Orchidaceae. In: Arditti J (ed) Orchid biology. Reviews and perspectives, vol III. Cornell Univ Press, London

Leong PC, Morris JP (1947) Available calcium in vegetable. Med J Malaya 1:289

Long C (2005) Swaziland's Flora-siSwati names and uses. SNTC Home Page

Mapunda LND (2007) Edible orchids in Makete District, the Southern Highlands of Tanzania: distribution, population and status. MSc dissertation, Uppsala University, Report No. 39

Morris B (1996) Chewa medicinal botany. A study of herbalism in Southern Malawi. Lit Verlag, Hamburg

Nyomora AMS (2005) Distribution and abundance of the edible orchids of the Southern Highlands of Tanzania. Tanz J Sci 31(1):45–54

Stewart J (1981) The genus Ansellia – orchids of Africa. Am Orchid Soc Bull 50(3):248–255

Stewart J, Campbell B (1970) Orchids of tropical Africa. A S Barnes & Co, New York

Stewart J, Campbell B (1996) Orchids of Kenya. St. Paul's Bibliographies, Winchester

Teoh ES, Teoh LKK (1999) Everywoman's book on menopause and the hormone replacement controversy. Times Books International, Singapore

Terashima H, Ichikawa M (2003) A comparative ethnobotany of the Mbuti and Efe hunter-gatherers in the Ituri forests, Democratic Republic of Congo. Afr Study Monogr 24(1, 2):1–168

Veldman S, de Boer H, Otieno J (2016) Species assessment in African orchid cake. Barcode applications, pp 10–11. https://uu.divaportal.org/smash/get/diva2:775683/FULLTEXT01.pdf

Veldman S, Gravendeel B, Otieno JN et al (2017) High throughput sequencing of African chikanda cake highlights conservation challenges in orchids. Biodivers Conserv 26(9):2029–2046

Veldman S, Kim SJ, van Andel TR et al (2018) Trade in Zambian edible orchids – DNA barcoding reveals use of unexpected orchid taxa for chikanda. Genes 9:595

Von Klein GH (1905) The medical features of the Ebers papyrus. Cornell University Library. Originally published in JAMA (1905)

Watt JM, Breyer-Brandwijk MG (1962) The medicinal and poisonous plants of Southern and Eastern Africa. E S Livingstone, Edinburgh

The Challenge: Orchid Conservation

18

Orchids in nature face several threats, the first of which is over-collection to supply (a) nurserymen who cater for growers of fancy plants flaunting them as status symbols or (b) drug dealers offering aphrodisiacs and miraculous cures. Habitat destruction is the other major threat. This may be caused (a) by human activity—lumbering; construction of dams, highways, housing and industrial sites; agriculture; cattle grazing, war; etc.—or (b) through natural disasters like fire, floods, soil erosion, earthquakes, global warming and alteration to the ecology (Bailes 2005) (Fig. 18.1).

Over-collection

During the season of the orchid craze in the nineteenth century, no jungle was safe from the plunder of orchid collectors dispatched by nurserymen who dreamt of making fortunes when their shipments arrived in Europe. At the second auction at Stevens Auction Rooms in London in the early 1850s, it was not uncommon for single orchid plants to sell above 10 pounds sterling, for example, *Angraecum caudatum* 18 guineas, *Angraecum eburneum* 18 guineas and *Vanda suavis* 17 guineas. A rare orchid could fetch several hundred guineas. On 4 May 1878, William Bull announced that he had received two consignments of orchids totalling 2 million plants! Nevertheless, Swinson (1970) commented that these 'imports did not reach their peak until

(Frederick) Sander was in full operation'. Astonishing.

A hundred years later the situation had not improved. When new *Paphiopedilums* were discovered in the karst forests and limestone hills of southern China and North Vietnam, eager suppliers shipped tens of thousands of plants to nurserymen in the United States and Europe. A CITES ban on the export and import of *Paphiopedilums* collected from the wild has stopped the pillage, and meanwhile China has embarked on ex situ conservation of the species. It remains to be seen whether a repeat of the orchid plunder will occur when Myanmar becomes an open country.

On the positive side, this historical over-collection brought orchids to the attention of the world and pleasure to millions. If it had not happened, there would be no orchid-growing hobby, and most beautiful orchids would still lie hidden in the jungles and remain unseen by most people living far from their habitats. Regulations forbidding the collection of orchids from the wild appear to go overboard when it even disallows private collection of orchids from fallen trees in timber concessions or ground orchids from sites destined to be flooded because of development (Hansen 2001).

As for medicinal orchids, fewer than 20 species are over-collected. Excessive demand for certain medicinal orchid species in China, India and Turkey prevails because some people in these

© Springer Nature Switzerland AG 2019

E. S. Teoh, *Orchids as Aphrodisiac, Medicine or Food*, https://doi.org/10.1007/978-3-030-18255-7_18

Fig. 18.1 *Coelogyne nitida* growing 15–20 m above the ground at Doi Inthanon (2500 m) in northern Thailand. Old trees provide the natural habitat for epiphytic orchids; hence deforestation poses a major threat to orchids. (©Teoh Eng Soon 2019. All Rights Reserved)

countries, subscribing to old traditions, consume such orchids on a regular basis. *Custom Records of China Imperial Customs* in 1884 indicate that in that year, excise was collected on 15 shipments of *Dendrobium* and 5 lots of *Baiji* (*Bletilla striata*) loaded at ports along the Yangtze. The first shipment of *Dendrobium* (probably *D. moniliforme*) arrived from Japan, the second from Annam (Vietnam) and the rest from the local provinces of Sichuan, Guangdong, Hubei, Jiangxi, Fujian, Zhejiang, Guangxi, Anhui, 'Chili' (?Hubei), Honan, Hunan and Shanxi (Hart 1884). China Imperial Records of 1888 also carried inventories of *Shihu* and *Tianma* (Braun 1888).

Chinese interest in the three herbs would definitely place these orchids at risk where it is not for the fact that many of the species involved can now be easily cultivated. During the mid-1970s over 100 tons of *Tianma* (*Gastrodia elata*) were collected annually from the wild because *Tianma* was the principal herb employed for the treatment of strokes and various neurological disorders. At that time China had to rely on its own produce to treat various ailments for the bulk of its population. When wild populations of *Tianma* became scarce, Chinese scientists learnt how to cultivate the parasitic orchid (Xu 1989; Xu and Guo 2000). Cultivated *Gastrodia elata* accounts for over 80% of commercial *Tianma* today, the best producers based at the salubrious province of Guizhou. Medicinal *Dendrobium* are cultivated, but demand continues to exceed supply from cultivated sources. Additionally, many herbalists consider wild plants, especially from classically renowned sources, to be better than cultivated varieties. Wild populations of medicinal *Dendrobium* in the neighbouring countries of Indochina, Thailand and Myanmar are being depleted because of the strong Chinese demand. *Baiji* (*Bletilla striata*) did not come under threat because it is easily cultivated.

Commercial exploitation of the forests in India has led to the disappearance of many species. In Arunachal Pradesh orchids are struggling to survive (Rao 2004). *Vanda tessellata* is disappearing from the subcontinent due to heavy deforestation of their host trees (Pandey et al. 2003). Medicinal orchids are harvested indiscriminately although trade in orchids is regulated under Schedule VI of the Wildlife Protection Act (1972). The situation is aggravated by cattle grazing, subsistence farming and other causes of habitat destruction.

The four rare orchid species that go into the production of *Astavarga* and *Eulophia spectabilis* (considered an aphrodisiac) have become scarce because of Indian demand. Furthermore, since several species are small and large numbers of plants are required to constitute a commercially significant supply, foraging for such plants would be conducted over a wide area, thus resulting in extensive species depletion. The situation is a repetition of the disastrous Turkish *salep* scenario (Sezik 1967). Being considered an aphrodisiac in India is a death sentence for an orchid in that country, and sometimes it may even pose a cross-border threat. On one occasion, a single shipment of *Flickingeria fugax* transported from Nepal to India consisted of hundred thousands of plants (Koopowitz 2001).

Although India has embarked on forest conservation, conservationists on the ground are worried that economic considerations may outweigh any desire to conserve orchids. In the Niyamgiri Hills of Orissa, 20 orchids are being used by the Dongaria Kandha tribes as medicinal remedies for various conditions, but the entire ecosystem of the hill areas has come under threat from mining operations (Dash et al. 2008).

Generally, however, medicinal orchids which feature as one of dozens of possible remedies employed by tribals and dwellers in remote villagers to treat simple symptoms such as cough, fever, insect bites, injuries, etc. are not in danger of being over-exploited. In fact, if an orchid is useful, villagers sometimes cultivate the plants in nearby forest to ensure a readily available supply (e.g. *Eria pannea*). Furthermore, this orchid would usually be only one of many possible remedies, and several alternative plants are often more readily available, either because they grow in the kitchen garden or occur in the bushes and forests that are within easy reach. Asian herbals seldom mention orchids because they are not so easily obtainable.

In the event that a miraculous cure is discovered in a particular orchid species, theoretically, it might put the orchid on the endangered list. However, Taiwan has shown that today, it is possible to mericlone most orchids and offer them at very reasonable prices. The jewel orchid, *Anoectochilus formosanus* (*Yaowang*) which is a popular herbal remedy, is available as either a dried herb or as living plants suitable for cultivation. In China, *Bletilla striata* is cultivated commercially to meet the demands for *Baiji*, and even the heterotrophic *Gastrodia elata* is now cultivated. Furthermore, when a potent phytochemical is identified, its structure can be determined, and methods can then be devised to synthesize the compound. Vanilla, malaxin (dihydroartemisin), gastrodin, denbinobin and many other compounds have been synthesized. The discovery of a useful chemical in an orchid would therefore not lead to its over-collection from the wild. What puts an orchid at risk is its continuing usage as a popular herbal remedy. Currently, this applies to *nobile*-type *Dendrobium*, but the problem can be easily rectified. Therefore, medicinal usage should not pose a problem.

Habitat Destruction

Population growth and accompanying demand for food, water, housing, electricity, transportation and other basic needs, coupled with extravagant consumption, human greed and misuse of power, are the main causes for the disappearance of primary forests and the destruction of orchid habitats. In many developing countries, extraction of timber followed by agricultural development or the construction of dams and cities has led to the loss of host trees and terrestrial orchid habitats. Such moves cannot be entirely prevented because humans must seek to improve their quality of life, but good governments should seek a common good and balance development with a need to preserve their own biosphere. Sadly, this is not always the case because many politicians only think of themselves and not their fellow countrymen and their country. Orchids

seldom feature as an item for consideration although, if truth be known, the orchid wealth of a country is an accurate index of that nation's ecological health.

Paphiopedilum dianthum, once thought to be endemic in China, was discovered in Vietnam in 2000, but logging soon led to the disappearance of the species in many of its newly discovered natural habitats (Averyanov et al. 2003). Likewise, *Vanda spathulata* previously very common at Veli near Trivandrum in Kerala State disappeared when the area was cleared and occupied by the Vikram Sarabhai Space Centre (Abraham and Vatsala 1981) (Fig. 18.2).

Forest Reserves

Numerous third-world countries have established forest reserves and national parks where commercial harvesting of timber and other forest products is prohibited in order to preserve the full range of its biodiversity and to ensure that this natural heritage is retained for posterity. However being poor or sometimes poorly governed, this is merely a show of official commitment, and favoured individuals or foreign companies are still being granted concessions to extract timber from inordinately large acreages, worse, even without enforcement to ensure that good forest management policies are being observed.

Malaysia and Indonesia have been accused of over-logging. Government figures state that Malaysia produces about 350 million cubic feet of timber annually and cuts an area of 128,000 acres. Timber is Malaysia's second largest export earner, after petroleum. Western environmentalists have criticized Malaysia for permitting its rainforests to be cut down. This is especially true in the Bornean states of Sabah and Sarawak, and it threatens the livelihood of tribal people. Ironically, Sabah, endowed with oil, gas and numerous natural resources, is one of the richest states in Malaysia, but it has a poverty rate of 23% which is four times the national average for Malaysia (www.undp.org. my). Meanwhile, unique orchid species such as *Phalaenopsis gigantea*, *Paraphalaenopsis denevei* and *Paphiopedilum rothschildianum*, although not

Fig. 18.2 *Dendrobium dixanthum* flourishing in deciduous forest at Chiang Dao (2000 m), northern Thailand. (©Teoh Eng Soon 2019. All Rights Reserved)

known to be medicinal, are no longer found in the forests of East Malaysia.

Timber is also a major export from Indonesia, the country with the most extensive forest in the world, after Brazil. Indonesia has about 122 million hectares of tropical hardwoods. Kalimantan and Sumatra, the two largest islands, between them have 55 million hectares of commercial value. At the end of 1980, 500 concessionaires, 70 foreign owned, were extracting timber from 49 million hectares of Indonesian forest (www. fao.org). The degree of over-exploitation may be appreciated by comparing this figure with the total biosphere reserves of India, only at 8.4 million hectares. In 1983, logging, slash-and-burn farming and prolonged draught resulted in a catastrophic fire in Kalimantan which burned for an entire year destroying 3.6 million hectares of tropical rainforest. The smog even hid the sun in Peninsular Malaysia and Singapore.

Among the various races, the Chinese has been the most concerned about leaving something behind for future generations. Chinese Buddhist texts encourage gratitude for ancestors seven generations removed (Cole 1998). What would we leave behind if we go on destroying forests at the present rate? Every year Nepal loses 2.3% of forests richest in orchid biodiversity (Subedi 2005). Indonesia loses 1.8 million hectares of forests annually (Budiharta et al. 2011).

Thailand has lost over 20% of its forests (2 million hectares) since 1990 (Bisson et al. 2003). When the history of the world is written in the twenty-third century by our descendants seven generations below, let it not be said that humans in the twentieth and twenty-first century were excessively consumerist and they were the most destructive of their environment.

Bhutan is probably the best preserved country in the world, thanks to the astonishing vision of its former king, King Jigme Singye Wangchuck, who sought to establish GNH (Gross National Happiness) as the index of a nation's success, in place of either GDP (gross domestic product) or PPP (purchasing power parity). Its population of 750,000 (figure corrected for year 2013) lives in an area of 38,394 sq km, 72% of which is under forest cover, 20% is under perpetual snow and only 8% is arable (Wangchuck 2006). In order to preserve this pristine orchid sanctuary which comprises almost the whole country, a conservation project has been started by Bhutan's National Biodiversity Centre (NBC) headquartered at Serbithang. The NBC aims to educate the public on the threats to orchids, protect the forests, promote ecotourism as an additional source of income, cultivate and propagate threatened or desirable orchids and reintroduce orchids in protected areas. Two orchid houses have been established at the National Botanic Gardens (Dalstrom et al. 2012) (Fig. 18.3).

Kasugayama Primeval Forest located above Nara City in Japan is possibly the oldest protected forest. Hunting and logging has been disallowed in this forest since 841 which is populated with tall cedars and numerous species of animals. Its footpaths connect various Shinto and Buddhist shrines which are well over a thousand years old, and this makes it an attractive tourist attraction. It is designated a World Heritage site.

China's massive infrastructure development, a population of 1.3 billion and its rapid economic expansion brought into focus the importance of careful national planning on conservation of the national landscape. Presently, no other country is making a greater effort at forest conservation. At the end of 1993, China had set up 766 nature reserves covering 66 million hectares. By 2007, the total number of nature reserves had increased to 2349, with acreage almost doubling to 150 million hectares, or slightly more than 15% of the territory. Over 1200 orchid species in approximately 174 genera occur in the country, distributed across a broad range of ecosystems, each of which need to be preserved if all species are to survive. Some endemic species are very restricted in their distribution, and it is important that such special habitats be protected. Where species are rare but have a wide distribution, it is essential to maintain contiguous or unfragmented forest ecosystems to conserve the total gene pool of the species. Soil erosion and destruction of loss of forest cover pose significant threats to geophytic and saxicolous species.

Many of the important orchid habitats are present in China's forest reserves: Xishuangbanna National Nature Reserve in southern Yunnan, Gaoligongshan National Nature Reserve in western Yunnan, Dinghushan National Nature Reserve in Guangdong, Jianfengling National Nature Reserve in Hainan, Wuyishan National Nature Reserve in Fujian and the Yachang Orchid Nature Reserve in Guangxi Province, Huanglong and Jiuzhaigou in Sichuan (Chen 1995; Jia 2007). The number of medicinal orchid species in the different nature reserves varies considerably, higher numbers occurring in warmer areas and in reserves that enjoy a wide range of microclimates. A count made on native orchids from Gaoligongshan Mountains at the western border of Yunnan shows that the massive Gaoligongshan National Nature Reserve harbours 108 medicinal species in its 405,549 hectares which reach up to 3916 m (Jin et al. 2009). Similar checks on their native flora indicate that at the Yachang Orchid Nature Reserve in Guangxi, there are 71 medicinal species, and Huanglong in Sichuan is home to 22 medicinal species (Liu et al. 2010; Pernier and Luo 2007; Teoh 2011). Chinese nature reserves shelter numerous subtropical and temperate species, many of which are endemic or subendemic.

According to Jia Jiansheng, the Deputy Director of the Department of Wildlife Conservation, State Forestry Administration of China, orchids

Fig. 18.3 *Cypripedium cordigerum* grows in alpine forest at 3000 m in Bhutan. Some of its natural habitats will disappear following the construction of more roads in the country. There is currently an ongoing effort to rescue the wild orchids. (©Teoh Eng Soon 2019. All Rights Reserved)

are an important component in the long-term project launched by the Chinese government to preserve wildlife. The project is tapping both on local and international expertise. Of particular importance are the new 22,000 hectare Yachang Orchid Nature Reserve in Guangxi Province established in 2005 which harbours 156 species of orchids, an ex situ conservation centre in Shenzhen, Guangdong Province and germplasm and seed banks focusing particularly on the resources of the southwestern region (Jia 2007). Seven orchid species including the medicinal *Changnienia amoena* Chien, *Dendrobium candidum* Wall ex Lindl., *Gastrodia elata* Bl. and *Phalaenopsis aphrodite* Rchb. f. were listed in the first volume of the *China Plant Red Data Book* which highlighted the rare and endangered species that required immediate protection (Fu 1992).

Although forests cover 64% of the land in the Korean Peninsula, 29 orchid species are listed as critically endangered in the country of which 11 are medicinal: *Calanthe discolor* forma *sieboldii, Cymbidium kanran, Cymbidium lancifolium, Cypripedium guttatum* var. *koreanum, C. japonicum, C. macranthos, Dendrobium moniliforme, Galeola septentrionalis, Goodyera repens, Gymnadenia conopsea* and *Liparis nervosa*. Habitats for some species are scarce and in the case of *Dendrobium moniliforme* limited to one or two. There are few plants in each habitat. *Bulbophyllum inconspicuum* occurs in small populations in fewer than ten places surveyed, and the species is also considered as endangered, although not critically so. Illegal collection has decimated the ornamental species. Careless over-collection of *Cremastra appendiculata* led to its habitat destruction and a rapid decline in plant population (Lee 2009). Fortunately, several of the medicinal species enjoy a widespread distribution either in Asia or even throughout the world, and from a global viewpoint, they would not be critically endangered. *Cremastra appendiculata*, for instance, is distributed across China to the Himalayas (Sikkim, Bhutan and Nepal) and could easily be reintroduced into Korea, albeit their genetic make-up might be different.

Numerous medicinal species that occur in Gaoligongshan are also present in northern Myanmar, Bhutan, Sikkim and Nepal, but in some places they are already under threat. Sharing information and plants between countries might be a way to conserve species that have become rare.

Much attention is given to the conservation of orchids in Australia and the Americas, to the extent that on an occasion Australian authorities even diverted a coastal highway in northern New South Wales to prevent annihilation of a colony of the endangered parasitic *Rhizanthella slateri* (Clements 2011). The American Orchid Society established the Orchid Conservation Program of the El Pahuma Orchid Reserve in Ecuador which seeks to protect over 200 species in 700 hectares of primary tropical montane forest (Glicenstein 2005). Its chairman, Leon Glicenstein, also undertook to plant 34 and 127 seedling plants of *Cypripedium acaule* in Cowles Bog, Indiana, in 2004 and 2005, respectively. When he checked on the bog in 2006, 35% of the plants were alive. Cherokee Indians used this orchid to treat nerve disorders (Hamel and Chiltoskey 1975), whereas Menomis employed it for male disorders. *Cypripedium* also enjoyed a place in the *United States Pharmacopeia* for the treatment of nerve disorders before the twentieth century.

In Situ Conservation

Planning for in situ conservation takes into consideration numerous factors, some of which are under human control, others not: size of conservation area, location, stability, scope for altitudinal migration, host trees, mycorrhiza, pollinators and accessibility. Ideally, sites selected should contain broad genetic banks of various orchids and be contiguous with other protected forests. It should not be prone to flooding, unlikely to be exploited for timber, agriculture or the building of dams. Untapped primary forests with high orchid density are best, but if already subjected to logging in the past, host trees must be present else they need to be reintroduced. Non-native host trees, for instance, *Samanea saman* (rain tree), have served as an excellent alternative host tree for epiphytic orchids in the Asian tropics. Pollinators are essential for maintain self-sustaining populations of orchids and genetic variation and evolution. Some conservationists have tried removing competitors from locations where rare orchids occur, much as one would do in one's own garden. Volunteers in New York on discovering a colony of the parasitic autumn coralroot (*Corallorhiza odontorhiza*) began clearing the site of invasive Japanese stiltgrass and black swallowwort (Taft 2013), a laborious task only applicable to locations in parks.

In situ conservation sometimes involves the reintroduction of an orchid to its original habitat or species relocation when the original habitat is slated for a large development project like the building of a dam. Should it be necessary to reintroduce some orchid species, it would be preferable if symbiotic mycorrhiza is reintroduced at the same time, although, as far as we know, this has not been attempted. In addition to producing sugars that enable orchid seeds to germination, mycorrhiza produces fungal elicitors that encourage cell division, seedling growth and synthesis of metabolites. Mycorrhiza also produces phytoalexins which protect the orchid against pathogens. Many scientists believe that with terrestrials, to achieve any degree of success, mycorrhiza symbionts are essential (Rasmussen and Whigham 1993; Chen et al. 2003; Bidartondo et al. 2004; Bidartondo and Read 2008; Hollick 2004; Preiss et al. 2010; Liu et al. 2010). Fortunately, this is usually not a monogamous relationship (Hadley 1970). *Dendrobium officinale* (syn. *Dendrobium catenatum*) in China's Guangxi Province associates simultaneously with five fungi species (Xing et al. 2013).

Designating areas for collection and preventing over-collection, the collection of plants before fruit set, are sensible ways to maintain the availability and quality of a desirable orchid. In Zambia, access to *chikanda* tubers is managed by the chief of each tribe who seasonally designates a harvestable *mambo* (wet meadow) within his chiefdom. Nevertheless in the central province of Serenje, harvesters

claimed that they were free to harvest tubers anywhere (Veldman et al. 2018).

Attempts to replant or leave behind young plants of terrestrial orchids in Turkey did not meet with much success because of indiscriminate over-collection. Education and enforcement are measures that need to be conducted for conservation efforts to prevail. Conservation efforts have to take into consideration the livelihood of rural folks, and with this in mind, an effort has been made to cultivate *Dendrobium catenatum* in karst and *danxia* forests with dual objectives in mind and to provide consistent, renewable source of the most desired species of *shihu* for the herb market and additional earnings for people living in these remote regions. This new beginning is meeting with some success (Liu et al. 2013). Similarly, there is a need to reintroduce *Dendrobium nobile* and other medicinal species en masse to Xin Yi in Guizhou Province before their pollinators disappear, and thereafter their harvesting should be regulated (Fig. 18.4).

An Indian farmer living in Sikkim or Himachal Pradesh could earn an excellent annual return provided he could carry out a proper cultivation of *Eulophia campestris* (Butola and Badola 2008) because this orchid is highly valued in the herb markets of northern India (Teoh 2011). In Himachal Pradesh, successful cultivation of *Dactylorhiza hatagirea* brought returns of Indian rupees 250,000–350,000 per hectare, far exceeding the returns from gathering the orchid from the wild (Butola and Badola 2006). There is no reason why farmers should not cultivate the orchid rather than resort to foraging for it in the wild. Education is the key.

Ex Situ Conservation

Botanic gardens serve to study and protect species plants, and many botanical gardens establish extensive nurseries to preserve their native plants. Ex situ conservation can sometimes be thwarted by political decisions. Some countries have lost their entire collection of botanical orchids due to appointment of directors with a cavalier attitude towards plants with which they are unfamiliar. It

Fig. 18.4 The rain tree (*Albizia saman*), native to South America, is now extensively naturalized throughout the tropics. It is home to many orchids. This old tree in Singapore almost at sea level is home to a large

is to be hoped that with more awareness and public interest in biodiversity and conservation, such mistakes will not be repeated.

Gene Banks

Gene banks have been established for commercially important *Vanilla*. Ex situ conservation and gene banks help to deal with the problems of severe drought, forest fires, flooding and epidemics which are factors beyond man's control allowing decimated species to be reintroduced into their original habitats. It is important to collect plants and seeds of medicinal orchids from as many locations as possible to ensure that no valuable genetic asset is omitted (Chen et al. 2006). Studies of their DNA should ensure that clones from different sources are indeed unique (Ding et al. 2002a, b; Li et al. 2005, 2008; Xu et al. 2010).

Vitrification has solved the problem of long-term seed storage by cryopreservation for many species (Thammasiri 2000, 2008, 2010; Vandrame and Faria 2011 MORE). Some terrestrial orchids that are difficult to germinate may need to be cryopreserved as green embryos. Alternatively studies are being undertaken to develop pretreatment solutions to deal with the problem of thick seed coats (Hirano et al. 2005; Lee 2011). Mycorrhiza-assisted germination and addition of growth factors may help. Raising seedlings with mycorrhiza boosts their survival when they are transferred to natural habitats (Hollick 2004; Yam 2010). Protocorms can also be vitrified and cryopreserved.

The cost of a physical set-up for gene banks is not prohibitive, but the effort requires co-operation, dedication, sustained effort and continuous support from institutions and governments. Countries which lack space for extensive conservation, and even countries with large nature reserves, can play a role by engaging in intensive conservation. That is to say that they could engage in intensive orchid research to select clones with superior medicinal yield; propagate desirable medicinal species; establish seed, embryo, pollen (Thammasiri 2000, 2008, 2010; Wood et al. 2000; Vendrame et al. 2007; Vendrame et al. 2008) and tissue banks; engage in developing genetic improvement; study mycorrhizal and bacterial relationships with orchids (Rasmussen and Whigham 1993; Compant et al. 2005; Liu et al. 2010; Chu et al. 2010; Tsavakelova et al. 2005; Sun et al. 2011); and produce such superior stock of medicinal plants in order to eliminate the necessity for collecting plants from the wild. In Taiwan, some cultivated varieties of *Anoectochilus formosanus* are recognized as the equivalent of the wild species (Chang 2007) and can fetch a similar price of US$100 per kilogram fresh weight.

The Queen Sirikit Botanic Gardens in Chiang Mai is engaged in ex situ conservation of native Thai orchids (Ongprasert 2012). At Mahidol University in Bangkok, Kanchit Thammasiri is developing efficient cryopreservation methods for efficient seed and protocorm storage and his is one of the 30 collaborating sites of the Orchid Seed Storage for Sustainable Use Programme of the Royal Botanic Gardens, Kew. Nevertheless, presently, only 28 out of Thailand's 1200 orchid species are in storage (Tobias 2012).

Concluding Remarks

Conservation of natural orchid species in situ, also ex situ, and cultivation of medicinal crops on a commercial scale are important activities that need to be promoted. When fur, seal skins and sandalwood became scarce in the nineteenth century, American traders started cultivating ginseng that fetched good prices in China. Silkworm is used to cultivate prized *Cordyceps*. *Bletilla striata* and *Gastrodia elata* are also extensively grown, and one hopes that in the near future, all

Fig. 18.4 (continued) community of *Bulbophyllum vaginatum* which flowers gregariously in response to a significant drop in temperature caused by tropical rainstorms. (©Teoh Eng Soon 2019. All Rights Reserved)

Fig. 18.5 *Coelogyne corymbosa* thriving as a lithophyte in Sikkim. It also survives as a terrestrial, but throughout northeastern India it is most commonly an epiphyte. (©Teoh Eng Soon 2019. All Rights Reserved)

medicinal *Dendrobium* will be cultivated and not plundered from the forests. It is a shame that poor villagers in many Asian countries are not taught and given the opportunity to grow medicinal orchids to provide abundant supply: instead, they merely eke out a meagre livelihood harvesting wild orchids. At the time of writing, some attempts are being made to rectify the situation, but the effort should be more widespread.

Conservationists dealing with orchids have focused their attention on species that are rare, critically endangered, endemic or attractive to collectors. Their efforts are laudable. However, conservation efforts are being directed to fewer than 10% of orchid species, whereas the proper focus of conservation should be biodiversity. Plant species that are unimportant today might well turn up to have an important role in the future (Fig. 18.5).

References

Abraham A, Vatsala P (1981) Introduction to orchids, with illustrations and descriptions of 150 South Indian orchids. TPGRI, Trivandrum

Averyanov L, Cribb P, Phan KL, Nguyen TH (2003) Slipper orchids of Vietnam. With an introduction to the flora of Vietnam. Royal Botanic Gardens, Kew

Bailes CP (2005) Orchids in Nepal, the conservation and development of a natural resource. Advisory report and recommendations. Richmond, Royal Botanic Gardens

Bidartondo MI, Burghardt B, Gebauer G et al (2004) Changing partners in the dark: isotopic and molecular evidence of ectomycorrhizal liaisons between forest orchids and trees. Proc R Soc 271(1500):1799–1806

Bidartondo MI, Read DJ (2008) Fungal specificity bottlenecks during orchid germination and development. Mol Ecol 17:3707–3716

Bisson J, Guiang ES, Walpole P, Tolentino D Jr (2003) Better governance critical to reversing forest degradation in Southeast Asia. Paper submitted to 12th World Forestry Congress, 2003, Quebec City, Canada

Braun R (1888) China. Imperial maritime customs II. Special Series No. 8. List of Medicines exported from Hankow and the other Yangtze Ports. Shanghai Inspector-General of Customs

Budiharta S, Widyatmoko D, Irawati et al (2011) The processes that threaten Indonesian plants. Flora Fauna Intern Oryx 45(2):172–179

Butola JS, Badola HK (2006) Growth, phenology and productivity of *Dactylorhiza hatagirea* (D. Don) Soo, a critically endangered medicinal orchid in Himalaya: domestication compared with wild. J Orchid Soc India 20:373–343

Butola JS, Badola HK (2008) Himalayan threatened medicinal plants and their conservation in Himachal Pradesh. J Trop Med Plants 9(1):125–135

Chang DCN (2007) The screening of orchid fungi (OMF) and the applications. In: Chen WH, Chen HH (eds) Orchid biotechnology. World Scientific, New Jersey

Chen SC (1995) Orchids and their conservation. Proc. 5th Asia Pacific orchid conference & show, Fukuoka, pp 15–18

Chen RR, Lin XG, Shi YQ (2003) Research advances on orchid mycorrhiza. Chin J Appl Environ Biol 9(1):97–101

Chen XM, Xiao SY, Guo SX (2006) Comparison of chemical composition between *Dendrobium candidum* and *Dendrobium nobile*. Zhongguo Yi Xue Ke Xue Yuan Xue Bao 28(4):524–529

Chu XL, Yang B, Gao L et al (2010) Species diversity of cultivable bacteria isolated from the roots of *Cymbidium faberi* Rolfe. Wuhan Zhiwuxue Yanjiu 28(2):199–205

Clements M (2011) Oral presentation. In: Programme World Orchid Conference, Singapore

Cole A (1998) Mothers and sons in Buddhism. Stanford University Press, Stanford

Compant S, Duffy B, Nowak J et al (2005) Use of plant growth promoting bacteria for biocontrol of plant diseases: principles, mechanisms of action and future prospects. Appl Environ Microbiol 71(9):4951–4959

Dalstrom S, Gyeltshen N, Gyelsshen C, Gyeltshen N (2012) Thunder dragon orchids. A conservation project in Bhutan designed to protect orchids. Orchids 81(8):494–497

Dash PK, Sahoo S, Bal S (2008) Ethnobotanical studies on orchids of Niyamgiri Hill Ranges, Orissa, India. Ethnobot Leaflets 12:70–78

Ding X, Xu L, Wang Z et al (2002a) Authentication of stems of *Dendrobium officinale* by rDNA ITS region sequences. Planta Med 68(2):191–192

Ding XY, Xu LS, Wang ZT et al (2002b) Molecular authentication of *Dendrobium chrysanthum*. Zhongguo Zhong Yao Za Zhi 27(6):407–411

Fu LK (1992) China Plant Red Data Book, rare and endangered plants. Science Press, Beijing

Glicenstein L (2005) Planning ahead. The AOS conservation committee works to protect orchids for future generations. Orchids 74(4):298–299

Hadley G (1970) Non-specificity of symbiotic infection in orchid mycorrhiza. New Phytol 69:1015

Hamel PB, Chiltoskey MU (1975) Cherokee plants their uses – a 400 year history. (Self published, address not stated, available through Amazon)

Hansen E (2001) Orchid fever. Methuen, New York

Hart R (1884) China imperial customs III. Misc. Series No. 17. List of Chinese medicines. Inspector General of Customs, Shanghai

Hirano T, Godo T, Mii M, Ishikawa K (2005) Cryopreservation of immature seeds of *Bletilla striata* by vitrification. Plant Cell Rep 23:534–539

Hollick SH (2004) Mycorrhizal specificity in endmic Western Australia terrestrial orchids (Tribe Diurideae): implications for conservation, PhD dissertation. University of Murdock, Australia

Jia JS (2007) The status of orchid conservation in China. Lankesteriana 7(1–2):48

Jin XH, Zhao XD, Shi XC (2009) Native orchids from Gaoligongshan Mountains, China. Science Press, Beijing

Koopowitz H (2001) Orchids and their conservation. Timber Press, Portland

Lee BC (ed) (2009) Rare plants data book of Korea. Korea National Arboretum, Gyeonggi-Do

Lee YI (2011) In vitro culture and germination of terrestrial Asian orchid seeds. In: Thorpe TA, Yeung EC (eds) Plant embryo culture. Methods and protocols. Humana Press Inc., Tootowa, NJ

Li T, Wang J, Lu Z (2005) Accurate identification of closely related *Dendrobium* species with multiple species specific gDNA probes. Biochem Biophys Methods 62(2):111–123

Li X, Ding X, Chu B, Zhou Q, Ding G, Gu S (2008) Genetic diversity analysis and conservation of the endangered Chinese endemic herb *Dendrobium officinale* Kimura et Migo (Orchidaceae) based on AFLP. Genetica 133(2):159–166

Liu H, Luo YB, Liu ZJ (2013) Using guided commercialized cultivation models to promote species conservation and sustainable utilization: an example from the Chinese medicinal orchids. Biodivers Sci 21 (1):132–135

Liu HX, Luo YB, Liu H (2010) Studies of mycorrhizal fungi of Chinese orchids and their role in orchid conservation in China – a review. Bot Rev 76:241–262

Ongprasert P (2012) Forest Management in Thailand. International Forestry Cooperation Office, Royal Forest Department, Ministry of Natural Resources and Environment, Thailand

Pandey NK, Joshi GC, Mudaiya RK et al (2003) Management and conservation of medicinal orchids of Kumaon and Garhwal Himalaya. J Econ Taxon Bot 27(1):114–116

Pernier H, Luo (2007) Orchids of Huanglong. Huanglong National Park, Sichuan

Preiss K, Adam IKU, Gabauer G (2010) Irradiance governs exploitation of fungi: fine-tuning of carbon gain by two partially myco-heterotrophic orchids. Proc R Soc Biol Sci Ser B 277(1686):1333–1336

Rao AN (2004) Medicinal orchid wealth of Arunachal Pradesh. News Lett Envis Node Indian Med Plants 1 (2):1–5

Rasmussen HN, Whigham DF (1993) Seed ecology of dust seeds in situ: a new study technique and its application in terrestrial orchids. Am J Bot 80: 1374–1378

Sezik E (1967) Turkiye'nin Salepgilleri Ticari Salep Cesitleri ve Ozellikle Mugla Salebi Uzerinde Arastirmalar. Doctoral Thesis. Istanbul Universitesi Eczacihk Fakultesinde (In Turkish). Summary in English Sezik E (1990)

Subedi A (2005): Orchids and sustainable livelihood: an initiative in Nepal Himalayas to manage globally threatened biodiversity. In: Raynal-Rogues A, Roguenant A, and Prat D: 18th world orchid conference proceedings, Djon, France. Paris: Naturalia Publications, pp 474–479

Sun L, Shao H, Liu L et al (2011) Diversity of siderophore-producing bacteria of *Cymbidium goeringii* roots. Weishengwu Xuebao 51(2):189–195

Swinson A (1970) Frederick Sander: the orchid king. Hodder & Stoughton, London

Taft D (2013) *Corallorhiza odontorhiza* restoration. Orchids 82(5):305

Teoh ES (2011) Medicinal orchids: the issue of conservation. Malay Orchid Rev 45:105–113

Thammasiri K (2000) Cryopreservation of seeds of a Thai orchid (*Doritis pulcherrima* Lindl.) by vitrification. Cryo-Letters 21:237–244

Thammasiri K (2008) Cryopreservation of some Thai orchid species. Acta Hortic 788:53–62

Thammasiri K (2010) Vitrification-based cryopreservation of *Grammatophyllum speciosum* protocorms. CryoLetters 31(4):347–357

Tobias P (2012) Conservation touches us all. Orchids 81:102–110

Tsavakelova A, Cherdyntseva TA, Netrusov AI (2005) Auxin production by bacteria associated with orchids. Mikrobiologica 74(1):55–62

Vandrame WA, Faria RT (2011) Phloroglucinol enhances recovery and survival of cryopreserved *Dendrobium nobile* protocorms. Sci Hortic 128(2):131–135

Veldman S, Kim SJ, van Andel TR et al (2018) Trade in Zambian edible orchids - DNA barcoding reveals use of unexpected orchid taxa for Chikanda. Genes 9 (12):595. Preprints. Creative Commons CC

Vendrame WA, Carvalho VS, Dias JMM (2007) In vitro germination and seedling development of cryopreserved *Dendrobium* hybrid mature seeds. Sci Hortic 114:188–193

Vendrame WA, Carvalho VS, Dias JMM, Maguire I (2008) Pollination of *Dendrobium* hybrids using cryopreserved pollen. Hortic Sci 43(1):264–267

Wangchuck ADW (2006) A portrait of Bhutan. Viking (Penguin), New Delhi

Wood CB, Pritchard HW, Millar AP (2000) Simultaneous preservation of orchid seed and its fungal symbiont using encapsulation-dehydration is dependent on moisture content and storage temperature. Cryo Letters 21 (2):125–136

Xing XK, Ma XT, Deng ZH et al (2013) Specificity and preference of mycorrhizal associations in two species of the genus Dendrobium(Orchidaceae). Mycorrhiza 23(4):317–324

Xu JT (1989) Studies on the life cycle of *Gastrodia elata*. Zhongguo Yi Xi Xue Ke Xue Yuan Xue Bao 11:237–241

Xu J, Guo S (2000) Retrospect on the research of the cultivation of gastrodia elata Bl., a rare traditional Chinese medicine. Chin Med J 113:686–692

Xu H, Ying Y, Wang ZT, Cheng KT (2010) Identification of *Dendrobium* species by dot blot hybridization assay. Biol Pharm Bull 33(4):665–668

Yam TW (2010) Conservation of the native orchids through seedling culture and re-introduction: a Singapore experience. Bot Rev 76(2):263–274